DATE DUE

Advances in
MARINE BIOLOGY

VOLUME 49

Restocking and Stock Enhancement of Marine Invertebrate Fisheries

By

JOHANN D. BELL*, PETER C. ROTHLISBERG[†],
JOHN L. MUNRO*, NEIL R. LONERAGAN[†],
WARWICK J. NASH*, ROBERT D. WARD[‡] AND
NEIL L. ANDREW[¶]

*The WorldFish Center, Penang, Malaysia
[†]CSIRO Marine Research, Cleveland, Australia
[‡]CSIRO Marine Research, Hobart, Australia
[¶]National Institute of Water & Atmospheric Research, Wellington, New Zealand

ELSEVIER

AMSTERDAM • BOSTON • HEIDELBERG • LONDON
NEW YORK • OXFORD • PARIS • SAN DIEGO
SAN FRANCISCO • SINGAPORE • SYDNEY • TOKYO
Academic Press is an imprint of Elsevier

Elsevier Academic Press
525 B Street, Suite 1900, San Diego, California 92101-4495, USA
84 Theobald's Road, London WC1X 8RR, UK

This book is printed on acid-free paper.

Permissions may be sought directly from Elsevier's Science & Technology Rights
Department in Oxford, UK: phone: (+44) 1865 843830, fax: (+44) 1865 853333,
E-mail: permissions@elsevier.com. You may also complete your request on-line
via the Elsevier homepage (http://elsevier.com), by selecting
"Support & Contact" then "Copyright and Permission" and then "Obtaining Permissions."

For all information on all Academic Press publications
visit our Web site at www.academicpress.com

ISBN-13 : 978-0-12-026149-9
ISBN-10 : 0-12-026149-9

PRINTED IN THE UNITED STATES OF AMERICA
05 06 07 08 09 9 8 7 6 5 4 3 2 1

CONTENTS

1. Introduction

2. Restocking Initiatives

3. Stock Enhancement Initiatives

4. Overview and Progress Towards a Responsible Approach

5. Lessons Learned

6. Management of Restocking and Stock Enhancement Programmes

7. Other Important Considerations for All Initiatives

8. Conclusions

PREFACE

Several marine fisheries are in jeopardy because of deterioration in habitats and chronic over-fishing. Other, better-managed, fisheries often fail to reach their full potential because the supply of juveniles regularly falls short of the carrying capacity (optimal stocking density) of supporting habitats. To help meet the growing demand for fish and shellfish, strong calls are now being made to restore marine capture fisheries and maximise their yields in sustainable ways. For some of these fisheries, additions of cultured juveniles to the population, redistribution of wild individuals and provision of new habitat can potentially assist the two main processes needed to restore and increase production where the causes of habitat degradation have been addressed. These processes are: (1) rebuilding spawning biomass of severely depleted populations to levels where the fishery can once again support regular harvests, a process we call 'restocking'; and (2) increasing productivity of operational fisheries by overcoming recruitment limitation, a process we call 'stock enhancement'. Releases of hatchery-reared juveniles and translocations of adults are being used for restocking initiatives. A broader range of interventions is being attempted for stock enhancement, including release of hatchery-reared juveniles; capture, culture and release of wild-caught juveniles; thinning and relocation of dense aggregations of 'spat'; and provision of additional habitat to increase settlement success.

In this review, we evaluate the status of restocking and stock enhancement initiatives for 11 groups/species of marine invertebrates. In doing so, we assess progress against the widespread call for a responsible practice. Overall, stock enhancement initiatives far outnumber restocking efforts. However, the goals of releasing cultured juveniles were not always well-defined, and it is now clear that the primary need of some target fisheries was restocking, not stock enhancement. It is still too early to tell whether the small number of restocking programmes will be successful. Stock enhancement has had mixed success. Large-scale releases of wild-caught scallop spat in Japan and New Zealand, involving up to billions of individuals per year, have delivered clear benefits, and the costs are now paid by the stakeholders. Releases of pond-reared shrimp in China in billions, and in Japan in hundreds of millions, have also increased catch, although operations in China have been much reduced since withdrawal of government support. Releases of tens of millions of sea urchins and abalone in Japan seem to have arrested declining yields in some cases but have not rebuilt catches to former levels. Many of the smaller-scale stock enhancement initiatives have had little effect on abundances of target species and failed to provide benefits exceeding the costs of producing juveniles.

We advocate a new approach to restocking and stock enhancement. This involves: (1) recognizing the need to apply these interventions to self-replenishing populations within a fishery, (2) assessing the status of each self-replenishing population and identifying whether restocking or stock enhancement is required; (3) using the simplest, reliable models to predict whether restocking or stock enhancement will add value to other measures available to achieve the relevant goal and (4) only proceeding with releases of cultured juveniles (or other interventions) where they are predicted to be beneficial. For release of cultured juveniles, efficient modelling will require sound information on the least cost of producing juveniles, and on rates of survival after release. There are no shortcuts: each self-replenishing population demands specific research and an objective, responsible approach. Indeed, a key lesson from this review is that there is no generic method for restocking and stock enhancement—release strategies developed for one species do not work for others, and those that succeed for a species at one place do not necessarily work elsewhere.

For marine fisheries, we conclude that restocking and stock enhancement have their greatest potential among the invertebrates. There are three reasons for this. First, many molluscs and echinoderms appear to have multiple, self-replenishing populations at surprisingly small spatial scales. This increases the likelihood that mass-releases of cultured juveniles will have an impact on population size. Second, the sessile/sedentary nature of many species means that juveniles can be released at densities that will maximize reproductive success in restocking programmes. Third, the shallow, inshore habitat of many species, combined with their lack of mobility, facilitates assessments of the success of the interventions and aids adaptive management.

Nevertheless, the challenges of restoring or increasing productivity for many marine invertebrate fisheries remain daunting. We trust that the lessons compiled here, and the new approach we advocate for assessing whether restocking and stock enhancement have a role to play, will assist managers to meet these challenges.

Johann D. Bell, John L. Munro and Warwick J. Nash
The WorldFish Center, Penang, Malaysia

Peter C. Rothlisberg and Neil R. Loneragan
CSIRO Marine Research, Cleveland, Australia

Robert D. Ward
CSIRO Marine Research, Hobart, Australia

Neil L. Andrew
National Institute of Water & Atmospheric Research,
Wellington, New Zealand

Chapter 1

Introduction

Reduced productivity from many of the world's coastal and marine capture fisheries is now a widespread cause of concern for governments (Valdimarsson and James, 2001; Hilborn *et al.*, 2003; Myers and Worm, 2003). The problems are by no means ubiquitous, because there are several well-managed stocks, but many large- and small-scale fisheries no longer provide yields close to their sustainable potential (Pauly *et al.*, 2002). The main problems fall into two broad categories. The first is degradation of the ecosystems that underpin the biological production of fisheries. This has occurred not only through developments on coasts and in catchments that reduce, modify or pollute the supporting habitats, but also through destructive methods of fishing. The second problem is over-fishing in all its forms (Pauly, 1987, 1988), often driven by technological and socioeconomic factors such as over-capacity, increased efficiency and lack of property rights (Charles, 2001; Pauly *et al.*, 2002).

The second of these problems, over-fishing, can be particularly acute for many of the commercially exploited marine invertebrates. The shallow distributions and sedentary or sessile life styles make many marine invertebrates accessible and easy to harvest. Their vulnerability is often exacerbated by the common mode of reproduction of marine invertebrates, broadcasting of gametes. Where fishing is intense, the distance between adults can be too great for external fertilization in sessile species and inhibit spawning in sedentary animals. Such 'Allee effects' are summarized by Caddy and Defeo (2003 and references therein).

Another consequence of the inshore distribution of many sessile or sedentary marine invertebrates is that stocks are often composed of meta-populations or multiple, self-recruiting populations (Stoner, 1997a; Shaklee and Bentzen, 1998; Botsford, 2001; Caddy and Defeo, 2003; Palumbi, 2003). Knowledge is accumulating rapidly about the factors selecting for self-recruitment among marine species (Strathmann *et al.*, 2002; Swearer *et al.*, 2002), often at surprisingly small scales, and the coastal features that promote such population structures (Caddy and Defeo, 2003; Largier, 2003). Although stock structures of most marine invertebrates have

ADVANCES IN MARINE BIOLOGY VOL 49
© 2005 Elsevier Ltd. All rights reserved

0065-2881-05 $35.00
DOI: 10.1016/S0065-2881(05)49001-X

yet to be assessed comprehensively, the accumulating evidence suggests that replenishment of over-fished populations may not occur easily from adjacent populations because of restricted exchange of propagules, particularly where local currents favour retention of larvae. Crustaceans and cephalopods can differ from other marine invertebrates in this regard, because their greater mobility means that the factors affecting exploitation and replenishment are more akin to those for finfish (Longhurst and Pauly, 1987). Overall, however, it comes as no surprise that the proportion of marine invertebrates that are over-fished or fully exploited exceeds the proportion of demersal and pelagic finfish with endangered stocks (Figure 1.1).

Several authors (Jackson *et al.*, 2001; Pauly *et al.*, 2002; Myers and Worm, 2003) have recently discussed the potential for rebuilding many of the world's capture fisheries. For some species, the scope for restoration is likely to exceed expectations, because the historical sizes of stocks are not well-documented. The basic steps needed to repair severely damaged fisheries are now well recognised; the quality and area of supporting habitats must be improved and fishing effort must be reduced. Both these measures can establish a larger spawning biomass capable of producing increased, if variable, harvests on a regular basis (Hall, 1999).

In addition to these basic steps, recent developments in hatchery technology, which have enabled the propagation of many species of coastal fish and invertebrates worldwide, could increase the productivity of selected capture fisheries through release of cultured juveniles into the wild. This approach

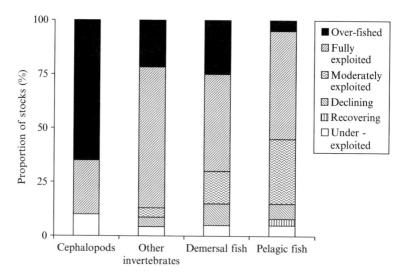

Figure 1.1 The proportion of stocks ranging from over-fished to under-exploited for four categories of marine fisheries (redrawn from Hall, 1999).

has been applied most widely in Japan, where 90 species of coastal fish and invertebrates have been released to augment wild stocks (Honma, 1993; Imamura, 1999). In general, however, this is a relatively new field of endeavour, and opinion remains divided on the potential, and the dangers, of this approach (Blankenship and Leber, 1995; Munro and Bell, 1997; Hilborn, 1998; Travis et al., 1998; Leber, 1999, 2002, 2004; Caddy and Defeo, 2003; Molony et al., 2003; Lorenzen 2005 and references therein). The potential for significantly enhancing large-scale, non-salmonid marine finfish stocks has been questioned because of the trivial contribution of hatchery releases relative to natural reproduction (Blaxter, 2000). On the other hand, the characteristics of invertebrate fisheries described previously make the processes of releasing cultured juveniles in restocking and stock enhancement programmes more feasible. Note that we draw a marked distinction here between 'restocking' and 'stock enhancement'. Although these terms both involve the release of juveniles to augment wild populations, and have often been used interchangeably together with terms like 'reseeding', 'sea/marine ranching', 'sea/marine farming' and 'culture-based fisheries' (Munro and Bell, 1997; Caddy and Defeo, 2003; Molony et al., 2003), the processes they describe have different goals (Manzi, 1990; Bell, 1999a, 2004; see also Hilborn, 1998; Travis et al., 1998; Lorenzen, 2005).

'Restocking' in the context of this review involves releasing cultured juveniles into the wild to restore the spawning biomass of a severely depleted fishery to a level where it can once again provide regular, substantial yields. Restocking should be considered when the time required for replenishment of the population by natural processes, under management measures such as a moratorium on fishing or the use of marine protected areas, is likely to be unacceptable to the fishing community. It will depend on knowing enough about the ecology of the species and nursery habitat, and survival of cultured juveniles in the wild, to indicate that mass-releases will 'fast-track' recovery of spawning biomass.

'Stock enhancement', on the other hand, is designed to increase the productivity of an operational fishery. This may be accomplished by augmenting biomass of spawning adults, but greater gains are possible by overcoming recruitment limitation (Munro and Bell, 1997; Grimes, 1998; Doherty, 1999; Lorenzen, 2005). Recruitment is considered to be limited when the natural supply of juveniles fails to reach the carrying capacity of the habitat. This often results from larval or juvenile mortality even when there are great numbers of breeding adults. Recruitment limitation is the rule rather than the exception among marine animals with pelagic larvae and arises for several reasons (reviewed by Thorson, 1950; Rumrill, 1990; see also Doherty, 1999 and references therein). For example, planktonic larval stages may perish en masse because of predation, starvation, or transport away from suitable nursery habitats or settlement sites. High post-settlement mortality can also occur when larvae settle in habitats where predators are abundant and shelter is limited. As a result of recruitment limitation, even unexploited or lightly fished habitats may not produce as many invertebrates as they could. Stock enhancement can redress this situation by augmenting the natural supply of juveniles to optimize production from a fishery (see also Malouf, 1989; Munro and Bell, 1997; Lorenzen, 2005).

The processes of restocking and stock enhancement should be sequential for any given species. Thus, severely over-fished populations could be restored by restocking, and then managed for optimum productivity through stock enhancement.

For both restocking and stock enhancement to be considered as possible management measures, there should be scope for mass-release of juveniles. Although aquaculture technology developed over the past couple of decades has paved the way for propagation of many marine species, there remain great differences in the potential for mass-production of cultured juveniles among marine invertebrates. These parallel the differences in propagation potential between some salmonids, particularly chum salmon, and marine fish with small pelagic eggs (Figure 1.2). Some invertebrates (e.g., scallops) can be collected easily in very large numbers (billions) as wild-settled juveniles ('spat') (Ventilla, 1982), whereas others (e.g., spiny lobsters) may be collected only in limited numbers. Many invertebrates (e.g., sea urchins, abalone, sea cucumbers, giant clams, lobsters) must be reared in hatcheries, where the expense of providing adequate shelter, food or surface area for settlement often restricts production. However, intensive pond rearing of shrimp can potentially produce large numbers of juveniles for stock enhancement (Liu, 1990). Overall, production of some marine invertebrates for stock enhancement programmes (e.g., scallops, shrimps) may be 1–3 orders of magnitude greater than production of non-salmonid marine fish (e.g., red sea bream and flounder) for similar purposes (Figure 1.2). Production at this scale indicates that restocking or stock enhancement has good potential for some populations of marine invertebrates.

Here we review restocking and stock enhancement initiatives for a wide variety of marine invertebrates (11 groups/species: giant clams, topshell, sea cucumbers, scallops, other bivalves, abalone, queen conch, shrimp, spiny lobsters, lobsters and sea urchins). We begin with those species/groups where restocking has been the primary object of releasing cultured juveniles (Chapter 2) and proceed to species where releases have been aimed mainly at stock enhancement (Chapter 3). This distinction has not always been straightforward, however, because stock enhancement has been attempted for some species when the primary need in hindsight was for restocking. To simplify matters, we have categorized the species/groups according to the original objective of releasing the juveniles, and then commented where we feel the initial goal was misdirected.

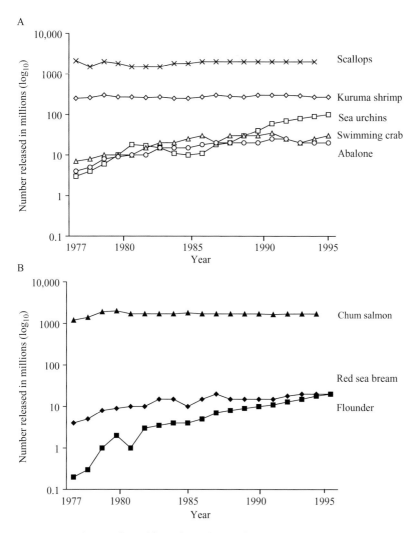

Figure 1.2 The number of juveniles released for the major species of (A) marine invertebrates and (B) fish in Japanese stock enhancement programmes (redrawn from Kitada, 1999).

In structuring this review of restocking and stock enhancement initiatives for marine invertebrates, we were influenced greatly by the landmark article by Blankenship and Leber (1995), in which they advocate a 'responsible approach' to marine stock enhancement. In general, they stress the need to: (1) consider releases of juveniles as just one of a range of management options, (2) evaluate all aspects of the process, (3) minimize negative impacts on wild stocks and (4) use adaptive management to improve cost-effectiveness. Accordingly, for each of the 11 groups/species, we present information to allow readers to gauge for themselves the progress that has been made toward the blueprint outlined by Blankenship and Leber (1995). For each group/species, we examine: the rationale for restocking or stock enhancement; methods for supplying the juveniles; defects in morphology or behaviour of the cultured animals; development of release strategies (including use of artificial habitat to enhance survival or productivity in some cases); precautions against loss of genetic diversity; risks of introducing diseases; development of systems for marking juveniles to assess the effectiveness of releases; status of the development of technology/release programmes; management regimes or requirements; current impediments to restocking or stock enhancement; and the need for further research. This information varies considerably among the various groups/species, depending on the purposes of the releases and the stage of development of restocking or stock enhancement programmes, so not all of these points are addressed for each group/species.

It is also important to note that we have not set out to address all the measures available to managers to replenish severely depleted stocks or enhance the productivity of operational fisheries. For example, marine protected areas are the subject of a large literature and are not considered here. Instead, this review is restricted to the more active forms of intervention, such as the mass-release of cultured juveniles and the modification or addition of habitats to increase survival of wild or released juveniles. Caddy and Defeo (2003) describe a broader range of approaches.

Where restocking or stock enhancement initiatives are underway for several species in a group, we have not always covered every species worldwide. Instead, we have sometimes reviewed initiatives on several species in one general area (e.g., bivalves in the United States), or one species at several locations (e.g., the sea cucumber *Holothuria scabra* in the Indo-Pacific), to report progress and issues typical for that group. Otherwise, Chapters 2 and 3 are based on past and current initiatives within the general distribution of the group/species.

The remainder of the review is in four parts. Chapter 4 compares and contrasts the approaches and outcomes across species and summarises general progress toward the components of a responsible approach outlined by Blankenship and Leber (1995). This has been done separately for the species/groups used in restocking and those used for stock enhancement initiatives. Chapter 5 outlines the major lessons that have emerged for the 11 groups/species. Chapter 6 focuses on the role of restocking and stock enhancement as part of a suite of management tools, outlines procedures for identifying when these interventions are likely to add value to other forms of management, and emphasises that restocking and stock enhancement are not the same processes and need to be managed differently. Chapter 7 deals with considerations for all restocking and stock enhancement initiatives, regardless of species, location and the methods used for producing juveniles. These are the need to: (i) measure success; (ii) maintain genetic diversity of wild stocks; (iii) reduce the risk of introducing diseases and (iv) minimise negative effects on the ecosystem. We conclude (Chapter 8) with an assessment of achievements, a proposal for integrating restocking and stock enhancement with other forms of fisheries management and the research issues that lie ahead to ensure that these interventions are implemented responsibly where they are deemed to be appropriate.

Chapter **2**

Restocking Initiatives

2.1. GIANT CLAMS

2.1.1. Background and rationale for restocking

The giant clams (Tridacnidae) of coral reefs in the Indo-Pacific are the largest bivalves that have ever lived, and their taxonomy, distribution, biology and ecology have aroused much interest in the past 30 years (see Munro, 1993a; Lucas, 1994; Mingoa-Licuanan and Gomez, 1996 for reviews). Apart from the size of the largest species, *Tridacna gigas*, which grows to a shell length (SL) of 1.4 m, and their symbiotic relationship with zooxanthellae (*Symbiodinium* spp.), giant clams have attracted attention because stocks of the larger species (*T. gigas*, *T. derasa*, *T. squamosa* and *Hippopus hippopus*) have been depleted seriously or extinguished within major parts of their range (Copland and Lucas, 1988; Munro, 1989; Lucas, 1994). Overfishing by foreign vessels to obtain the highly prized adductor muscle was the main cause of reduction in stocks (Munro, 1989). Domestic harvesting for subsistence and commerce has also been significant in Asia (Mingoa-Licuanan and Gomez, 1996), the Pacific (McKoy, 1980; Lewis *et al.*, 1988) and Japan (Murakoshi, 1986).

In response to the decline in stocks, giant clams were listed under Appendix II of the Convention on International Trade in Endangered Species (CITES) (Wells *et al.*, 1983). However, it soon became apparent that protection by CITES alone would not bring about restoration of some depleted populations, because densities had been reduced to the point where successful reproduction was improbable. Replenishment from 'neighbouring' (often distant) stocks was also unlikely because of the relatively short duration of the larval stage (Bell, 1999b and references therein).

In the early 1980s, several institutions embarked on a concerted effort to restore stocks of giant clams by propagating juveniles and placing them in coral reef habitats (Figure 2.1). These institutions were: the Okinawa Prefectural Fisheries Experimental Station, the University of Papua New Guinea, the Micronesian Mariculture Demonstration Center, the Australian Centre for International Agricultural Research, the Marine Science Institute at the University of Philippines and the WorldFish Center.

2.1.2. Supply of juveniles

The production of juvenile giant clams for placement on coral reefs has not been straightforward. Settled juveniles are relatively easy to rear, because the eggs are comparatively large (∼90 μm) and the larvae require little or no nutrition until they are inoculated with zooxanthellae (Heslinga *et al.*, 1990;

A

B

C

Figure 2.1 (A) Broodstock of the giant clam *Tridacna gigas* used for hatchery operations in Solomon Islands (photo: M. McCoy). (B) Village rearing of juvenile giant clams in protected cages in Solomon Islands; cage covers have been removed to show the clams (photo: M. McCoy). (C) Cultured *Tridacna gigas* growing in the intertidal zone at Orpheus Island, Australia (photo: J. Lucas).

Braley, 1992; Lucas, 1994). However, the process of growing juveniles until they are large enough to escape predation is extraordinarily slow compared with other marine invertebrates (see Section 2.1.4). The juveniles must be retained in land-based nurseries for about 9 mo, until they reach a size of 20–25 mm SL, before they are robust enough to be transferred to ocean nurseries. This stage of the rearing process can be shortened by several weeks by optimising the application of dissolved inorganic nitrogen to fertilize the zooxanthellae (Grice and Bell, 1997, 1999), but there is still a bottleneck.

2.1.3. Defects caused by rearing

There is great variation in the growth of juvenile giant clams during the land-based nursery phase. This variation can be attributed in part to variable fitness of individuals, although much of it is also undoubtedly caused by localised crowding in the tanks. However, even if unfit juveniles survive the land-based nursery phase, they are unlikely to emerge from the ocean nursery, so such individuals are generally weeded out before they reach the release size.

2.1.4. Release strategies

The preparation of giant clams for restocking in the wild does not stop at the end of the land-based nursery phase. Giant clams placed in the ocean suffer heavy mortality from carnivorous gastropods (*Cymatium* spp. and Pyramidellidae), boring sponges, crabs, flatworms, triggerfish (Balistidae), rays (Myliobatidae) and turtles (Heslinga *et al.*, 1990; Govan, 1995). For the larger species (*Tridacna gigas* and *T. derasa*), further husbandry for a period of ~3 yr is needed before the young clams attain a size refuge from predation.

Wire mesh cages are effective in protecting several species of giant clams during the period when they are vulnerable to predators (Heslinga *et al.*, 1990; Calumpong, 1992; Bell *et al.*, 1997a; Table 2.1). Survival of the two smallest species, *Tridacna crocea* and *T. maxima*, which burrow into the substratum, can be improved by drilling into dead colonies of coral (*Porites* spp.) and placing the seed clams directly into the holes (Murakoshi, 1986). *Tridacna gigas* has also been reared successfully in the intertidal zone in Australia, where predation is lower than in subtidal habitats (Lucas, 1994).

In short, rearing giant clams in land-based nurseries and sea cages to prepare them for restocking does not differ from the methods for farming these species. The ~4 yr needed to rear giant clams to the size where they are no longer vulnerable to predation makes production of 'juveniles' for

Table 2.1 Survival of five species of giant clams (Tridacnidae) during initial phases of growing on in wire-mesh sea cages used to prepare the animals for restocking (at least 36 mo is needed for giant clams to attain an effective release size).

Species	No. of sites	Duration of growing on (mo)	Mean % survival (± SD)	References
Tridacna gigas	32	10	54.0 ± 18.6	Bell *et al.* (1997b)
T. derasa	11	24	92.2 ± 9.1	Hart *et al.* (1998)
T. squamosa	10	8	66.6 ± 18.3	Foyle *et al.* (1997)
T. maxima	11	19	38.9 ± 16.6	Hart *et al.* (1998)
T. crocea	11	17	39.0 ± 22.6	Hart *et al.* (1998)

restocking a laborious and expensive process. The best prospect for reducing the financial burden involved in growing giant clams for restoring populations is to link restocking to farming (see Section 2.1.8).

2.1.5. Minimising genetic impacts

The fact that populations of giant clams have been so drastically reduced in many places provides scope for hatchery-reared animals to have a great influence on the genetic diversity of remnant wild stocks. However, it is difficult to produce giant clams with gene frequencies comparable to the wild population from which they were derived (Figure 2.2). This problem is caused by the low proportion of adults that can be induced to shed eggs (Benzie, 1993a; Benzie and Williams, 1996) but can be combated by maintaining large numbers of wild broodstock, replacing broodstock regularly, maximising the number of matings during each spawning and releasing several cohorts, each derived from different parents (Munro, 1993).

Comparisons of genetic diversity among populations of wild giant clams show that stocks in relative proximity to one another are not necessarily closely related. There are major barriers to gene flow among the central and western Pacific, and between Micronesia and the Philippines, Australia's Great Barrier Reef (GBR) and the Solomon Islands (Ablan *et al.*, 1993; Benzie, 1993a; Benzie and Williams, 1997; Figure 2.3). The most comprehensive data have been collected for *Tridacna maxima*, but data for *T. gigas* and *T. derasa* are consistent with this pattern. There is also high genetic divergence of *T. squamosa* between the east and west coasts of Thailand (Kittiwattanawong *et al.*, 2001). Such information has been used to reduce the impact of restocking giant clams on the genetic diversity of wild stocks where it has been necessary to import 'seed' to rebuild severely depleted populations (see Munro 1989 for the list of such situations). For example,

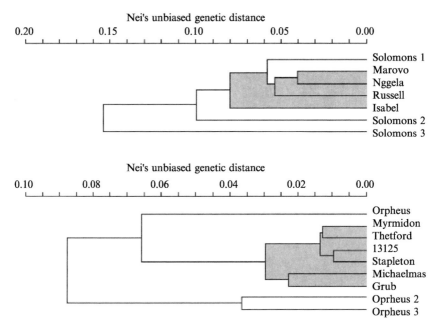

Figure 2.2   Genetic differences among cultured batches of the giant clam *Tridacna gigas* relative to each other and to the natural (shaded) populations from which they were derived in the Solomon Islands and near Orpheus Island, Australia (redrawn from Benzie, 1993a).

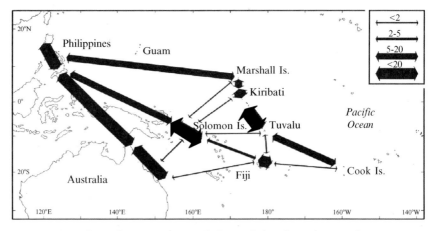

Figure 2.3   Gene flow among populations of the giant clam *Tridacna maxima*. Thickness of arrows represents different levels of dispersal, given by the average number of migrants per generation (redrawn from Benzie and Williams, 1997).

T. gigas used for restocking in the Philippines were derived from the Great
Barrier Reef and the Solomon Islands, not Micronesia.

2.1.6. Disease risks and other environmental impacts

Although giant clams have relatively few parasites and pathogens
(Humphrey, 1988), *Rickettsia* sp. has been associated with mortalities
(Norton *et al.*, 1993a). Other bacteria (including *Vibrio* spp.) and protozoan
parasites (e.g., *Perkinsus* sp. and *Marteilia* sp.), which cause serious losses in
other bivalves, have also been found in giant clams (Humphrey, 1988;
Norton *et al.*, 1993a).

The potential for transferring pathogens, and also gastropod predators,
within and among countries during restocking programmes has been recognised. For example, giant clams dispatched from Palau were routinely placed
in quarantine for 30 d (Heslinga *et al.*, 1990). Despite such precautions,
Cymatium spp. and pyramidellids were transferred occasionally to several
countries (Eldredge, 1994). Consequently, improved protocols have been
developed for the transfer of giant clams (Norton *et al.*, 1993b; Humphrey,
1995). Larvae are now exported before being inoculated with zooxanthellae.
This reduces the potential for transferring parasites and predators associated
with larger clams and reduces contamination from the tissues of other giant
clams used to supply zooxanthellae. Protocols for the exporting country
emphasize the use of quarantine facilities to produce larvae, isolation of
batches, maintenance of water quality and hygiene, and regular testing to
detect pathogens. Protocols for the importing country are similar, except that
more frequent tests are recommended (i.e., at least twice over 3 mo).

2.1.7. Status of restocking initiatives

The concerted effort to produce juveniles has led to placement of giant clams
on coral reefs in 16 countries, although survival of restocked clams has
varied widely within and among nations (Bell, 1999b). The main causes of
mortality were predators, storms, poaching and lack of husbandry. Small
clams were more vulnerable to predators and storms, and larger clams were
targeted by poachers. Despite the large number of restocking programmes,
there is no firm evidence that restocked giant clams have contributed to the
next generation. In part, this reflects the lack of appropriate sampling to
detect juveniles. Recruitment of *Tridacna crocea* occurred in Okinawa,
Japan, after large numbers of the cultured clams had been placed on coral
reefs (Murakoshi, 1986). However, in the absence of a genetic marker for

restocked clams, one cannot rule out recruitment from remnant wild populations living in reserves.

In general, enthusiasm for restocking giant clams has waned since the late 1980s and early 1990s, because it is an expensive, long-term exercise. In several Small Island Developing States (SIDS), restocking stopped after release of one or two cohorts, so well-planned protocols for maintaining genetic diversity have not been implemented. Exceptions are found in the Philippines and Japan, where restocking of giant clams has maintained momentum. The Philippines has developed a responsible programme for restoring stocks of *Tridacna gigas*. Because of the extremely low abundance of this species nationwide, seven cohorts of juveniles were imported from relatively closely related populations in Australia and the Solomon Islands. By 2002, seven batches of F_1 progeny had been raised from the imported animals and placed at 24 protected sites (Mingoa-Licuanan and Gomez, 2002).

In Japan, >2 million *Tridacna crocea* have been transplanted at 37 sites in Okinawa since 1977, with a mean survival of ~50% after 4 yr, and more than 500,000 *T. squamosa* have also been restocked at 22 sites (see Bell, 1999b and references therein). It is important to note, however, that the projects in Japan are more akin to stock enhancement than restocking (Murakoshi, 1986).

2.1.8. Problems to overcome

The two immediate obstacles to restocking giant clams are the high cost of rearing clams until they are large enough to escape predation and the poaching of transplanted clams. Poaching can be countered by placing giant clams in well-managed marine protected areas (Munro, 1993b; Braley, 1996; Mingoa-Licuanan and Gomez, 1996) or under the surveillance of resort owners and villagers who have an interest in safeguarding them. The high cost of producing clams is not easy to circumvent. Bell (1999b) describes a model that should reduce the cost by linking restocking to farming. In this model, giant clam farmers are required or encouraged to rear a small proportion (<5%) of their clams to the size where they will survive when placed on coral reefs. The model has at least four benefits: (1) outlays by governments would be restricted to the costs of seed clams, compliance and distribution of the mature clams; (2) multiple cohorts would eventually be transplanted to a wide variety of sites; (3) awareness of the need to conserve wild stocks would be increased among giant clam farmers and coastal communities and (4) there would be a repository of 'wild' broodstock for use by the farming industry.

A potential weakness of the model is that the genetic diversity of wild stocks might be diluted by the outplanted 'farm animals' (Benzie, 1993b;

Newkirk, 1993). This problem could be overcome by legislating to ensure that clams retained by farmers for restocking are derived from wild stock. The model also assumes that SIDS have the resources to purchase seed from a private hatchery. In countries where this is beyond the means of the government, development agencies would have to be encouraged to meet these costs (Bell, 1999c; Bartley *et al.*, 2004).

2.1.9. Management

Even when restocking can be linked to farming and development agencies assist SIDS with the production of seed, the relatively low numbers of giant clams that can be produced for restocking, and the long generation times for the larger species (9 yr in the case of *Tridacna gigas*), augur against rapid recovery of stocks. This places the onus on managers of coral reef resources to safeguard the remnant stocks of giant clams, particularly where they still occur at densities that enable them to reproduce. Many Pacific Islands have banned the export of wild giant clams unless they have been produced through aquaculture, in keeping with CITES. However, as the human populations of the islands increase, other measures will be needed to reverse the decline of the remnant stocks and to support the restocking initiatives. Such measures may include the creation of more marine protected areas, legislation to restrict harvests to subsistence or traditional uses or, in particularly severe cases, a complete moratorium on taking giant clams until populations recover.

2.1.10. Future research

The limited number of giant clams that can be produced for restocking highlights the need to maximize the reproductive contribution of each individual. To achieve this goal for each species of giant clam, Bell (1999b) believes that answers to the following questions are needed:

1. Which habitats provide the best rates of survival and growth for each species of giant clam?
2. How many giant clams are needed at each site to produce a spawning population size that yields genetically diverse progeny?
3. How should giant clams be arrayed at transplant sites to optimise the tradeoff between fertilization success and dispersal of gametes from the site?
4. Where should aggregations of giant clams be located to maximise the likelihood their propagules encounter suitable substrata for settlement?

The answers to many of these questions will depend on the use of the genetic markers described by Macaranas (1993) to identify the progeny of restocked giant clams and will only be answered after many decades of sustained effort.

2.2. TOPSHELL

2.2.1. Background and rationale for restocking

The large topshell *Trochus niloticus*, also commonly known as trochus from its generic name, is an herbivorous gastropod associated with coral reefs. The nacre of this species has been in demand for buttons since the early 1900s, and the species is heavily fished throughout much of the tropical Indo-Pacific (Nash, 1993). Global harvests of trochus approximate 6,000 tonnes per annum (Heslinga *et al.*, 1984; Eldredge, 1994) but are highly variable from year to year for individual countries (Figure 2.4).

Despite its occurrence throughout much of the Indo-Pacific, the natural distribution of trochus is patchy. The short duration (3–5 d) of the larval phase and the isolation of many islands and reefs have apparently prevented dispersal of juveniles to some suitable coral reef habitats within the overall range of the species. However, this situation has now changed, because adult trochus have been translocated to almost 60 places in the Pacific (Gillett, 1986, 1993; Eldredge, 1994; Isa *et al.*, 1997), including many outside the natural range of the species (Figure 2.5). Trochus have also been introduced occasionally as cultured juveniles (Hoffschir, 1990). Many of the introductions have been successful, for example, in the Cook Islands (Sims, 1985; Nash *et al.*, 1995) (Table 2.2) and French Polynesia (Doumenge, 1973). Viable fisheries have been established after an initial moratorium on harvesting of up to 24 yr, with little apparent impact on other species of commercial importance (Gillett, 1986; Nash, 1993).

The shallow distribution of trochus, its predictable occurrence on the windward side of reefs and movement to high points on the substratum to spawn (Nash, 1993) make the species both easy to collect and vulnerable to over-fishing (Heslinga *et al.*, 1984). Trochus fisheries are managed variously: by size limits, quotas, short open seasons, long-term closures and reserves (Heslinga *et al.*, 1984; Nash, 1993; Amos, 1997; Foale, 1997; Purcell, 2004a). Quotas taken during a short open season can be an effective form of management, because the product is non-perishable and can be stored and sold progressively, if needed, to obtain the best prices. Stocks can also recover relatively rapidly through implementation of long-term closures or reserves, because juveniles, which are highly cryptic (Smith, 1987; Castell

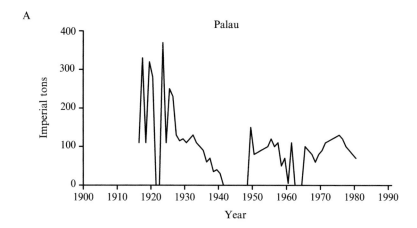

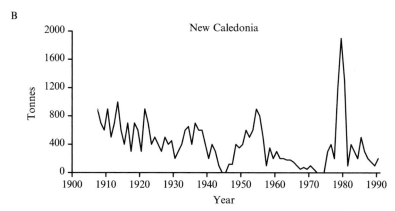

Figure 2.4 (A) Landings of topshell *Trochus niloticus* as live weight from Palau between 1915 and 1979 (redrawn from Heslinga, 1981a). (B) Exports of trochus shell from New Caledonia between 1907 and 1990 (redrawn from Etaix-Bonnin and Fao, 1997). Note that low harvests in New Caledonia in 1941–1945 and 1961–1973 were not caused by over-fishing but were the effects of World War II and the 'nickel boom', respectively.

et al., 1996), are often present in reasonable numbers and reach sexual maturity within a further 1–2 yr (Lincoln-Smith *et al.*, 2000).

In many instances, however, stocks have been fished down to unproductive levels (Heslinga and Hillman, 1981; Heslinga *et al.*, 1984; Nash, 1993; Crowe *et al.*, 1997). Consequently, there have been concerted efforts in Japan, Australia, Indonesia and several Pacific Islands to rebuild populations of trochus through release of cultured juveniles and translocation of adults (Nash, 1988; Crowe *et al.*, 1997; Purcell and Lee, 2001; Crowe

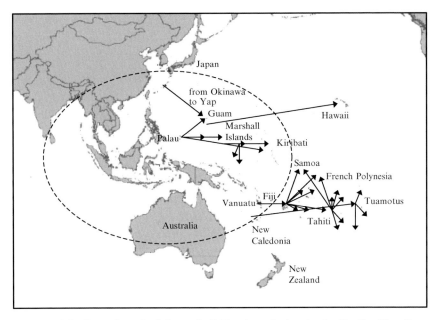

Figure 2.5 Transfers of adult topshell *Trochus niloticus* in the Pacific. The direction of transfers between countries is indicated only once even though numerous separate transfers may have occurred (based partly on information from Eldredge, 1994). The natural range of topshell is within the dashed-line ellipse.

et al., 2002). In some places (e.g., Okinawa, Japan) the boundaries between restocking and stock enhancement, as defined in Chapter 1, are somewhat blurred. On balance, however, most of the initiatives can be regarded as restocking.

2.2.2. Supply of juveniles

Trochus can be mass-produced relatively easily in hatcheries (Heslinga and Hillman, 1981; Isa *et al.*, 1997; Lee and Lynch, 1997). Spawning of wild broodstock is induced by raising water temperature or by irradiating the water with ultraviolet light (UV). The fertilized eggs develop into lecithotrophic (non-feeding) larvae and settle onto hard surfaces covered in diatoms within 3–5 d. They grow to a size of 3–4 mm in 3 mo by feeding on a variety of epiphytic algae, at which point they are robust enough to handle (Isa *et al.*, 1997).

The cost of providing a large surface area for grazing, and the time required to rear juveniles to a size where they escape most predation, can

Table 2.2 Estimates of abundance of topshell *Trochus niloticus* and harvests landed on Aitutaki Atoll, Cook Islands, from 1980–1987 after introduction of the species in 1957 (based on information in Sims, 1988).

Feature of fishery	Year					
	1980–1981	1983	1984	1985	1986	1987
Population estimate	470,000	336,000	339,000	305,000	360,000	385,000
Duration of harvest	15 mo	3 mo	12 d	3 d	No harvest	2 harvests of 24 h
Harvest (tonnes)	~200	35.7	45.7	27.0	No harvest	Day 1: 12.0 Day 2: 33.1
Harvest ratio	No accurate harvest data	31.0%	49.8%	26.5%	No harvest	Day 1: 11.5% Day 2: 36.0%
CPUE (kg/person/d)	No accurate harvest data	7	13	36	No harvest	Day 1: 63 Day 2: 141
Remaining stock	Unknown	232,000	170,000	224,000	360,000	217,000

be considerable (Isa *et al.*, 1997). These impediments, and the general failure of releases involving trochus of <30 mm (Crowe *et al.*, 2002, 1997; Purcell and Lee, 2001), have led to three separate ways of restocking trochus: release of relatively large numbers of small (1–4 mm) juveniles at comparatively low cost; release of juveniles grown on to >40 mm in sea cages to help overcome high rates of predation and translocation of adults (Section 2.2.4). The production of juveniles >40 mm in hatcheries is the most expensive of these options, but the costs can be lowered substantially by using trochus as grazers to reduce the growth of epilithic algae in tanks used to rear giant clams (Clarke *et al.*, 2003) and then transferring them to sea cages, including those used for growing on giant clams (Amos and Purcell, 2003; Clarke *et al.*, 2003). Purcell (2001a) has also shown that trochus can be reared effectively in sea-cage monocultures once they reach a size of 12 mm.

2.2.3. Defects caused by rearing

Juvenile trochus reared in hatchery tanks made from fibreglass have thinner shells than wild individuals of the same size (C. Lee, personal communication 1995). Because predation by fish, crabs, stomatopods and carnivorous gastropods are the main sources of mortality of released trochus (Vermeij, 1976; Nash, 1993; Castell, 1996; Isa *et al.*, 1997), this defect in shell strength could potentially reduce the effectiveness of restocking. The problem has been partly solved by rearing juveniles in tanks with coral rubble 'preseeded' with benthic diatoms to provide the snails with a source of calcium carbonate and additional food supply (Lee, 1997).

Purcell (2002) has identified another difference in shell structure between wild and cultured trochus that has implications for restocking; the spines on the lateral edge of each whorl of wild shells <30 mm are missing in hatchery-reared animals >15 mm (Figure 2.6). The spines are thought to help trochus deter predators by enabling them to 'lock' into crevices. The spines are eventually lost after the juveniles make the transition from living in holes and crevices to grazing on the surface. The absence of spines in cultured trochus >15 mm suggests that hatchery-reared individuals may have lost their hiding adaptations prematurely, which could be a factor in the high mortality of released cultured shells <30 mm described in Section 2.2.4.

Hatchery-reared trochus seem to have few obvious behavioural defects that make them vulnerable to predation. For example, Castell and Sweetman (1997) found that both cultured and wild trochus exuded white mucus that inhibited predation by a carnivorous gastropod. Heslinga (1981a) noted that cultured trochus had the same nocturnal feeding pattern as wild specimens even in the presence of abundant food, and Clarke *et al.*

Figure 2.6   (A) Juvenile wild and cultured topshell *Trochus niloticus* from Western Australia showing loss of protective whorls in the cultured specimens (photo: S. Purcell). (B) A tagged, cultured juvenile topshell on coral rubble in a protective cage in Western Australia (photo: S. Purcell). (C) Adult topshell, with the meat removed, ready for sale (photo: A. Mercier).

(2003) observed that cultured trochus released on a reef flat behaved in a similar way to wild individuals. A behavioural problem that might affect survival is the inability of cultured trochus to right themselves quickly if dislodged from the substratum. In the laboratory, hatchery-reared snails could not restore themselves to an upright position when displaced, whereas 90% of individuals released for 8 d recovered normal body position within 5 min of being disturbed (Isa et al., 1997). This indicates that cultured trochus must learn 'righting' behaviour and may be susceptible to predation if dislodged by rough seas shortly after release.

2.2.4. Release strategies

Considerable effort has been devoted to the release of cultured trochus <30 mm. Habitat preferences of juvenile wild trochus have been studied (Heslinga, 1981b; Castell, 1997; Colquhoun, 2001), and hatchery-reared animals at a range of sizes have been outplanted experimentally into structurally complex intertidal and subtidal habitats in five countries (Castell, 1996; Crowe et al., 1997, 2002; Isa et al., 1997). In these experiments, there was a general trend of better survival with increasing shell size; nevertheless, it is overwhelmingly evident that the number of trochus that survived to reproductive age rarely justified the necessary investment in hatchery rearing. The experiments also revealed that survival of trochus <30 mm was better in habitats exposed at low tide than in tide pools that harbour predators, and that individuals cannot survive if displaced onto sand during rough weather. The disappointing results with juveniles <30 mm have led to the three alternative strategies for restocking trochus mentioned earlier: release of juveniles of 1–4 mm, release of juveniles >40 mm, and translocation of adults.

The release of very small juveniles has been proposed as a means of avoiding the high costs involved in rearing the snails to a size at which they survive well (>35–40 mm) (Crowe et al., 2002) and overcoming the defects that can occur from long-term hatchery culture. It has been tested in Western Australia and shows enough promise to warrant further investigation. After 12 mo, there was a significant but modest increase in the number of juvenile trochus at restocked locations relative to controls (S. Purcell, personal communication 2001). This method depends on the release of very large numbers to ensure that enough of them find suitable shelter and survive to adulthood. Multiple sites with releases of 100,000 per site at densities of ~5 m^{-2}, and protection from predators during the first few days, are recommended for this strategy (S. Purcell, personal communication 2001).

Growing of juveniles to >40 mm is attractive when the costs of production can be reduced substantially by rearing trochus together with giant clams (Clarke *et al.*, 2003), or in cages (Amos and Purcell, 2003). The survival of cultured trochus of ~40 mm released onto reef crests was 76–85% after 4 wk (Clarke *et al.*, 2003) and 36–47% after 6 mo (Purcell *et al.*, 2004). Furthermore, survival was not dependent on the structural complexity of the habitat (Clarke *et al.*, 2003). Because released animals grow at the rate of ~2 mm mo^{-1}, and trochus reach sexual maturity at 55–60 mm (Isa *et al.*, 1997), there is a strong expectation that this strategy will deliver reasonable rates of survival to reproductive age.

The translocation of trochus to restock existing or previous fisheries, as opposed to creation of new fisheries (Gillett, 1986, 1993; Eldredge, 1994), has been limited to Western Australia (Purcell and Lee, 2001) and Vanuatu (M. Amos, personal communication 2001). In Western Australia, adult trochus were held in pens for 3 mo in an attempt to increase spawning success before being allowed to disperse. However, there was little evidence that the abundance of juveniles at sites with broodstock increased relative to natural variation in abundance of juveniles at control sites after 18 mo (S. Purcell, personal communication 2001). In contrast, there was a significant effect of releasing broodstock in Vanuatu. After 14 mo, the average density of juveniles at sites where broodstock were released was 775 ha^{-1}, whereas it was 0–50 ha^{-1} at control sites (S. Purcell, personal communication, 2001). The difference between the two studies was attributed to variation in current regimes, with conditions in Vanuatu being far more conducive to retention of larvae.

The advantages of translocating adults, compared with the release of cultured juveniles, for restocking trochus are clearly backed up by the number of cases where lucrative fisheries have been created by introducing trochus and implementing an initial long-term moratorium on harvesting (Table 2.2). In short, a breeding population can be established in the wild immediately and responsibly if sufficient adults can be obtained from nearby sources. On the other hand, released cultured juveniles may not survive to adulthood in adequate numbers, especially when released at a size of 1–4 mm. Thus, the main reason for restocking with hatchery-reared juveniles would be to overcome any difficulty in obtaining sufficient adults for an initial effective population size.

2.2.5. Use of artificial habitat

In Japan, artificial habitats have been used in an attempt to improve the survival of released trochus. Rectangular concrete nurseries <10 m^2, with depths ranging from 0–60 cm, were constructed on reef flats exposed at low tide to receive cultured juveniles (Kubo *et al.*, 1991). The production of algae in the nurseries was 2–10 times greater than in land-based tanks, and the numbers of predators visiting the enclosures were low. However, other herbivorous gastropods and echinoderm competitors quickly inhabited the structures. The trochus were also at high risk of being washed out of the nurseries by strong wave action. The latter problem was overcome by fitting complex plastic and concrete structures inside the nurseries (Kubo *et al.*, 1992).

2.2.6. Minimising genetic impacts

Genetic relationships were studied in six species within the family Trochidae on the Australian Great Barrier Reef, during which *Trochus niloticus* from two locations were compared (Borsa and Benzie, 1993). Otherwise, little is known about the population structure of trochus. However, considering the scale and frequency of translocations (Figure 2.5), there are likely to be many similarities among stocks of trochus in the Indo-Pacific. The limited number of individuals used to establish populations (e.g., 40 in the case of French Polynesia; Cheneson, 1997) suggests that some of the founded stocks should have limited genetic diversity. The way that trochus have been spread to many locations greatly reduces the likelihood of adverse genetic effects from any future restocking of such populations.

The existence of good records for many of the translocations (Gillett, 1986; Eldredge, 1994) provides the opportunity to look at the effect of founding population size on genetic diversity at multiple sites. Apart from the considerable academic value of such research, a thorough investigation of stock structure throughout the range of the species would enable resource managers to identify particular populations associated with favourable attributes such as growth rate and shell quality. This may promote measures to maintain any existing advantages of populations in particular countries or selective breeding for future releases of trochus in those countries where it has already been introduced.

2.2.7. Disease risks and other environmental impacts

No diseases of juvenile or adult trochus have been reported (Lee and Amos, 1997). However, the shells of trochus are colonized by a wide variety of epiphytic algae and boring and encrusting invertebrates (Nash, 1993). Thus, transfer of animals from one location to another, and from nurseries to the wild, is a potential biosecurity risk unless appropriate measures are taken to destroy the commensal organisms.

None of the introductions of trochus seem to have had a detrimental environmental impact, with the possible exception of the one at Aitutaki Atoll, Cook Islands, where trochus may have competed with the closely related herbivorous gastropod, *Turbo setosus* (Sims, 1985; Gillett, 1986). However, Nash (1993) casts doubt on this possibility because *T. setosus* was still common at Aitutaki Atoll in 1992, some 35 years after the introduction of trochus.

2.2.8. Systems for marking juveniles to assess effectiveness of releases

Trochus released in small-scale restocking experiments have often been marked by gluing plastic tags to the shell or placing marks on the outer or inner nacre (Castell *et al.*, 1996, 2003; Crowe *et al.*, 1997) (Figure 2.6). However, Castell *et al.* (1996) have demonstrated that censuses under-estimate survival of trochus tagged in these ways by 19–30% because of the highly cryptic nature of the juveniles. To overcome this problem, Crowe *et al.* (2001) used a marking system similar to that developed by Glazer *et al.* (1997b) for queen conch (see Chapter 3, Section 3.4.7). Aluminium tags were attached to the shells of trochus as small as 16 mm and recovered using an underwater metal detector. The system was effective for finding cryptic juveniles beneath coral rubble and had no effect on the survival of the animals. The limitations of the system are that the detector has to be within 8 cm of the animal and the tags had an effective operational life of no more than 3 mo, because they became dislodged frequently after that length of time.

The preoccupation with physical tags to assess the survival of released trochus must change if the relative merits of the three release strategies described in Section 2.2.4 are to be evaluated. Instead, efforts are needed to identify a simple genetic mark that will enable assessment of the contribution of released, or translocated, trochus to population size. The fact that F_1 and F_2 trochus have been produced relatively easily in hatcheries (Heslinga, 1981a; Lee and Amos, 1997) makes selection for a conspicuous attribute (e.g., altered shell colour pattern) a possible option for creating a simple genetic tag.

2.2.9. Status of restocking initiatives

The considerable literature on experiments to restock trochus in five countries, including the release of 132,000 individuals in Okinawa in 1990 (Isa *et al.*, 1997), does not provide any evidence that released hatchery-reared juveniles have contributed to the next generation. This may be due to the lack of follow-up surveys or to the inability to discern progeny from released and wild animals. On the other hand, the restocking of 300 adults at two sites in Vanuatu (M. Amos, personal communication, 2001) increased recruitment of juveniles in a way that is unlikely to have been by chance alone and is in line with the outcome of numerous introductions of trochus to create fisheries in the Pacific. This indicates that restocking with adults may be the most successful method for trochus, although it does not work equally well at all locations (see Section 2.2.4).

2.2.10. Management

Apart from the need for a complete moratorium on fishing until densities reach levels where they can provide substantial and sustainable yields, management of restocked trochus fisheries should not differ greatly from the measures listed in Section 2.2.1 for unaugmented stocks. Because fishing for trochus (and sea cucumbers, Section 2.3.1) is among the few options to obtain cash for many coastal communities in the Pacific, the resource should be managed in a socially harmonious way that spreads the benefits widely throughout the community (Amos, 1997; Tuara, 1997). Amos (1997) and Purcell (2004) identify three key elements for managing restocked trochus fisheries: (1) an initial moratorium (probably ≥ 5 yr), (2) dedication of one or more sanctuaries to maintain the supply of larvae to fishing areas and (3) short fishing periods to coincide with the financial needs of communities. The duration of the fishing periods could be regulated in either of two ways. First, 'slot' size limits can be used to conserve at least 30% of egg production (Nash, 1993). In this type of regulation, both maximum and minimum size limits would be set, a minimum size to ensure that trochus spawn before capture and a maximum size to gain the benefit of the greater egg production of larger snails (Nash, 1993). This measure also eliminates the landing of older trochus, which typically have poor shell quality because of boring organisms. Second, by setting a total allowable catch (e.g., 60% biomass) above a certain size limit (Preston, 1997; Tuara, 1997) based on standard methods for estimating population size (Nash, 1993; Cheneson, 1997).

Translocation of adults from restocked areas to nearby areas that do not receive larvae, a method proposed for sea cucumbers (Section 2.3.9), would seem to be particularly suitable for trochus, because their larvae have limited dispersal. Indeed, viable trochus fisheries have been created from the transfer of as few as 40 animals (Cheneson, 1997).

Amos (1997) has raised concerns for management about establishment of factories that process trochus shells into button blanks before export. Although this would seem to bring further economic benefit to SIDS, the minimum numbers of shells needed by competing factories has promoted over-exploitation of stocks in Vanuatu. Clearly, construction of button factories should not be encouraged unless they can be viable under the constraint of intermittent harvests that stem from the short open seasons recommended previously.

Our conclusion that restocking of trochus is best accomplished by the translocation of adults provides an important focus for fisheries managers. It should be possible to avoid the costs associated with restocking simply by protecting remnant trochus until numbers rebuild. Restocking by moving adults or by release of cultured juveniles can then be restricted to locations without viable reproductive populations.

2.2.11. Future research

If it is possible to identify sites where larvae will be retained, the most critical questions for optimizing restocking of trochus through movement of adults or release of cultured juveniles are: (1) How many adults are needed to rebuild the population to productive levels in a reasonable time? (2) How many very small juveniles must be released to achieve the desired number of adults? (3) Do some habitats bestow better survival, faster growth and increased egg production on trochus? and (4) What is the typical dispersal distance of trochus larvae? The answer to this last question will reveal the appropriate scale for restocking activities and establishment of fishing reserves.

2.3. SEA CUCUMBERS

2.3.1. Background and rationale for restocking

Exploitation of sea cucumbers is based largely on high demand for the boiled and dried body wall of these animals in China, where the processed product, known as 'beche-de-mer' or 'trepang', is highly prized for its curative and medicinal properties (Preston, 1993; Chen, 2003, 2004). The body and viscera of some species are also favoured in Japan and Korea (Yanagisawa, 1996). Although more than 1400 species of sea cucumbers have been described, fisheries concentrate mainly on the 20–30 species from the Indo-Pacific that have thick body walls (Conand, 1990; Conand and Byrne, 1993; Holland, 1994; Rowe and Gates, 1995; Hamel et al., 2001). For the purposes of this review, we have chosen to highlight the situation in the Pacific.

Fishing for beche-de-mer is one of the oldest and most widespread forms of commerce in the Pacific Islands (Conand and Byrne, 1993; Holland, 1994). It has brought considerable benefits to SIDS, because harvesting, processing and storage of sea cucumbers is easy and requires no specialised equipment or refrigeration (Preston, 1993). In fact, beche-de-mer provides the most revenue from non-finfish marine resources for several Pacific nations (e.g., the Solomon Islands, Richards *et al.*, 1994) and is one of the few sources of income for some coastal communities.

Despite the importance of beche-de-mer in the Pacific, the fishery no longer yields its full potential because of over-exploitation. The beche-de-mer fishery in this region has followed a 'boom and bust' cycle (Battaglene and Bell, 2004 and references therein; Figure 2.7). In many places, the initial phase of exploitation reduced the resource to the point at which there was little capacity for replenishment, and it often took 3–4 decades before stocks recovered to levels at which profitable harvests were again possible. The long time frames needed for recovery of beche-de-mer fisheries are because sea cucumbers can be harvested to very low levels, and there is often limited scope for replenishment from elsewhere because of the isolation of islands and the relatively short larval periods (Battaglene and Bell, 2004). In cases of extreme over-fishing in remote locations, stocks have been virtually extinguished. For example, surveys in Chuuk (Truk) in the northern Pacific in 1988 found only two individuals of one edible species of sea cucumber 50 years after thousands of tons were exported to Japan in the late 1930s (Richmond, 1997).

Artisanal fisheries tend to over-harvest sea cucumbers because of easy access to the resource and high prices (Conand and Byrne, 1993; Preston, 1993; Holland, 1994). Over-exploitation has been exacerbated by lack of good information for managers. The number of species targeted, the remote and artisanal nature of the fishery and the difficulty in distinguishing between some sea cucumbers once they are processed into beche-de-mer makes it difficult for SIDS to collect those data (e.g., catch-per-unit-effort, growth, mortality) normally used to manage marine fisheries on a sustainable basis (Shelley, 1986; Conand and Byrne, 1993; Lokani, 1995). Often, the only available data on species composition and size frequency of wild stocks are those provided by exporters (Adams, 1992; Holland, 1994; Lokani, 1995).

As a result, management regimes, or recommendations for management, for the beche-de-mer fishery in many Pacific nations have been based on scant knowledge of the size at sexual maturity for the numerous species and measures aimed broadly at preventing 'recruitment' and 'growth' over-fishing (Adams, 1992; Holland, 1994; Lokani, 1995). These measures include size limits, prohibition on collection by divers using compressed air, closed seasons and marine protected areas. Such regulations have been of limited success (Battagelene and Bell, 2004 and references therein) mainly because

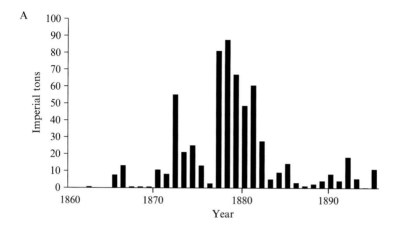

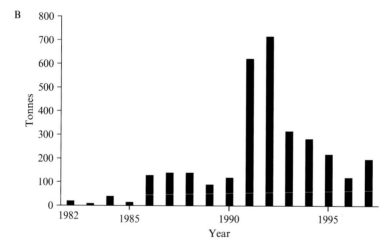

Figure 2.7 Variation in export of beche-de-mer from Solomon Islands: (A) 1854–1896 and (B) 1982–1997 (redrawn from Battaglene and Bell, 2004).

SIDS often lack the financial and human resources to enforce them. The diverse social and coastal tenure systems of many Pacific nations also complicate the application of regulations (Conand and Sloan, 1989; Lokani, 1995; Aswani, 1997).

   Provided SIDS can increase their capacity for management of inshore marine resources, there is now general recognition that the fastest way to restore the productivity of sea cucumber fisheries in certain situations could be to rebuild the stocks by releasing cultured juveniles (Battaglene and Bell, 2004). This approach has been inspired to some extent by the knowledge that

Apostichopus japonicus has been produced and released in stock enhance-ment programmes in Japan (Yanagisawa, 1996, 1998) and that it has been mass-produced in hatcheries in China for many years (Lovatelli *et al.*, 2004).

The sandfish *Holothuria scabra* (Figure 2.8) has been identified as the species most suitable for restocking in the Pacific because of the relative ease of producing juveniles and because it is among the most valuable of the tropical sea cucumbers (Battaglene and Bell, 1999, 2004). Consequently, most of this section refers to information for *H. scabra*. The stock enhance-ment of *A. japonicus* in Japan, and the major farming and 'put-and-take' sea ranching activities for *A. japonicus* in China, are covered briefly in Section 2.3.11.

2.3.2. Supply of juveniles

The hatchery methods for propagating sandfish pioneered by James *et al.* (1994) have been modified by Battaglene (1999), Battaglene *et al.* (1999) and Pitt and Duy (2004) to provide reliable technology for producing juvenile sandfish. The larvae can be reared on a variety of microalgae routinely available in hatcheries. The juveniles can be grown to a size suitable for release in the wild ($\sim$60 mm and 15 g) in 4 mo with addition of little or no food by stimulating the availability of organic matter, diatoms and bacteria (Pitt, 2001; Battaglene and Bell, 2004; Pitt and Duy, 2004; and references therein). The weak points in the process are the reliance on wild or non-acclimated broodstock, with the result that larvae cannot be produced year-round in all areas, and the high levels of mortality at first feeding and settlement. Scaling-up production of juvenile *Holothuria scabra* to rear the hundreds of thousands or millions of individuals needed for effective restocking programmes has yet to be achieved.

Scaling-up the production of juveniles is perhaps the most challenging part of the work remaining to be done. At a putative release size of 15 g, and with stocking densities that do not inhibit growth ($\sim$225 g m^{-2}), a large surface area of tanks or ponds is required to culture the necessary numbers of juveniles. Battaglene and Bell (2004) have suggested that this may be possible by combining the culture of sandfish and shrimp, thereby saving the costs of constructing and managing rearing ponds. This may be a possibility even for farms dedicated to rearing *Penaeus monodon,* because juvenile sandfish can tolerate salinities as low as 20 ppt (Mercier *et al.*, 1999). In addition to producing sandfish 'seed' for restocking, such 'polyculture' may have other advantages for the shrimp farmers. In particular, the feeding and burrowing behaviour of sandfish may help assimilate excess nutrients and bacteria toward the end of the production cycle, when eutrophication of pond water is a common problem. The potential for such polyculture should be

Figure 2.8 (A) Fisherman with his catch of sea cucumbers in New Caledonia (photo: S. Purcell). (B) An adult sandfish *Holothuria scabra* (photo: S. Purcell). (C) Processing sandfish in New Caledonia by drying them on racks (photo: S. Purcell). (D) Seagrass meadows in New Caledonia, a suitable release habitat for juvenile sandfish (photo: S. Purcell).

revealed by experiments to determine whether there are any negative interactions between shrimp and sandfish. A significant decrease in growth of
sandfish because of buildup of ammonia-N can occur when they are reared
with the shrimp *Litopenaeus stylirostris* in laboratory tanks (S. Purcell,
personal communication 2005). Pitt *et al.* (2004) have shown potential for
combining the culture of sandfish and the shrimp *Penaeus monodon.* They
found no adverse effects of co-culture on the shrimp, although in a minority
of cases shrimp did prey on the sandfish when water quality was poor and
the sea cucumbers became stressed. Their experimental treatments were
unreplicated, however, and so the results must be regarded as preliminary.
Overall, more research is needed to determine whether these results hold true
over a wider range of sizes for both sandfish and shrimp and during growing
on in commercial-scale earthen shrimp ponds.

   The alternative to ponds for rearing cultured juveniles is a farming method
developed in India and Indonesia, where sandfish are held in pens and
cages on the seabed and fed various waste products including chicken
manure and bottom sludge from shrimp ponds (Muliani, 1993; James,
1994, 1996). However, such methods would make it difficult to mass-
produce the very large numbers of individuals needed for restocking
programmes.

2.3.3. Defects caused by rearing

Little is known at this stage about the consequences of rearing *Holothuria
scabra* in hatcheries, because observations on the survival of cultured juveniles in the wild are rare. Dance *et al.* (2003) reported intense predation of
juveniles in one of two habitats (see Section 2.3.4) and questioned whether
the production of saponins by the sea cucumbers, presumed to deter
predators, may have been affected by rearing conditions in the hatchery.

   Cultured sandfish do not seem to have important behavioural deficits. The
juveniles have a strong diel burrowing cycle (Mercier *et al.*, 1999) and, when
released into seawater ponds with a good food supply, attained a size of
400 g and produced eggs within 12 mo of release (Pitt and Duy, 2004).

2.3.4. Release strategies

Research is underway to determine optimal release strategies for juvenile
cultured sandfish by identification of the habitats, times, sizes and stocking
densities that result in the greatest proportion of released animals surviving
to reproduce (Purcell *et al.*, 2002). A concerted effort to learn about
the ecology of juvenile sandfish in preparation for such experiments is

summarised in the review of the biology and ecology of sandfish by Hamel *et al.* (2001). We now know that sandfish use seagrass beds as nursery habitats (Figure 2.8); pentactula larvae settle preferentially on seagrass blades and remain there for about a month before descending to the substratum. Once they reach a size of 10–40 mm, juvenile sandfish start diel burrowing behaviour, burrowing around sunrise and emerging in the evening. The juveniles actively select organically rich sediments and are tolerant of salinities as low as 20 ppt. Daily movement is limited, averaging <1 m d^{-1}.

In a preliminary experiment on factors affecting survival of released sandfish, Dance *et al.* (2003) placed juvenile cultured sandfish with a mean size of 35 mm on soft substrata near mangrove-seagrass, and lagoonal coral reef flat habitats in the Solomon Islands. Mean recovery at the mangrove seagrass sites was 95–100% 1 h after release and 70% 3 d later. At the coral reef flat sites, however, mean recovery was as low as 37.5% after 1 h, and 100% loss, which was attributed to predation by fish, occurred in two of the three releases within 48 h. Dance *et al.* (2003) also demonstrated that survival of juvenile sandfish was improved significantly by caging in 8-mm mesh. However, they did not continue the experiments for long enough to determine whether high rates of predation occurred for larger individuals in the coral reef flat habitat once the cages were removed.

A recent study by Purcell *et al.* (2005) has shown that stress caused by handling, and long duration times during transport of juveniles to release sites, can inhibit burrowing behaviour of cultured sandfish on release. Protection from predators until the normal diel burrowing habit is established may be necessary.

2.3.5. Minimising genetic impacts

Genetic analysis of sandfish from several locations in Bali, Indonesia; the Great Barrier Reef and Torres Strait in Australia; the Solomon Islands and New Caledonia have revealed significant differences in genetic diversity among populations (Uthicke and Benzie, 2001; Uthicke and Purcell, 2004). Remarkably, one of the two populations studied in the Solomon Islands was more closely related to a population in Australia than it was to the other one in the Solomon Islands, only 20 km away. Significant variation has also been recorded at relatively small spatial scales (60–130 km) in New Caledonia (Uthicke and Purcell, 2004). However, at the scale of sites, Uthicke and Benzie (1999) found no significant variation in gene frequencies between sandfish from shallow and deep water or between black and brown colour morphs.

On the basis of the distinct genetic structuring of sandfish populations, Battaglene and Bell (2004) have stressed the importance of releasing cultured juveniles in the same areas as broodstock are collected. They also highlight the problems of small effective population sizes when sandfish are induced to spawn in hatcheries (as few as 10% of individuals shed gametes) and propose the same measures listed in Section 2.1.5 for conserving the genetic diversity of wild stocks during restocking programmes.

2.3.6. Disease risks and other environmental impacts

In the Pacific, the only disease found during the hatchery production of *Holothuria scabra* has been the occasional occurrence of skin peeling in 3-mo-old individuals (Battaglene and Bell, 2004). The cause of this disease, which is brought on by overcrowding, is thought to be ciliated protozoa. There is no known treatment for the skin peeling condition, but it does not seem to be highly contagious and can be avoided through good husbandry. In Madagascar, however, a contagious disease caused by a bacterial infection resulted in mortality of hatchery-reared sandfish within 3 d (Eeckhaut *et al.*, 2004). The symptoms first appear as a white spot on the integument close to the cloacal aperture, which then spreads rapidly.

Battaglene and Bell (2004) have recommended the following quarantine procedures before release of juvenile sandfish: (1) removal of damaged or sick animals during the culture period; (2) placement of juveniles in tanks with hard surfaces, flow-though seawater and aeration for 48 h before release to evacuate intestines of sand, combined with regular removal of sand and faecal material from the tanks; (3) examination of a subsample of individuals for external parasites and diseases before release and (4) removal of any algae or invertebrates associated with juveniles taken from quarantine tanks to avoid translocation of other organisms to the wild.

No other environmental impacts have been identified for the release of cultured sandfish. Indeed, a major reason why restocking is being assessed as a potential management tool for this species is that increased numbers of sea cucumbers are unlikely to have detrimental effects on other important fauna. The two factors that support this assertion are that sandfish occur naturally at very high densities, e.g., 13,500, 6,000 and 2,562 ha^{-1}, for Papua New Guinea, New Caledonia and Torres Strait, respectively (Shelley, 1986; Conand, 1990; Lokani *et al.*, 1996), and that they feed low in the food chain, burrowing in mud and sand to process organic matter, bacteria, diatoms, cyanophyceans and foraminiferans (Wiedemeyer, 1992; Mercier *et al.*, 1999). Sea cucumbers in general are believed to play a beneficial role for assemblages associated with soft substrata through bio-turbation of sediments (Hammond, 1982; Massin, 1982; Wiedemeyer, 1992; Mercier *et al.*, 1999).

2.3.7. Systems for marking juveniles to assess effectiveness of releases

Sea cucumbers have so far proved difficult to tag, although Conand (1989) has reported tag retention rates for species with thick body walls that are probably good enough for estimating short- to medium-term survival rates for larger size classes. The suitability of physical tags for estimating survival of released, cultured juveniles still remains to be demonstrated. However, even if a physical mark can be developed, it would not ultimately be appropriate for assessing the contribution of released sandfish to restoration of over-fished stocks. A genetic tag, capable of being retained through several generations, is needed for this purpose. Dependence on the use of wild broodstock for production of juveniles rules out selection of individuals with rare alleles to create a genetic marker. The only simple genetic tag for sandfish would be the use of the black morph in breeding programmes. However, Battaglene and Bell (2004) list some potential problems with this approach: black adults are not common in the wild so accumulation of sufficient broodstock may be difficult, and because the pattern of inheritance for the black morph is unknown, there is no guarantee that progeny derived from released animals will be easy to detect. Alternatively, it might be possible to use a rare mitochondrial DNA marker (see Chapter 7, Section 7.1).

Because of the time and cost required to identify genetic tags for sandfish, it may be more prudent to rely on other methods to assess the effects of releases on restoration of stocks. The application of sampling programmes designed to assess impacts, developed by Underwood (1992, 1995), to restocking and stock enhancement programmes is described by Munro and Bell (1997), Battaglene and Bell (2004) and Purcell (2004b).

2.3.8. Status of restocking initiatives

Development of technology for restocking sandfish has progressed to the point where basic methods for propagating juveniles are available but, as outlined in Section 2.3.2 and 2.3.4, methods are now required to scale-up the production of juveniles cost-effectively and for releasing cultured juveniles in the wild at high rates of survival.

Strehlow (2004) has used a socio-economic approach to evaluate the financial benefit of a restocking programme in Vietnam. Using estimates for the cost of producing juvenile sandfish in a hatchery and different rates of survival for released animals and their progeny, he modeled the costs and time needed for recovery of a local fishery (assumed to be at a negligible level). The model allowed fishing within certain limits. Under a scenario in which 15% of released sea cucumbers survive and with conservative levels of spawning by released animals and their progeny before capture, re-establishment of a local population approaching 1 million animals is estimated to take ~10 yr.

2.3.9. Management

Battaglene and Bell (2004) stress the need for complete protection of released sandfish and the remnant wild stock until their progeny have replenished the population to the point at which substantial harvests can be sustained each year. Where resources are only sufficient to produce enough juveniles for releases in one area for each self-replenishing population and dispersal of their larvae is predicted to cover only a small part of the available habitat, Battaglene and Bell (2004) advocate translocation of some progeny from the restocked area to depleted areas within the range of the population (Figure 2.9). Such a scheme would spread sandfish to many areas of available

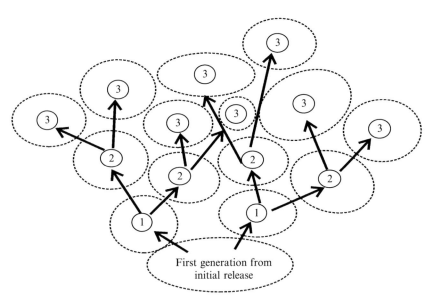

Figure 2.9 Scheme for translocation of the progeny of sea cucumbers from restocked to depleted areas to distribute juveniles to all suitable habitats. Numbers refer to the generation derived from an initial release of cultured animals. Broken lines denote area of larval dispersal of progeny of restocked sea cucumbers (redrawn from Battaglene and Bell, 2004).

habitat in three generations if two translocations of progeny were made from each location.

2.3.10. Future research

According to Battaglene and Bell (2004), the remaining major challenges for restocking of sandfish are: (1) learning how to scale-up the rearing techniques to produce the hundreds of thousands or millions of juveniles needed to have an impact on the abundance of wild populations, (2) developing effective strategies for releasing cultured juveniles and (3) assessing the economic viability and cost-benefit of restocking.

The preliminary work on polyculture of *Holothuria scabra* in shrimp ponds (see Section 2.3.2) indicates that it may be possible to produce juveniles *enmasse* cost-effectively in this way. However, much additional work is needed to determine the optimal sizes of sandfish and shrimp to grow together to determine whether there are any deleterious interactions between the species and to identify any synergies in the process that may be advantageous to shrimp farmers. In the event that polyculture with shrimp is not possible, development of more intensive feeding regimes during the nursery phase will be needed to reduce the pond area required for large-scale production.

Research on the best habitats for releasing sandfish is underway. The imperative is to do well-designed field experiments to identify the optimal size at release, stocking density and time for release. Further work on determining whether provision of shelters for released juveniles increases survival would also be beneficial.

Assessment of the viability and cost-effectiveness of restocking programmes for sandfish will require careful comparisons of the costs of producing juveniles, their rates of survival in the wild and the relative merits of this way of increasing spawning biomass compared with other forms of management.

2.3.11. Stock enhancement, farming and sea ranching of *Apostichopus japonicus*

In contrast to the initiatives underway in the Indo-Pacific to develop methods for restocking *Holothuria scabra,* cultured juveniles of *Apostichopus japonicus* are routinely released in stock enhancement programmes in Japan. The total output from the 17 hatcheries producing *A. japonicus* exceeds 2 million juveniles of 8–10 mm per year (Yanagisawa, 1998). However, total hatchery production is limited compared with other marine invertebrates in Japan (Imamura, 1999) because of the relatively high cost of rearing juveniles using powdered algal diets.

In Japan, the juveniles are released on artificial habitat because they are highly vulnerable to predation. Reefs with a good cover of seaweed are preferred (Yanagisawa, 1998). Survival rates of 15% have been obtained for juveniles of 13–16 mm after 6 mo by placing them in purpose-built concrete enclosures with artificial algae (Ito *et al.*, 1994), but comparable survival was also observed for juveniles of 2.5 mm released on 'artificial rock' (Yanagisawa, 1998). The direct cost of producing juveniles has not yet been integrated with the costs of labour and hatchery facilities in Japan to determine whether the additional harvest stemming from released *Apostichopus japonicus* results in a net economic benefit. However, the great value that can be added to *A. japonicus* by salting the entrails to produce 'konowaba' provides considerable potential for stock enhancement of this species to be profitable in Japan.

China has taken a different, highly successful approach to increasing the production of *Apostichopus japonicus*. This is well described by Chen (2004) and other chapters in the volume edited by Lovatelli *et al.* (2004), and summarised briefly in the following. Two systems are used for producing these sea cucumbers in China: farming in earthen ponds and pens in the sea, and 'put-and-take' sea ranching. For farming, hatchery-reared juveniles are grown in converted shrimp ponds bordering the Yellow Sea in northern China. The ponds are usually 1–4 ha in size and prepared by installing numerous small (~3 m diameter) stone reefs to provide suitable habitat for *A. japonicus*, which is normally associated with hard substrata. Juveniles are placed into the ponds at a size of 20 mm and a stocking density of 10–15,000 ha^{-1}. The rate of survival at harvest size 12–18 mo later is 20–30%.

For the sea ranching operations, larger stone reefs are constructed in coastal areas leased from the government by private companies. Juveniles of 20–50 mm are placed on reefs by divers and then harvested when they reach market size 18 mo later. This activity accounts for 75% of total production of *A. japonicus* in some provinces but does not constitute either restocking or stock enhancement. As mentioned previously, it is mainly a put-and-take operation. Nevertheless, it is likely that some spawning occurs before all the sea cucumbers are removed from the reefs where they were placed. This presumably helps to maintain the wild population in the vicinity of the sea ranching operations. Sea ranching has also taken the pressure off the wild fishery; several reserves have been established to protect wild stocks, and Chen (2004) reports that the biomass of wild populations is now 'close to optimal'.

In recent years, there has been a staggering increase in the production of *Apostichopus japonicus* in China. More than 1 billion juveniles are produced in hatcheries each year, and total harvests in 2003 were estimated at 6,750 t dry weight (~200,000 t wet weight), which approximates the total harvest of sea cucumbers from the rest of the world (Chen, 2004).

However, the intensity of production is now causing concern that emerging disease problems may intensify, resulting in serious economic losses. A concerted effort is currently underway to isolate and identify the various bacteria, fungi and parasites involved (Wang *et al.*, 2004). Interestingly, the price for processed *Apostichopus japonicus* in China has increased along with production. This has been attributed to greater demand for this traditionally prized product from an increasingly affluent population.

Chapter 3

Stock Enhancement Initiatives

ADVANCES IN MARINE BIOLOGY VOL 49   0065-2881/05 $35.00
DOI: 10.1016/S0065-2881(05)49003-3

3.1. SCALLOPS

3.1.1. Background and rationale for stock enhancement

Trawling or dredging for scallops (Pectinidae) in shallow coastal and marine waters supports important fisheries worldwide (Shumway, 1991). Ten species have yielded harvests >5,000 tonnes wet weight per year, with yields from several species exceeding this by 1–2 orders of magnitude (Shumway and Castagna, 1994). The countries with important scallop fisheries are Japan, the United States, Canada, China, Norway, Chile, Mexico, Argentina, United Kingdom, France, New Zealand and Australia (Shumway, 1991).

Scallop fisheries have two things in common. First, there is great inter-annual variation in the natural supply and successful settlement of juveniles. This leads to large differences in production from year to year, because the supply of juveniles regularly falls short of the carrying capacity of the habitat (Orensanz et al., 1991; Robinson, 1993; Dao et al., 1999; Kitada, 1999). The historical production of *Patinopecten yessoensis* from Mutsu Bay in Honshu, Japan, before the development of aquaculture, is a typical example of this variation (Figure 3.1). Second, there has often been severe over-fishing (Ventilla, 1982; Caddy, 1989; Arbuckle and Drummond, 2000). Once again, the fishery for *Patinopecten yessoensis* in northern Japan provides a pertinent example: production in Sarufutsu, Hokkaido, steadily declined from >20,000 y^{-1} in the early 1900s to relatively trivial catches from 1955–1973 (Figure 3.1).

The good news for several scallop fisheries, however, is that a cost-effective method has been developed to stabilize production at levels that can exceed historical catches. The technology was developed in Japan and represents the most successful form of stock enhancement currently practiced for marine invertebrates. It had its origin in the 1930s, when trials commenced to collect scallop spat on cedar twigs suspended from nets. These methods progressed to the use of synthetic collectors in the 1960s and finally to the major breakthrough, an 'onion bag' to surround the collector with fine mesh, allowing spat to enter and settle but retaining them when they detach at a size of ~10 mm (Ventilla, 1982). This innovation improved the catch rate of spat dramatically and made the culture of scallops economically viable. The spat were used both for growing on to market size in hanging culture and for 'sowing' on the seabed to restore and then enhance the capture fisheries

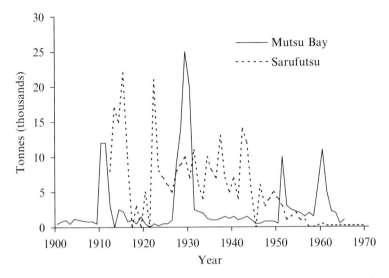

Figure 3.1 Production of the scallop *Patinopecten yessoensis* at two locations in Japan: Mutsu Bay, Honshu (data from Aoyama, 1989) and Sarufutsu, Hokkaido (data from Masuda and Tsukamoto, 1998) before stock enhancement and/or hanging culture.

(Ventilla, 1982; Masuda and Tsukamoto, 1998; Kitada, 1999 and references therein). This basic technique has now been applied in many countries (Shumway, 1991), although in some areas where wild spat cannot be collected cost-effectively (e.g., the United States and France [Tettelbach and Wenczel, 1993; Shumway and Castagna, 1994; Dao *et al.*, 1999]), seed for restocking and stock enhancement are produced in hatcheries. In such cases, the scale of stock enhancement is limited. For example, in France, the largest single releases have involved production of 5 million postlarvae resulting in a harvest of 100 tonnes of live scallops (Dao *et al.*, 1999).

Stock enhancement of scallops stands out among marine invertebrates for the nature and scale of the operations. Efficient and sustainable ways have been developed for managing the enhanced resource by co-operatives and the private sector. Japan and New Zealand are leading the world in stock enhancement of scallops, and the following summary draws heavily on examples from these two countries.

3.1.2. Supply of juveniles

Most stock enhancement programmes for scallops use the Japanese 'onion bag' collecting system in which fine-meshed bags containing monofilament gill net or a similar material are suspended from longlines (Figures 3.2

and 3.3) to obtain seed for release in the wild. The deployment of these spat collectors is often a finely-tuned process. In Mutsu Bay, Japan, for example, the main settlement peak is predicted by careful monitoring of the gonadosomatic index of mature scallops, water temperatures, and the size–frequency distribution, density and distribution of the planktonic veliger larvae (Ventilla, 1982; Ito, 1990). Spatfalls in Japan vary enormously (from <1,000 to >100,000 spat per bag); 1,500 spat per bag is considered

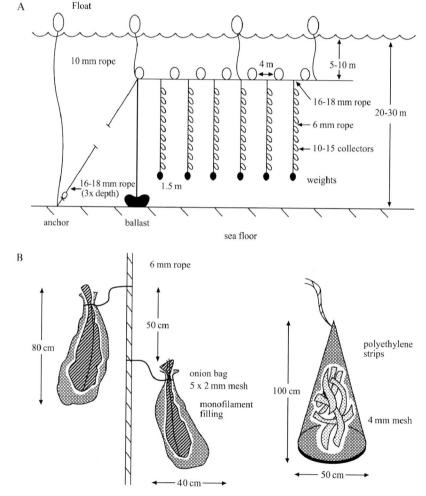

Figure 3.2 Basic designs for (A) long-lines used to suspend spat collectors for scallops and (B) 'onion bag' systems for collecting spat (redrawn from Ventilla, 1982).

Figure 3.3 (A) A typical pectinid scallop *Pecten fumatus* from Australia (Photo: R. Kuiter), (B) spat collectors within 'onion bags' (photo: M. Kurata), (C) juvenile scallops growing in a lantern net (photo: M. Kurata) and (D) scallops *Patinopecten yessoensis* caught by dredging in Hokaiddo, Japan (photo: M. Kurata).

optimal. The spat are removed from collectors at 3–5 mm and held in 'pearl' or 'lantern' nets (Figure 3.3) in intermediate culture until they reach ~30 mm (Ventilla, 1982; Aoyama, 1989). They are then moved over periods up to 50 h from good sources of spatfall to good growing areas, which may be hundreds of kilometers away. Survival rates during transport are 80–90%, provided temperatures are kept constant below 8°C, and the spat remain moist (Ventilla, 1982). The intermediate culture phase takes ~10 mo (Table 3.1), and survival is generally ~90% (Ventilla, 1982). The juvenile scallops are then ready for release on the seabed. In New Zealand, spat collectors are also deployed when data from monitoring plankton indicate that larvae are ready to settle. However, there is no intermediate culture of spat in New Zealand because of fouling problems (Bull, 1994). Instead, they are retained in the collecting bags until they reach 20–30 mm and then placed on the substratum (K. Drummond, personal communication 2001).

3.1.3. Defects caused by rearing

A major advantage of using wild spat for stock enhancement of scallops is that the seed have no inherent morphological, behavioural or genetic deficits. The one possible exception is that ~10% of seeded *Patinopecten yessoensis* harvested from the seabed in Japan have deformed shells (Ventilla, 1982). This is apparently a consequence of intermediate culture of juveniles, because 30% of scallops retained permanently in hanging culture also have this problem. The shell deformation has little effect on survival, which exceeds 90% in hanging culture, and minimal effect on marketing, because much of the crop is sold as boiled or dried meat (Ventilla, 1982).

Concerns have been raised about the quality of hatchery-reared juvenile scallops in Europe, particularly the strength of their shells (Haugum et al., 1999) and their ability to avoid predators by swimming and burrowing in the sediment (Fleury et al., 1996, 1997). Haugum et al. (1999) demonstrated that the crab *Cancer pagurus* was able to crush a significantly greater number of hatchery-reared than wild scallops of 45–65 mm. Fleury et al. (1996, 1997) recognized two main effects on the 'vitality' of juveniles: conditions in the hatchery and stress associated with transport to the release site. An effective measure of vitality for juvenile scallops has yet to be developed; however, Fleury et al. (1997) suggest that the adenylic energy charge in smooth muscle, which represents available energy reserves, is a possible measure. They have demonstrated that it decreases consistently for 3 days after release of hatchery-reared scallops of all sizes. To ensure that cultured juveniles have sufficient energy reserves to evade predators on release, Fleury et al.

Table 3.1 Main stages in the development, collection and growing of the scallop *Patinopecten yessoensis* in stock enhancement programmes in northern Japan (after Aoyama, 1989).

			Stage				
	Spawning	Larvae/trochophore	Attached spat	Benthic spat	Juvenile	Young	Adult
Month	March	April	May	July	Oct.–Dec.	May	Oct.–Dec.
Age			40 d	4 mo	8–10 mo		20–22 mo
Shell length			300 μm	10 mm	30 mm	6 cm	10 cm
		Spat collection	← →				
				Intermediate rearing ← →			
						Bottom culture ←	

(1997) recommend acclimation of spat at release sites before transfer to the bottom. Overall, however, survival of wild and hatchery-reared scallops in Europe is comparable (Dao *et al.*, 1999), indicating that transport and handling of seed before release has a more pronounced effect on survival than origin of the spat. The effects of transport and handling on survival have also been recognized in New Zealand (see Section 3.1.4).

3.1.4. Release strategies

There are three basic strategies for release of scallop spat captured on collectors: natural release, direct release and release after intermediate culture (Robinson, 1993). Natural release involves use of collectors without onion bags so that spat drop off when the byssus deteriorates at a shell length of ~10 mm. The resulting juveniles are then harvested from the seabed after about 8 mo and transported to areas designated for stock enhancement. An average rate of survival of 15% has been reported for this method in New Zealand (Bull, 1994), and higher densities of scallops were observed near experimental sites in Canada where this technique was used (Robinson, 1993). Spat can also fall from the outsides of 'onion bags', longline ropes, etc. Indeed, large concentrations of small scallops on the seabed at spat collection sites have proved to be a significant additional source of seed in New Zealand (K. Drummond, personal communication 2001). Direct release involves the use of collectors with onion bags and deliberate placement of the spat directly on the seabed once they reach a size where they are less vulnerable to predation. This strategy has been necessary in New Zealand, because heavy fouling of lantern nets has made intermediate culture impractical. Survival rates for direct release in New Zealand have been estimated at 15% (Bull, 1989), but this relatively low survival is offset to some extent by considerable savings in labour and equipment. The intermediate culture of spat in lantern nets is the only strategy that provides good control over all factors considered to be important for releasing scallops, namely size of seed, time of release, release habitat and stocking density (Fleury *et al.*, 1997).

Stock enhancement programmes for scallops in Japan use intermediate culture, not only because many of the steps involved are the same as those required for large-scale production of scallops in hanging culture but also because size at release has a marked effect on survival (Ventilla, 1982; Robinson, 1993 and references therein). After 30 years of releasing juvenile scallops in Japan, many lessons have been learned about how to maximize survival and growth. Initially, in Mutsu Bay, for example, scallops were sown on the seabed at densities of 6–10 m^{-2}, with another 6 shells m^{-2} in hanging culture above them (Ventilla, 1982). However, 'unexplained' mass mortalities and a series of toxic red tides between 1975 and 1980 wreaked havoc with production, which slumped by >50% in the late 1970s (Ventilla, 1982). Eventually, maximum stocking densities were set at 5–6 scallops m^{-2} for both hanging culture and stock enhancement on the seabed combined (Ventilla, 1982; Aoyama, 1989). This is now standard practice in northern Japan and is comparable to maximum densities used in Europe (Dao *et al.*, 1999), although 3–4 scallops m^{-2} are used as a stocking density in New Zealand (K. Drummond, personal communication 2001). Stocking at densities >5–6 scallops m^{-2} in Japan reduces growth rate and extends the time to harvest past the preferred 2.5–3.5 yr (Ventilla, 1982). Extensive experience in Japan has also shown that acceptable rates of survival depend on releasing scallops at a size of ~30 mm and that growth and survival are best at depths <40 m on sediments with a low mud content (Ventilla, 1982). The importance of selecting suitable substrata has also been stressed by Thomson *et al.* (1995) and Barbeau *et al.* (1996); otherwise scallops may disperse or die because of sedimentation or wave action.

Release strategies vary slightly among stock enhancement programmes using different species of scallops. For example, a relatively small seeding size is used for *Pecten novaezelandiae* in New Zealand. However, several programmes have recognized that the major factors affecting survival of scallops during release are handling and transport to the release site; scallops are more sensitive to temperature, exposure to air and handling stress than other bivalves (Dao *et al.*, 1999). This applies to transfer of spat from collectors to lantern nets for further growth (as described in Section 3.1.2) and transport of seed before release. For instance, full immersion of juvenile *Pecten novaezelandiae* from the time they are removed from onion bags until they are placed on the seabed has improved survival by 50% and reduced the number of spat that need to be released (Arbuckle and Metzger, 2000).

Removal of predators is another widespread strategy used to enhance survival of released scallops. Predation is the major cause of mortality in wild and seeded scallops (Lake *et al.*, 1987; Minchin, 1991; Barbeau *et al.*, 1994; Nadeau and Cliché, 1998a) and varies significantly among sites (Barbeau *et al.*, 1996). Predators can also trigger an escape swimming response in scallops, causing them to disperse (Peterson *et al.*, 1982), and dispersal can be as great as 10 km yr^{-1} in some species (Posgay, 1981). Indeed, abundance and persistence of predators are prime factors determining the suitability of sites for stock enhancement of scallops (see Section 3.1.8). In Japan and New Zealand, starfish are the major predators (Ventilla, 1982), whereas crabs cause the most problems in Europe (Minchin, 1991; Fleury *et al.*, 1997; Strand *et al.*, 2004), and both starfish and crabs prey on scallops in the United States and Canada (Elner and Jamieson, 1979; Tettelbach and Wenczel, 1993; Barbeau *et al.*, 1994; Hatcher *et al.*, 1996; Nadeau and Cliché, 1998a).

Starfish are removed by dredging before release of scallops in Japan. This method is effective at reducing densities of up to 300 ha^{-1} by five-fold and has been promoted by incentive schemes organized by fishery co-operatives (Ventilla, 1982). Predation by crabs can be combated by trapping the crabs (Minchin, 1991; Spencer, 1992; Fleury et al., 1997), releasing scallops during periods of low water temperature when the predators are less active (Barbeau et al., 1996), placing scallops within fenced areas to exclude crabs (Strand et al., 2004) and rearing seed to larger sizes (Fleury et al., 1997). The latter technique is far more effective against starfish than crabs (Barbeau et al., 1996). Attempts to protect scallop broodstock in cages for a restoration project off Long Island, New York, were only partially successful, because starfish everted their stomachs through the mesh and preyed on the enclosed adults (Tettelbach and Wenczel, 1993).

3.1.5. Minimising genetic impacts

In Japan, a study on *Patinopecten yessoensis* by Fujio and von Brand (1991) supports the view that the effects of stock enhancement on genetic diversity are reduced when wild juveniles are used. They confirmed that the Okhotsk coast of Hokkaido was a mixing zone for spat from different populations and that juveniles collected from this area were representative of wild stocks. They also demonstrated that the proportion of homozygotes in seeded populations was high initially, but decreased over time, and was much lower 2–3 yr after release of juveniles. From this, we infer that although spat collection methods allow 'unfit' homozygotes to survive initially, selective mechanisms that favour survival of heterozygotes in scallops (Beaumont and Zouros, 1991) are not interrupted by the stock enhancement process, and normal levels of heterozygosity occur by the time scallops reach spawning size.

Although there are no problems using spat collected locally for stock enhancement of scallops, important population attributes may be lost in species with restricted gene flow (e.g., *Chlamys opercularis* in Europe; Beaumont, 1982) if spat are moved among locations. In the worst-case scenario, introgression of genes from animals introduced into the local population may affect important adaptations such as synchrony of the reproductive cycle with larval food supply. Thus, scallops imported from another region may not spawn at the best times for larval survival and therefore may contribute little to replenishment. The latter scenario is supported by the work of Mackie and Ansell (1993) and Cochard and Devauchelle (1993), who showed that differences in reproductive cycles between populations of *Pecten maximus* in Europe were maintained even when juveniles were transplanted to other locations. Petersen et al. (1996)

also noted that geographic variation in spawning times among populations of *Argopecten irradians* in North Carolina may render transplanted
scallops unfit for stock enhancement or restocking programmes.

Interestingly, electrophoretic techniques failed to detect important genetic
differences among European populations of *Pecten maximus* (Beaumont
et al., 1993), highlighting the need to use a large number of loci for such analyses
and to confirm any apparent lack of gene flow from mitochondrial or nuclear
DNA analyses with observations of important biological functions (see Section
7.2). This general caution also applies to scallop spat produced in hatcheries;
juveniles should be released where the broodstock were collected. This caution
does not apply to species with high levels of gene flow among populations (e.g.,
Placopecten magellanicus on the east coast of North America [Beaumont and
Zouros, 1991] and *Chlamys islandica* in Norway and the North Atlantic
[Fevolden, 1989]) or to locations where spat from different stocks mix, like
the one in Japan described by Fujio and von Brand (1991).

The introductions of *Argopecten irradians* from the United States to
China, *Patinopecten yessoensis* from Japan to China, France and Ireland,
and *Pecten maximus* from France to Chile (Mortensen, 1999) raises the
possibility of hybridization between introduced and closely related species.
If hybrids are viable, there may be a risk of breaking down co-adapted gene
complexes, as described previously. Such consequences will be related
mainly to genetic differences among populations and the relative sizes of
the source and recipient populations (Beaumont, 2000).

Although some of the studies cited here indicate that homogeneous populations of scallops can occur at substantial spatial scales, this is certainly not
always the case. For example, Peterson *et al.* (1996) used transplants of
spawners, and monitoring of settlement and recruitment, to provide reasonable evidence for self-replenishing populations of *Argopecten irradians* on a
basin scale within sounds in North Carolina. Such small-scale spatial structuring of scallops will have important implications for management (see
Chapter 6) and needs to be confirmed by the genetic approaches mentioned
previously.

3.1.6. Disease risks and other environmental problems

Although use of wild scallop seed in many countries precludes transfer of
pathogens from hatcheries, neither wild nor released scallops are immune
from organisms that can destroy or impair them. Mortensen (1999) briefly
describes the mechanisms that spread deleterious organisms among scallops,
some regulations designed to minimize such events and other factors that
should be considered to reduce risks further.

The most serious causes of mortality have been attributed to *Rickettsia*-like prokaryotic organisms that result in myodegeneration (Gulka *et al.*, 1983), although these are not always pathogenic, and to the 'brown tide' chrysophyte *Aureococcus anophagefferens* (Tettelbach and Wenczel, 1993). The *Rickettsia*-like organism caused 100% mortality of *Placopecten magellanicus* in Narragansett Bay, Rhode Island, in 1979–1980, and brown tide resulted in death of up to 95% of *Argopecten irradians* at various sites in Long Island Sound, New York, in 1985 (Tettelbach and Wenczel, 1993). Dinoflagellate blooms (red tides) also rendered large harvests of scallops unsalable in Japan during the late 1970s because of paralytic shellfish poison and other toxins (Ventilla, 1982). *Dinophysis fortii, Gonyaulax catanella* and *Protogonyaulax tamarensis* were the causative organisms (Ueda et al., 1982; Ventilla, 1982). During these red tides, toxin levels exceeded the safety level of 4 mouse units g^{-1} in all parts of the scallop except the adductor muscle (Ueda et al., 1982), which could be sold if boiled (Ventilla, 1982). The red tides were attributed to runoff of nutrient-enriched water from rice fields but also coincided with periods when scallops were overstocked at many locations in northern Japan because of the combined densities of shells in hanging culture and on the seabed (Section 3.1.4).

A Sacculina-like parasitic barnacle was reported from *Patinopecten yessoensis* by Ventilla (1982) and by Elston and Burge (1984), but this was later shown to be a copepod parasite (Nagasawa *et al.,* 1988). The only way to arrest infestation by the copepod is to destroy affected scallops before the larvae are released into the plankton. The shell burrowing polychaete *Polydora ciliata* also poses problems for *Patinopecten yessoensis* in Japan (Ventilla, 1982). It is associated with muddy or fine sandy-muddy substrata and has damaged the shells of 70–80% of scallops sown in such habitats. Tettelbach and Wenczel (1993) indicate that *Polydora* spp. are also significant pests in some parts of the United States and have been known to infest and weaken the shells in entire populations of *Argopecten irradians*.

Clearly, translocations of spat from collecting to growing areas, transfers of adults within their natural distribution and the introductions of scallops described in Section 3.1.5 have potential to spread organisms that affect scallops. Movements of scallops must therefore be made responsibly (Sindermann, 1993; Blankenship and Leber, 1995). Although the risks can never be reduced to zero, Sindermann (1993) has described procedures that can minimize them. The transfer of the parasitic protozoon *Perkinsus karlssoni* to populations of *Argopecten irradians* in Canada from the Unites States is a case in point; diligent histopathological examination of scallops transferred to Canada, followed by three generations of quarantine, all failed to detect the parasite in granulomas. However, it was present and appeared in scallops cultured in open water 10 yr later (Sindermann, 1993 and references therein).

3.1.7. Systems for marking juveniles to assess effectiveness of releases

Determining the contribution of released juveniles to total harvest of scallops has been relatively easy because a stress band is laid down in cultured shells at seeding, and evidence of fouling during intermediate culture persists on both valves (Dao *et al.*, 1999). These natural tags have made development of a mark for individuals released in stock enhancement programmes unnecessary. In the few programmes where scallops have been released to restore stocks, e.g., in Long Island, New York, (Tettelbach *et al.*, 1997), genetic variation between the wild and cultured scallops has been used to demonstrate that ~25% of recruitment was derived from hatchery-reared animals (Krause, 1992). Identification of the genes involved in determining shell colour, and the heredity of these genes (Adamkewicz and Castagna, 1988), is one way of providing a simple genetic tag. The contribution of restocking to spawning biomass can then be estimated by comparing gene frequencies for shell colour in the population before, and several generations after, release of hatchery-reared juveniles selected to have significantly different gene frequencies consisting of dominant alleles. Care will be needed, however, in any such hatchery breeding programmes to ensure that the fitness and genetic diversity of released scallops is not altered in other ways.

3.1.8. Status of stock enhancement initiatives

Collection and intermediate culture of scallop spat until they reach a size refuge is a prime example of how aquaculture can increase productivity of marine resources on a sustainable basis. The techniques developed in Japan have resulted in a more stable industry in Hokkaido (Kitada and Fujishima, 1997) that now releases 3 billion scallop seed per year and produces yields of ~250,000 t yr^{-1}, which are about three times greater than the maximum catches before enhancement (Figure 3.4). The capture of spat also supports yields from hanging culture that exceed 200,000 t yr^{-1} (Kitada, 1999). The returns from stock enhancement in Japan are not based on outstanding rates of survival; only 25–30% of released juveniles survive to harvest. The overall recovery rate is even lower at 20% (Ventilla, 1982), because dredging recaptures 80% of the seeded scallops. In France, the commercial fleet has recovered released scallops at rates varying from 5–36%, with an overall average of 25% (Dao *et al.*, 1999). In New Zealand, however, rates of survival now routinely exceed 50% as a result of the combined effects of improved handling and harvesting after 2 years (Arbuckle and Metzger, 2000).

Despite the outstanding success of scallop stock enhancement programmes in some countries, the release of wild-caught juveniles has not

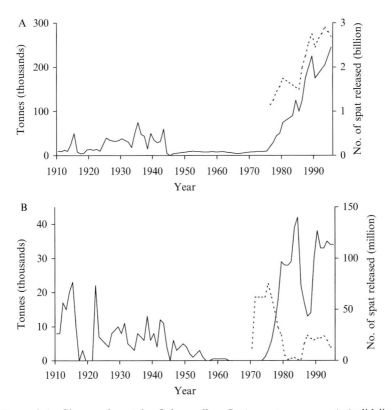

Figure 3.4 Changes in catch of the scallop *Patinopecten yessoensis* (solid lines) in (A) the whole of Hokkaido (redrawn from Kitada, 1999) and (B) Sarufutsu, Hokkaido (redrawn from Masuda and Tsukamoto, 1998) since the release of cultured juveniles on the seabed (dashed lines).

always resulted in improved fisheries. In New Zealand, for example, direct release of spat from collectors has been successful on the South Island but has not enhanced stocks effectively on the North Island. Attempts on the North Island failed for two reasons: (1) larvae produced by released scallops were not retained in the area, resulting in poor catches of spat, and (2) the released scallops had very high rates of predation (K. Drummond, personal communication 2001). Similarly, in Australia, where initial trials resulted in relatively low mortality of released juveniles (Thomson, 1992), larger-scale application of the technology from Japan struggled to increase yields from the capture fishery. In this case, poor survival of spat because of smothering by sediments (Thomson *et al.*, 1995) and high rates of predation by crabs (C. Gardner, personal communication 2001) were the problems. Stock enhancement has also faced difficulties in Canada, where survival rates of 1–6% in commercial-scale releases (Nadeau and Cliche, 1998b) have not matched previous rates of 40% in experimental trials (Hatcher *et al.*, 1996). Nadeau and Cliche (1998b) remain optimistic that improvements in release strategies, including removal of starfish predators, will raise survival to the 30% level required for economic viability (Cliche *et al.*, 1997). Clearly, reliable data on abundance of predators, and the retention of larvae, are of paramount importance in assessing the potential for stock enhancement of scallops at each location. Although retention of larvae is not critical if spat can be obtained elsewhere and the fishery is to be managed on a put-and-take basis, greater yields do occur where larvae are entrained by local currents and fishing is managed to ensure ample opportunities for spawning before harvest (see Section 3.1.9).

The ideal of increased, stable catches has not always been achieved, even for the most successful scallop stock enhancement programmes. For example, yield varies considerably among co-operatives in Japan (Fujita and Mori, 1990; Kitada and Fujishima, 1997), and seeding 'failures' occurred in New Zealand in 1993 and 1994, leading to a three-fold decline in landings in 1996 compared with 1994 (Arbuckle and Drummond, 2000). Also, the general increase in yields since the first harvest of released scallops in New Zealand in 1986 has yet to reach the peak catch in 1975 (Arbuckle and Drummond, 2000).

3.1.9. Management

Harvests from scallop stock enhancement programmes have been improved by: (1) stocking at densities that maximize the yield of acceptable market-sized shells, (2) interventions to reduce predation and (3) rotational fishing (Ventilla, 1982). The rotation system permits only a subset of the suitable areas to be fished in any one year and allows scallops in unfished areas to contribute to natural replenishment. The concept of rotating harvests for scallops dates back to 1895 in Japan (Masuda and Tsukamoto, 1998). Rotation stabilizes production, albeit at lower levels, even without enhancement (Arbuckle and Metzger, 2000; Drummond, 2004). However, the number of areas involved, and the interval between harvests, must be adapted to each location for the system to be effective. For example, Masuda and Tsukamoto (1998) report that landings from the Tokoro area of Hokkaido did not improve until the number of areas was increased to four, with the occasional use of a fifth 'reserve' area. Kitada (1999) illustrates how harvests from enhanced scallop fisheries in Japan managed by rotation often exceed predicted yields and concludes that the excess is due to natural replenishment, not improved survival of released juveniles. Interestingly, initial large-scale releases to restore stocks followed by rotating harvests does not yield the same catches as regular seeding combined with rotation. This is evident from the fishery in Sarufutsu, Hokkaido, where maximum yields decreased by 25–65% between 1984 and 1989 after cessation of seeding from 1981–1986 (Figure 3.4). Maximum harvest levels were subsequently restored by release of a reduced number of seed.

The management of *Pecten novazelandiae* in the South Island of New Zealand has developed in a similar way. The area licensed for commercial, recreational and customary scallop fishing in Tasman and Golden Bays is routinely divided into nine fishing sectors and some permanent reserves, with the number and location of sectors open to fishing varying each year depending on the status of the resource (Arbuckle and Metzger, 2000; Drummond, 2004). Adaptive management is used to fine-tune the seeding operations. Annual resource surveys determine how many seed are needed to reach carrying capacity; then deployment of spat collectors is adjusted to produce the required number of spat. Improvements in survival of seed because of full immersion during handling and transport means that this process now requires the release of fewer spat. Over the past decade, a reduction in the number of spat required, from 1.2 billion to 300 million y^{-1}, has been achieved (Bull, 1994; Arbuckle and Metzger, 2000).

Stock enhancement of scallops has worked best where fishermen have taken responsibility for all essential operations. In Japan, this is done by co-operatives, which formed quickly, based on those already existing for farming seaweed and oysters (Ventilla, 1982; Masuda and Tsukamoto, 1998). The Kawauchi co-operative, one of 12 in Mutsu Bay, provides a typical example. In 1982, it had access to 17 km of shoreline and produced 3,000 tonnes of scallops from 25 million shells in hanging culture and an additional 200 million shells sown on the seabed (Ventilla, 1982). On the South Island of New Zealand, the government has endorsed transfer of all aspects of scallop enhancement and some aspects of management to the Challenger Scallop Enhancement Company Limited (Challenger), subject to approval of Challenger's annual management plan (Arbuckle and Drummond, 2000). The keys to success of this arrangement are inclusion of the indigenous Maori people as at least 20% shareholders in the commercial fishery and allocation of the approved catch for the year in proportion to individual transferable quotas (ITQs) held by shareholders. Thus, ITQ holders are the direct beneficiaries of investments they make to improve yields. These investments are funded by a levy of ~20% of individual earnings per year, which covers costs such as collection and release of juveniles, annual resource surveys, scientific advice and development of the annual management plan. Further details about Challenger are provided by Arbuckle and Drummond (2000) and Arbuckle and Metzger (2000).

3.1.10. Marketing

Although scallops are a luxury food item, and there are many ways to market them, including boiled, dried, frozen, fresh and live products (Ventilla, 1982), the great increase in scallop production in Japan has been accompanied by a decrease in price (Masuda and Tsukamoto, 1998; Kitada, 1999). The challenge is to maintain good levels of profitability as production continues to increase through well-organized stock enhancement programmes. The Sarufutsu co-operative in Hokkaido has met this challenge by wholesaling their crop and constructing the freezing, boiling and drying processing facilities needed to add value to their product (Masuda and Tsukamoto, 1998). Concerted efforts to broaden the popularity of scallops within Japan have also expanded the market considerably.

3.1.11. Problems to overcome

Except where predation and larval retention are problems, the main impediments to stock enhancement of scallops are social and legal, not biological. Although fisheries for scallops are compatible with several other uses of waterways including trap, gillnet and recreational fisheries, commercial shipping and recreational boating, they conflict with trawling for other species and with aquaculture (Dao *et al.*, 1999; Arbuckle and Drummond, 2000). Thus, a major impediment to stock enhancement of scallops in many countries is access to the large areas of seabed needed to create viable fisheries (Fleury *et al.*, 1997; Dao *et al.*, 1999). The size of the area needed is indicated by the successful fisheries in Japan, which rotate harvests among several areas up to $\sim$20 km^2 each (Ventilla, 1982; Masuda and Tsukamoto, 1998). In New Zealand, the total area available to Challenger for scallop enhancement is 1400 km^2. Even if priority use of extensive areas of suitable scallop habitat can be obtained, these rights can be contended legally. This has occurred in the area used by Challenger in New Zealand (Arbuckle and Drummond, 2000), where the rights allocated to Challenger have now been disputed by mussel farmers.

A less complicated but real impediment to creation of enhanced scallop fisheries is that fishermen must forego harvests over large areas while rotational fishing grounds are established and seeded. Dao *et al.* (1999) indicate that this will not be easy to achieve in Europe.

3.1.12. Future research

Now that proven methods exist for augmenting the supply of juvenile scallops, and for stabilizing harvests at the carrying capacity of the habitat, the major research task is to identify appropriate locations for successful application of the technology. Assuming that the social and legal impediments to securing the large areas needed for scallop fisheries can be overcome, transfer of technology for management by stock enhancement depends on finding positive answers to the following questions for each site: (1) Can sufficient spat be collected, or produced in hatcheries, and raised in intermediate culture cost-effectively for large-scale increases in production? (2) Do released juveniles survive predation to market size at a viable rate (usually >25%)? (3) If not, can predation be reduced at reasonable cost? (4) Do local currents retain larvae to replenish the stock?

3.2. OTHER BIVALVES

3.2.1. Background and rationale for stock enhancement

Apart from cultured oysters (Ostreidae) and mussels (Mytilidae) that are reared in purpose-built farming systems, and which are not the subject of this review, many bivalves, including 'wild' oysters, support commercial, recreational and traditional fisheries (FAO, 2002). These are represented well by the clam and oyster fisheries in the United States so we have limited our review to that area of the world.

Clams and oysters have supported large fisheries on the east coast of the United States. The fishery for hard clams *Mercenaria mercenaria* on Long Island has been particularly significant, accounting for 45% of the national catch in the 1970s (Malouf, 1989; Manzi, 1990). However, over-fishing of hard clams has occurred, resulting in demands to restore stocks to former levels and large public 'seeding' efforts in Massachusetts and New York (Malouf, 1989; Manzi, 1990; Walton and Walton, 2001). Decreases in catch rates of hard clams and soft-shell clams *Mya arenaria* elsewhere on the east coast have also resulted in stock enhancement programmes (MacKenzie, 1989; Beal, 1991; Rice *et al.*, 2000; Arnold, 2001), as well as calls for other forms of management (e.g., rotational closures [Peterson, 2002]).

Historical harvests of the oyster *Crassostrea virginica* have dwarfed other bivalves, totalling 160 million pounds (>70,000 t) of meat per annum in the early 1900s (MacKenzie, 1996). The upper Chesapeake Bay was a major area of production, with landings peaking at 615,000 t of live shell in 1884 (Rothschild *et al.*, 1994). Other important oyster-producing areas on the east coast were New Haven Harbour, Connecticut; Delaware Bay, New Jersey; James River, Virginia; Apalachicola Bay, Florida; and several estuaries in Louisiana (MacKenzie, 1996). Regrettably, the yield of oysters has fallen to ~25% of historical levels (MacKenzie, 1996). In some of the major oyster areas (e.g., Chesapeake Bay) production is now a mere shadow of former levels, with landings at ~1% of peak historical levels (Rothschild *et al.*, 1994; Hargis and Haven, 1999; Wesson *et al.*, 1999) (Figure 3.5). Ironically, the over-fishing of oysters, which now has to be addressed by restoring areas where juveniles can settle in large numbers, was in part brought about by a form of stock enhancement. Young oysters and dead shells were removed from oyster reefs and spread in other areas to allow the oysters to grow at a faster rate. This removal was carried out to the point where the hard substrata needed to 'harness' the settlement of spat each year was destroyed (see Section 3.2.5).

Significant fisheries for clams have also been established on the west coast of the United States, where 95% of commercial clam production traditionally comes from the State of Washington (Schink *et al.*, 1983), although landings have varied substantially in the past 50–100 years (Figure 3.6). In particular, there has been a decline in commercial harvest of razor clams *Siliqua patula*, an increase in production of geoduck clams *Panopea abrupta* resulting from discovery of extensive subtidal stocks in the late 1960s and creation of fisheries by accidental introduction of the Manila clam *Venerupis japonica* (Chew, 1989; Westley *et al.*, 1990; Lipton *et al.*, 1992).

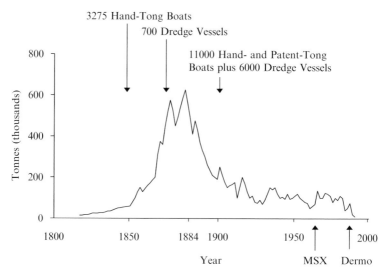

Figure 3.5 Historical landings of the oyster *Crassostrea virginica* from the Maryland portion of Chesapeake Bay, showing changes in fishing effort and the onset of the diseases, MSX and Dermo (redrawn from Rothschild *et al.*, 1994). Details of the different types of fishing gear and boats are given by MacKenzie (1996) and Luckenbach *et al.* (1999).

Although landings of razor clams had decreased substantially by the 1940s from peak harvests in 1915, and regulations had to be introduced to conserve stocks (Schink *et al.*, 1983), the subsequent decline in production of this species was not related only to over-fishing. Competition for razor clams as crab bait, and inexpensive canned clams from the east coast, caused a collapse in the local canning industry. The outcome was a switch from a commercial to a recreational fishery for razor clams (Schink *et al.*, 1983; Figure 3.6).

Marked decreases in settlement ('set') of juvenile razor clams on one third of clam beds because of intense harvest, inconsistent natural reproduction and disease led to interventions to increase catches on public beaches (Westley *et al.*, 1990). For geoduck clams, the impetus for stock enhancement was to overcome the slow rates of natural replenishment (15–60 yr) of this large (>4 kg), long-lived (100 yr) species (Figure 3.7) to provide more individuals for harvesting from intertidal beaches open to the public (Westley *et al.,* 1990; Beattie, 1992; Beattie and Blake, 1999 and references therein). However, after discovery of large subtidal stocks in Puget Sound, stock enhancement was also investigated as a way of doubling the conservative harvest levels of 2,300 t yr^{-1} (1% of adult stock) implemented to sustain commercial yields (Beattie, 1992). Measures were also developed to enhance production of the popular introduced Manila clam (Chew, 1989; Westley *et al.*, 1990).

Stock enhancement techniques used for clams in the United States stand out because of their focus on recreational fisheries, use of customized shelter, movements of wild animals and modification of the habitat. On the other hand, interventions to improve harvests of oysters have been aimed at industry and now depend heavily on restoration of habitat needed for the successful settlement of postlarval oysters (MacKenzie, 1996; Luckenbach *et al.*, 1999).

The practice of 'planting' seed clams and oysters on the substratum makes it somewhat difficult to distinguish between stock enhancement and farming. For the purposes of this review, we consider that stock enhancement, rather than farming, is the objective if: (1) the shellfish are not enclosed fully throughout the production cycle, (2) seed are placed in locations or microhabitats where they occur naturally and (3) natural replenishment accounts for a substantial proportion of the harvest over and above placement of seed. We regard transfer of seed outside the natural reproductive range of a species as creation of fisheries, not stock enhancement.

The main features and outcomes of various stock enhancement initiatives for clams and oysters in the United States are summarized below, illustrated primarily with examples from hard- and soft-shell clams and oysters on the east coast, and razor, geoduck and Manila clams on the west coast. There is a vast literature on both hard clams (see Kraeuter and Castagna, 2001 and

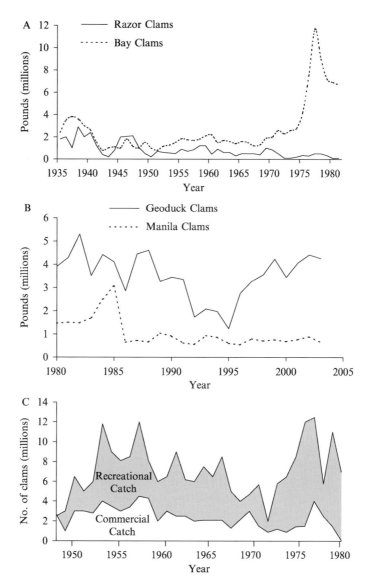

Figure 3.6 Clam catches from the state of Washington, United States. (A) Razor clams *Siliqua patula* and bay clams (13 species, dominated by geoduck clams *Panopea abrupta*) from 1935–1981. (B) Manila clams *Venerupis japonica* and geoduck clams from 1980–2003. (C) Razor clams taken by recreational and commercial fishermen from 1949–1980. (Redrawn from data in Schink *et al.*, 1983; and data supplied by Washington Department of Fish and Wildlife).

Figure 3.7 (A) The hard clam *Mercenaria mercenaria* (photo: National Oceanic and Atmospheric Administration of the United States. Central Library Photo Collection). (B) Geoduck clams *Panopea abrupta* from Washington State, United States. (photo: D. Rothaus). (C) Geoduck clam *in situ*—all that can be seen are the inhalant and exhalant siphons (photo: D. Rothaus). (D) An oyster reef in Chesapeake Bay, United States, exposed at low tide (photo: Virginia Institute of Marine Science). (E) Transferring oysters to a larger vessel in Chesapeake Bay (photo: Virginia Institute of Marine Science).

references herein) and oysters (Kennedy *et al.*, 1996; MacKenzie, 1996; Luckenbach *et al.*, 1999 and references therein). It is difficult to do justice to this literature in the broad context of this review, and we encourage those interested in learning more about management of these species to read more widely from the references quoted throughout this section.

3.2.2. Supply of juveniles

Four methods have been used in the United States in attempts to increase productivity of clam and/or oyster fisheries: transplantation of naturally occurring seed; release of juveniles reared in hatcheries; translocation of broodstock; and provision of additional habitat for settling juveniles. For many species (e.g., oysters, hard clams and Manila clams) more than one method is used to enhance stocks.

Transplanting juveniles that have settled naturally to areas where recruitment is low but growth and survival are good has long been part of the industry for oysters, and fisheries for hard- and soft-shell clams, on the east coast (MacKenzie, 1989, 1996; Malouf, 1989; Menzel, 1989; Manzi, 1990; Leard *et al.,* 1999 and references therein). This method has also been used for stock enhancement of razor clams on the west coast, where abundant natural seed are harvested from subtidal beds inaccessible to the public at a size of 5–15 mm and replanted on sparsely populated beaches open to clam diggers (Rickard and Newman, 1988). Such transplantations have two benefits: they reduce the often extreme densities that occur after a good 'set', thus reducing stunting and high rates of mortality, and they permit production in areas that do not receive many settling larvae.

Generalised hatchery methods for production of the millions of juveniles needed for stock enhancement are now routine for popular species of clams in the United States (Loosanoff and Davis, 1963; Chew, 1989; Malouf, 1989; Menzel, 1989; Beattie, 1992; MacFarlane, 1998 and references therein). The only exception is the geoduck clam, which requires specially designed nurseries with sand substratum, because small juveniles depend on collecting food and detritus from the surface of the sediment during the epifaunal phase after metamorphosis (Beattie, 1992). Although hatchery methods are now well established for several species, the cost of rearing clams to the size needed for adequate survival when unprotected in the wild limits the number of seed available for stock enhancement programmes (Malouf, 1989). This is particularly true for hard clams, which should be grown to at least 20–25 mm before release (see Section 3.2.4.1).

A recent development to overcome this high cost has been the mass-release of hundreds of millions of larval hard clams into a subbasin of the Indian River lagoon, an area with high larval retention (Arnold, 2001). It remains to be seen whether localised, careful placement of larvae will overcome the failures of similar approaches with lobsters (see Section 3.7) and finfish (Blankenship and Leber, 1995). Efforts so far have resulted in larval densities of hard clams as great as any on record (Arnold, 2001), although settlement success has yet to be quantified (Arnold *et al.,* 2002).

Translocation of spawners to enhance production has been limited mainly to hard clams in places like Great South Bay, Long Island (Kassner and Malouf, 1982; Manzi, 1990 and references therein), and in Florida (Arnold, 2001). The assumptions behind this method are that: (1) concentration of spawners at higher densities will increase fertilization success and the supply of larvae (Arnold et al., 2002), and (2) if spawning happens to be asynchronous with the local population, larvae will be available for settlement over longer periods. A variation on this theme is the 'relay' of large quantities (>1,000 t yr^{-1}) of hard clams from closed (polluted) areas of New York to certified depuration waters (Hendrickson et al., 1988; Rice et al., 2000). Apart from making more clams available for safe consumption, this activity aims to increase the density of spawning stock in receiving waters. Relaying of shells from areas where harvesting is prohibited (because of pollution) to approved areas is also a regular practice in the oyster industry (MacKenzie, 1996; Leard et al., 1999). In this case, however, it is done mainly to meet the conditions necessary for marketing oysters under the National Shellfish Sanitation Program (see Section 3.2.10).

The methods for enhancing recruitment by providing additional (artificial) habitat are covered separately in Section 3.2.5.

3.2.3. Defects caused by rearing

Malouf (1989) has issued a general caution that seed clams produced in hatcheries should not be assumed to be of the same quality as wild juveniles. This problem has received relatively little scrutiny, although Velasquez (1992) reported that hatchery-reared geoduck clams had great variation in shell thickness, resulting in 7–60% shell breakage, depending on the production batch and harvesting method. The condition of the shell in juvenile geoduck has a significant effect on short-term predation by crabs and *Nassarius* spp. gastropods, and the ability of the clam to burrow (Beattie, 1992; Velasquez, 1992). In this particular case, however, any deficit in short-term fitness of cultured geoduck caused by thin shells has a limited effect on the success of stock enhancement; it pales into insignificance compared with the subsequent rates of overall predation in subtidal habitats during the first year (see Section 3.2.4).

3.2.4. Release strategies

Redistribution of oyster seed from reefs (also known as beds or bars) with high densities of juveniles, to areas for growing on, has been a feature of the industry since the mid-1800s (Kennedy and Sanford, 1999; Leard et al.,

1999). Firm substrata covered with a few centimetres of soft mud are preferred. If the mud is too deep, bivalve shells, gravel or sand are applied to make the bottom firmer. Stocking densities range from ~20–60 t ha^{-1} (Leard *et al.*, 1999). Beds are subdivided into grids to facilitate uniform distribution of seed. Since the onset of the diseases MSX and Dermo (see Section 3.2.7), seed have been distributed in areas of estuaries where salinities are too low to be tolerated by these pathogens.

Juvenile razor clams collected from deeper water are planted on surf beaches by pouring them from buckets ahead of advancing waves. This is done at night on an incoming tide to avoid desiccation (Westley *et al.*, 1990). The vast numbers of seed released apparently 'swamp' the effects of predation.

Juvenile Manila clams are placed on beaches with a low slope where moderate currents and wave action favour retention of larvae, supply oxygen and food for fast growth and remove waste products (Chew, 1989).

Intertidal, low-energy, muddy-sandy habitats in shallow bays are the preferred areas for 'planting' hard clams (Kraeuter and Castagna, 1989). This species can be placed at a range of tidal levels, although there is a tradeoff between better survival and slower growth with increasing height above the low-water line (Kraeuter and Castagna, 1989).

Predation is widely regarded as the dominant factor limiting survival. Unprotected clams often have rates of mortality >95% (Chew, 1989; Gibbons and Blogoslawski, 1989; Kraeuter and Castagna, 1989; Marelli and Arnold, 1996; Kraeuter, 2001; Walton and Walton, 2001 and references therein). Among the many predators of clams, crabs are the most potent and have been known to eat >100 hard clams per day of the small seed size (Arnold, 1984; Gibbons and Blogoslawski, 1989). Starfish and carnivorous gastropods can also be a major problem. Removal of starfish and gastropods, and a range of protective structures (e.g., mesh covers, cages, gravel barriers) and chemicals (e.g., lime, copper compounds, brine solutions, poison baits) have been used to reduce predation on clams and oysters (Chew, 1989; Kraeuter and Castagna, 1989; MacKenzie, 1989, 1996; Beal, 1993a; Rice *et al.*, 2000). Some of the chemical methods have now been prohibited (MacKenzie, 1989).

The development of anti-predation measures for geoduck clams, which have rates of survival of <1% in subtidal habitats after 2 yr (Beattie, 1992), provides an interesting example of the steps needed to protect seeded clams. The solution to intense predation by crabs was found during experiments using PVC tubes in the intertidal zone. Survival of juveniles, planted in partially buried tubes covered with netting, over a 9-mo period increased from 21–67% (Beattie, 1992). The lessons learned from intertidal experiments were then transferred to the subtidal zone, and survival increased to 12% (Beattie, 1992). However, deployment and retrieval of the thousands of tubes necessary to protect the number of juveniles required to double the sub-tidal catch through stock enhancement presented formidable problems (Beattie, 1992).

Other factors important to growth and survival of seeded clams are: (1) release size, (2) type of substratum, (3) time of year, (4) abundance and composition of other infauna and (5) stocking density (Kraeuter and Castagna, 1989; Beal, 1991, 1993a; Beattie, 1992; Peterson et al., 1995).

3.2.4.1. Release size

Given the high rates of predation, survival of cultured juvenile clams placed in the wild is related strongly to size at release. For example, survival of soft-shell clams of 3–17 mm shell height (SH) from low intertidal sites in Maine increased significantly with size (Beal, 1991, 1993a; Beal and Kraus, 1991). In one of these experiments, survival varied from 70% for clams with a mean size of ~12 mm SH to 15% for those of 5 mm SH (Beal, 1991). Typically, <5% of the recovered clams had shells undamaged by predators. Such results indicate that successful release of juvenile clams depends on identifying the threshold size at which predation is lessened. For hard clams, this size is generally 20–25 mm (Arnold, 1984; Kraeuter and Castagna, 1989; Peterson et al., 1995; Arnold et al., 2002), although some predation can still be expected from the blue crab, Callinectes sapidus until the clams reach ~ 40 mm (Kraeuter and Castagna, 1980; Arnold, 1984).

3.2.4.2. Type of substratum

If released clams are not protected from predators until they reach a size refuge, they have high rates of mortality. Identifying habitats that provide adequate protection is, therefore, critical to achieving acceptable rates of survival. In laboratory experiments, hard clams were more vulnerable to predation by blue crabs in sand and sandy-muddy substrata than in crushed oyster shell or gravel (Arnold, 1984). Similar results have been obtained from field experiments. For example, juvenile hard clams of 16–22 mm released in seagrass beds survived in significantly greater proportions than those placed on sand; and clams planted in oyster-shell hash had significantly greater mean rates of survival than those released onto sandy-muddy substrata (Peterson et al., 1995). These findings concur with experiments on predation rates by crabs foraging in different substrata, which showed that success was significantly greater on sand than on a sand/shell mix (Sponaugle and Lawton, 1990).

3.2.4.3. Time of year

The time of year when seed clams are placed into habitats is important for both hard and soft-shell clams. Field experiments in Maine with *Mya arenaria* have shown repeatedly that hatchery-reared juveniles placed on intertidal flats at the end of the first growing season have poor (typically <30%) over-winter survival when ice scours the top few centimetres of mud (Beal, 1993b). To overcome this problem, clams are held in deeper water during winter and placed on the flats the following spring.

In the case of hard clams, the recommendation is to plant juveniles from late autumn to early winter (Peterson, 1990; Peterson *et al.*, 1995; MacFarlane, 1998) after water temperatures decline and crab predators enter winter dormancy. Because the clams grow well throughout winter in the southeast United States, there is no penalty of delayed growth for this strategy. Moreover, clams are less vulnerable to crabs the following spring because of their larger size (Peterson *et al.*, 1995 and references therein).

3.2.4.4. Abundance and composition of other bivalves and benthos

The literature on interactions between planted clams and other fauna is one-sided, dealing mainly with effects of other species on growth and successful recruitment of clams. For example, Kraeuter and Castagna (1989) provide a range of examples where growth and survival of hard clams has been impeded by co-occurring macro-invertebrates. Reasonable attention has been given, however, to ingestion by clams of their own larvae or larvae of other bivalves. There are conflicting reports on the effects of this 'predation' on populations (Kraeuter, 2001 and references therein).

There is evidence that the presence of at least one species of clam, *Gemma gemma*, can enhance settlement of *Mercenaria mercenaria*. Ahn *et al.* (1993) showed that the physical structure provided by small gem clams, and the way these clams modified the sediment, significantly increased the numbers of hard clams that settled. They concluded that dense assemblages of small suspension feeders could enhance settlement of another bivalve species.

There is little information on effects of increased densities of bivalves, resulting from the release of cultured juveniles, on other benthic species. However, Commito and Boncavage (1989) found that abundance of oligo-chaetes was correlated positively with the density of mussels, *Mytilus edulis*, and that it declined by 50% when the mussels were removed.

3.2.4.5. Stocking density

Dense plantings of bivalves can result in competition for food and reduced growth rates (Kraeuter and Castagna, 1989; Kraeuter, 2001). It is difficult, however, to specify a recommended stocking density for each species. Experiments need to be done at each location where clams are released to establish optimum stocking density. This is because food availability, and therefore the threshold density below which growth will not be affected substantially, will vary among sites because of differences in primary productivity, water flow, sediment characteristics, intertidal exposure, etc.

Surprisingly, relatively high stocking densities have resulted in improved survival when hard clams have been placed in trays subtidally and intertidally (Eldredge *et al.*, 1979). However, these conditions are not typical of those where clams are placed unprotected directly on the sediment in large-scale release programmes. Low stocking densities are recommended for such releases. For example, Peterson *et al.* (1995) cited a stocking density of 1 individual m^{-2} as one of the main reasons why hard clams planted in North Carolina had an average survival rate of 35% over a 14-mo period. They concluded that low-density planting of seed clams is a viable method of reducing predation and maximizing returns. This is supported by the experiments of Sponaugle and Lawton (1990), who showed that crab predation on hard clams was lowest when clams occurred at low densities in heterogeneous substrata.

3.2.5. Use of artificial habitats

The provision of additional habitat has had its greatest application in the oyster fisheries on the east coast of the United States. As early as the 1800s, oystermen realised that *Crassostrea virginica* needed hard substrata to settle successfully. This launched the widespread practice of spreading 'cultch' to provide the necessary surfaces for settlement. Initially, dead oyster shells collected at shucking plants or mined from ancient buried reefs were used for cultch. However, because the shells were also in demand for other purposes (e.g., to build roads or burned to make lime), other materials have been used for cultch. These include dredged shells of the clam *Rangia cuneata,* limestone, gravel, crushed concrete and coal fly-ash aggregate or gypsum stabilized with cement (Haywood *et al.*, 1999; Leard *et al.*, 1999). Settlement of oysters is generally greater on limestone and cement-stabilized gypsum than on the other non-oyster materials (Haywood *et al.*, 1999). Cultch has been used extensively and application of >20,000 m^3 at a site is common. It is usually dispersed from deck barges using high-pressure 'water canons' (Figure 3.8). In recent years, the practice has been to spread cultch near reefs

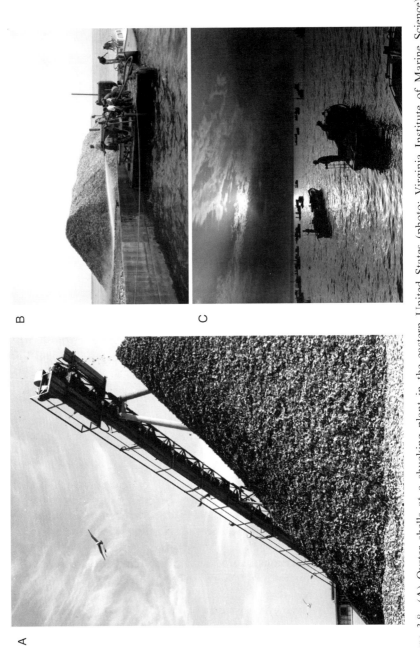

Figure 3.8 (A) Oyster shells at a shucking plant in the eastern United States (photo: Virginia Institute of Marine Science). (B) Spreading oyster shell 'cultch' from a barge in Chesapeake Bay, United States. (photo: Virginia Institute of Marine Science). (C) Oystermen using hand tongs to collect *Crassostrea virginica* in Chesapeake Bay (photo: Virginia Institute of Marine Science).

that were productive in the past. Cultch is also used to provide better conditions for growing seed oysters transplanted to other sites (Leard *et al.*, 1999).

Old shells and other materials are also being used to rehabilitate oyster reefs. Restoration of oyster reefs is necessary, because removal of large oysters for market, and seed for transplanting for growing on, has destroyed the fragile nature of the reefs (Rothschild *et al.*, 1994). In many cases, oyster reefs have been totally obliterated (Hargis and Haven, 1999).

Interestingly, initial harvests of oysters from reefs were seen as beneficial; beds in Chesapeake Bay were more extensive 30 years after the fishery began. Apparently, the early harvests broke up the tightly consolidated virgin reefs, reducing competition for food among adult oysters and providing increased areas for larval settlement (Kennedy and Sandford, 1999). As harvesting increased, particularly after the introduction of dredges and hydraulic equipment, the vertical profile of oyster reefs was greatly reduced and often disappeared altogether (Rothschild *et al.*, 1994). As a result, many formerly productive areas are now covered by silt and no longer provide suitable substrata for settlement of oysters (Kennedy and Sandford, 1999). It is ironic that some of the earliest attempts at stock enhancement—the removal of cultch and seed from oyster reefs to produce more oysters elsewhere—has been partly responsible for the demise of the fishery.

Restoration of oyster reefs will not be easy because they have been built up over thousands of years (Hargis, 1999). Undisturbed oyster reefs have been likened to coral reefs; they have increased in height with rises in sea level and provide the dominant physical structure where they occur (Hargis, 1999; Hargis and Haven, 1999) (Figure 3.9). The value of oyster reefs as habitat for fish and macro-invertebrates is now also recognised (Breitburg, 1999; Posey *et al.*, 1999). Consequently, considerable attention is being given to learning how to rebuild reefs at their original locations. These efforts are designed to restore the ecosystem benefits of reefs and provide the basis for future oyster production (Eggleston, 1999; Whitlatch and Osman, 1999). Key issues are re-creating the original vertical profile and dimensions of reefs and identifying which ones to rehabilitate first to maximize the rate of replenishment of meta-populations of *Crassostrea virginica*. There is also evidence that aggregating broodstock on reconstructed reefs may accelerate restoration of oyster populations in areas where local circulation promotes retention of larvae (Southworth and Mann, 1998).

Habitats have also been created or modified to enhance settlement of clams. In the case of Manila clams, this process involves increasing the sand-mud-gravel substratum needed for successful settlement of this species by spreading a 25–200 mm layer of gravel (5–75 mm dia) on previously poor habitat at low tide (Chew, 1989; Westley *et al.*, 1990). The gravel is applied at sites where densities of larvae are high and predators low and where strong

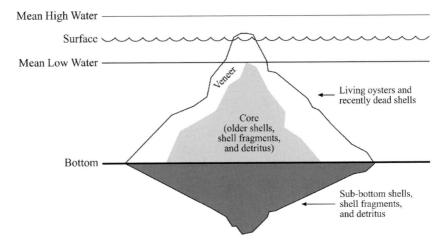

Figure 3.9   Historical structure of undisturbed oyster reefs in Chesapeake Bay, United States (redrawn from Hargis, 1999).

wave action, sedimentation or eutrophication will not render the artificial substrata unsuitable for settlement (Westley *et al.*, 1990). Brush or sandbags have also been placed on beaches to create barriers and form eddies to enhance the 'set' of Manila clams (Chew, 1989).

3.2.6. Minimising genetic impacts

Much of the research on clam genetics in the Unites States has focused on protocols for selective breeding of captive stock to improve growth during culture (Humphrey and Crenshaw, 1989 and references therein). There has also been an emphasis on maintaining heterozygosity, because it is correlated positively with growth (Dillon and Manzi, 1987). Such knowledge is generally not applicable to restocking and stock enhancement programmes, where the objectives are to maintain the natural gene frequencies of wild populations. This has not always been the approach, however, and some programmes have aimed to produce 'improved' hard clams for restoration and stock enhancement of wild populations (Manzi *et al.*, 1988).

Population genetics of hard clams has focused mainly on determining whether *Mercenaria mercenaria* and *M. campechiensis* are different species, and on determining the population structure of *M. mercenaria*. The existence of the two species has now been accepted, although there is a substantial area of Florida where they hybridize (Dillon and Manzi, 1989; Bert and Arnold, 1995; Arnold *et al.*, 1996; Hilbish, 2001). Populations of *M. mercenaria* from

Massachusetts to Florida were not strongly differentiated using allozymes, suggesting that they might be interconnected by gene flow (Humphrey and Crenshaw, 1989). North of Massachusetts, however, populations of *M. mercenaria* become increasingly fragmented. Hilbish (2001) has suggested that these northern populations may be extinguished and recolonized more frequently. For the oyster *Crassostrea virginica* there is a distinct genetic break between the Atlantic coast and the Gulf of Mexico (Reeb and Avise, 1990; Hare *et al.*, 1996; Hare and Avise, 1998).

It remains to be determined whether population genetic tools are able to detect finer-scale differences in population structure for these species. An important part of this endeavour will be gauging the extent to which gene flow among populations from different areas represents potential for replenishment or whether very closely related populations of clams and oysters in some places are functionally isolated, as is evident for scallops in Europe (see Section 3.1.5). Hilbish (2001) infers that populations of hard clams are probably generally localized and that other methods of stock delineation, in addition to genetic approaches, will be needed to define populations for management purposes. The multivariate morphometric analysis of the soft-shell clam *Mya arenaria* by Amaratunga and Misra (1989), in which they delineated at least three populations in the region of Nova Scotia–New Brunswick, provides an example of how other methods can help identify the spatial population structure of clams (see also Chapter 7, Section 7.2).

3.2.7. Disease risks

The wide range of pests, parasites and diseases affecting clams during hatchery production, and the measures used to combat these problems, are described comprehensively by Gibbons and Blogoslawski (1989) and Ford (2001). We repeat just a few of their points for general interest: (1) clams generally have fewer disease problems than oysters; (2) great care is needed to avoid spreading pathogens among hatcheries through translocation of broodstock and (3) the risk of diseases occurring in hatcheries can be lowered substantially through reduction of stress, provision of proper sanitation, high-quality water and algal food and a constant environment. Given the wide range of prokaryotic organisms implicated in disease outbreaks, and the fact that juvenile hard, geoduck and soft-shell clams for stock enhancement are derived mainly from hatcheries, vigilance is needed to ensure that such diseases are not spread to wild stocks or atypical hosts. The risk of transferring diseases to, or contracting them from, other species needs to be taken seriously. Clam hatcheries have fallen foul of the latter scenario; the bacterium *Vibrio anguillarum,* which has caused rapid mortality in clam embryos, is normally a pathogen of eels (Gibbons and Blogoslawski, 1989).

Wiegardt and Bourne (1989) also highlight the role of past introductions of bivalves in introducing pathogens to the west coast of the United States. They advocate careful control of translocations and introductions to reduce the risks of disease. In Long Island, New York, regulations have been introduced to specify locations of hatcheries supplying hard clam seed in an effort to lower incidence of disease.

The risk of losses to clam stock enhancement programmes varies because some species are more susceptible to disease than others (for example, hard, geoduck and Manila clams are more resistant than soft-shell clams) (Chew, 1989; Kraeuter and Castagna, 1989; Beattie, 1992; Ford, 2001). Hard clams in Florida have, however, exhibited unusually high levels of gonadal neoplasia, a tumourous condition in which abnormal germ cells arise within the follicles and ducts, causing a debilitating and chronic reduction in production of gametes (Bert *et al.,* 1993 and references therein). The incidence of gonadal neoplasia is greater in transplanted hard clams (see Section 3.2.9).

There has been massive mortality of oysters in Delaware and Chesapeake Bays since the 1950s because of the diseases MSX (caused by the protozoan *Haplosporidium nelsoni*) and Dermo (the protozoan *Perkinsus marinus*) (Andrews, 1996; MacKenzie, 1996 and references therein). By 1959, 50% of oysters on the seed beds and 90–95% of oysters on leased growing areas in Delaware Bay had died from MSX (Ford and Haskin, 1982). The transfer of oyster seed among estuaries on the east coast has increased the spread of MSX and Dermo. Effective ways of controlling these diseases have yet to be found, although both pathogens do not tolerate salinities below 8–9 ppt. Fortuitously, neither disease makes the oysters unsafe for human consumption.

A related issue is that oysters from the east coast of the United States have been transferred to the west coast and to Europe. Such transfers have rarely resulted in successful establishment of commercially viable oyster populations (Lipton *et al.,* 1992; MacKenzie, 1996). However, some bivalves and gastropods associated with *Crassostrea virginica* are now established on the west coast. At least two gastropods introduced to Europe with *C. virginica,* the filter-feeding Atlantic slippersnail *Crepidula fornicata* and the carnivorous Atlantic oyster drill *Urosalpinx cinerea,* are now serious pests (Carlton and Mann, 1996, cited in MacKenzie, 1996).

Clams also share some of the problems affecting scallops (see Section 3.1.6). For example, paralytic shellfish poisoning, caused by *Gonyaulax* sp., occasionally closes recreational fisheries for intertidal razor clams on the west coast (Schink *et al.,* 1983).

3.2.8. Systems for marking juveniles to assess effectiveness of releases

The need to assess the contribution of released clams to recreational and commercial fisheries in South Carolina led to development of a practical tag for cultured hard clams. The mark is conferred by cross-breeding local specimens with those exhibiting the distinctive 'notata' shell colouration (Manzi, 1990). Eighty percent of hatchery-reared seed have the 'notata' markings compared with 1% of wild stock, providing a simple way of estimating the number of released clams in the population. The identification of released soft-shell clams has proved to be even more straightforward because hatchery-reared shells develop distinct marks once transplanted (Beal, 1993a). Other simple methods for mass-marking bivalves are also available (e.g., staining the shells with iron [Fe]) (Koshikawa et al., 1997).

3.2.9. Status of stock enhancement initiatives

Malouf (1989) laments the lack of data on survival of hatchery-reared hard clams until they reach minimum harvest size and their contribution to natural replenishment. He surmises that the efforts of several towns on Long Island to release 2–3 million juvenile hard clams per year were driven by the need 'to do something' to increase catches. Alarmingly, he demonstrates that even if survival of the seed clams was as high as 50%, these stock enhancement programmes would have contributed only 2–4% to annual yields. He concludes that the contribution of stock enhancement to hard clam fisheries at the scale 2–3 million seed per town was insignificant and that it would need to be increased by an order of magnitude to account for 25% of landings. A more recent study by Walton and Walton (2001) confirms that hatchery releases of 1–2 million hard clams per year are still used by at least 23 managers in Massachusetts and New York, because they are perceived to benefit the fishery. Mean survival to market size was estimated by the managers to be $\sim$50% at best, although many of these assessments lacked rigour. Clearly, the cost–benefit of such releases needs to be evaluated quantitatively.

Peterson et al. (1995) provide a more optimistic assessment from stock enhancement trials with hard clams in North Carolina. They used a strategy that combined low stocking density (1 m^{-2}), relatively large seed (14–22 mm) and winter releases in protective habitat to achieve 35% survival (Section 3.2.4). Their assessment resulted in a projected return of $1.43–$2.33 for every dollar invested. It remains to be seen, however, whether sufficient seed can be obtained from hatcheries to implement stock enhancement programmes for hard clams at the scale required to benefit fishermen.

The outcomes of transplanting adult clams to increase local abundance of subsequent year classes have not been promising. Manzi (1990), drawing on the earlier work of McHugh (1981), demonstrated that transplantation of 125,000 spawning hard clams to Great South Bay, Long Island, contributed a trivial 0.0005% to total annual larval production. He does not dismiss the possibility that transplants of broodstock can be effective but implies that the scale of the process, and the location, will affect success. In another trial, densities of adult hard clams in Florida were increased 25-fold by transplanting clams into sanctuaries for spawners (Arnold, 2001). However, more than 50% of the clams died shortly after transplantation, because of unusually low salinities and difficulties in burrowing (Arnold *et al.*, 2002). The problems did not stop there; the incidence of gonadal neoplasia, which reduces reproductive capacity (see Section 3.2.7), was much greater in transplanted than in undisturbed clams, and most of the clams died within 12 mo (Arnold *et al.*, 2002). Clearly, a sound understanding of the habitat requirements of hard clams, and the size of the natural population, is fundamental to determining the contribution that can be made by increasing the numbers of spawners.

Few advances have been made in methods for growing *Crassostrea virginica* in the eastern United States for many years. The basic techniques for transferring seed from settlement areas to growing areas developed in the 1800s are still in use. Unfortunately, as mentioned in Sections 3.2.1 and 3.2.5, the size of the industry is much diminished because of elimination of many of the reefs needed to provide substrata for settling larvae, siltation that has rendered much of the remaining settlement surfaces unsuitable, and the diseases MSX and Dermo. This is nowhere more evident than in Chesapeake Bay (Figure 3.5). The rehabilitation efforts described in Section 3.2.5 will hopefully restore some of the former production, although much of the potential growing area is still affected by MSX and Dermo.

Little information is available on the success of transplanting juvenile razor clams to surf beaches on the west coast except that release of >90 million juveniles in Washington State in 1985 (Rickard and Newman, 1988) was regarded as 'clearly successful and cost-effective' (Westley *et al.*, 1990). The factor most likely to affect the success of this form of stock enhancement is the regularity of good 'sets' of juvenile clams. Large numbers of seed have to be available for transplanting at least once every 3 yr for this method to increase production substantially (Westley *et al.*, 1990).

Stock enhancement of geoduck clams on public intertidal habitat in Washington State is believed to have potential to increase recreational catch by 50% (Beattie, 1992). Consequently, releases were implemented with assistance from volunteer groups (Stuller, 1995; Beattie and Blake, 1999). The PVC tube method (see Section 3.2.4), or a variation of it, has also been adopted recently by the commercial sector. As a result, culture of geoduck clams in Washington has increased rapidly. For example, shellfish companies had invested US$15 million by 1999, with expected returns of US $400,000 ha^{-1} (Beattie and Blake, 1999). Development of commercial intertidal culture of geoduck clams in Washington, in the presence of the substantial subtidal resource, is presumably driven by catch limits and high prices. Government support for subtidal stock enhancement of this species was terminated in the early 1990s because of the apparently insurmountable problem of predation on the released clams (Beattie, 1992). However, predation has now been overcome in Canada, where commercial companies in British Columbia have developed 'planting machines' that place juveniles under protective netting in leased areas (Beattie and Blake, 1999).

Only small quantities of hatchery-reared Manila clams have been planted on beaches in Washington State, because the industry relies mainly on natural replenishment of this introduced species (Chew, 1989). Yields of Manila clams settling on gravel are estimated to be >16 t ha^{-1}, and can be made 3 yr after modification of the substratum (Chew, 1989).

3.2.10. Management

Few special provisions have been introduced for clams on the east coast of North America since the implementation of stock enhancement. Stocks continue to be managed by measures such as size limits, closures and quotas. In some areas (e.g., Rhode Island) clams located behind the 'pollution' line are relayed to areas open to fishing to augment the exploited stock (Section 3.2.2).

According to Malouf (1989), the most important decision concerning stock enhancement of hard clam fisheries is not how to fine-tune management to extract the maximum benefit from the cost of enhancement, but whether planting of juveniles is an appropriate management tool in the first place. He draws attention to the high cost of planting seed, which often consumes 50% of funds available for management of clams. Given the high fecundity and survival of wild adults, he questions whether the financial resources would not be used more effectively in other ways, such as establishment of reserves and reassessment of quotas. Malouf (1989) concludes, however, that stock enhancement does have potential to achieve two goals within integrated management programmes for clam fisheries: (1) reestablishment of self-sustaining clam populations in relatively small areas and (2) maintenance of intense recreational harvesting in areas where the community meets the cost of planting seed. Rice et al. (2000) describe a recent example of the latter situation in Rhode Island. Arnold et al. (2002) also acknowledge that stock enhancement of hard clams using any of the three strategies they tested, transplanting spawners, planting juveniles and releasing larvae, is fraught with difficulties and unlikely to be economically viable. They conclude that such interventions are only likely to be useful in the form of restocking programmes to restore a collapsed population.

The management of oyster fisheries has centred around maintaining the areas where seed are collected, leasing growing areas and regulating the harvests of oysters from public and private beds. Legislation includes: size limits designed to deliver the most desirable product and prevent overharvesting; closed seasons to protect sufficient oysters for replenishment; and gear; and operating restrictions to control the rate of harvest and limit damage to sensitive oyster reefs (MacKenzie, 1996; Leard *et al.*, 1999). The latter management measure is illustrated by the fishery in Delaware Bay. There, the state of New Jersey introduced the 'rough cull law' to ensure that dead shells were returned to the water during collection of seed and adult oysters to provide sufficient substrata for settling larvae. This law mandated that no more than 15% of material removed from the beds could be shell (MacKenzie, 1996).

Much attention has also been given to regulations to ensure that harvested oysters are fit for human consumption. The National Shellfish Sanitation Program (NSSP) was established in 1925. The NSSP issues uniform sanitation standards for growing, harvesting, processing and shipping shellfish. These standards are incorporated into state laws and regulations. States conduct sanitary surveys of harvest areas and catchments and then classify growing areas as: approved areas, conditionally approved areas, restricted areas and prohibited areas. Shellfish may be harvested from approved areas at any time. Oysters taken from restricted areas need to be purified in a depuration facility or relayed for a period (usually at least 15 d) to approved areas before sale. Details of regulations for the other classified areas are provided by MacKenzie (1996).

There are few specific management regulations for enhanced stocks of clams on the west coast of United States. These fisheries continue to be managed by the suite of established measures, including pre- and post-harvest surveys, lease areas, catch limits and recreational permits (L. Blankenship, personal communication 2003).

3.2.11. Future research

It is now clear that one of the most likely uses of hatchery releases of hard clams, or transplantations of broodstock, will be to restore small populations that have collapsed. Further research is needed, using methods in addition to population genetics, to identify these small, self-replenishing populations. For those that have been over-fished severely, the imperative is to determine whether releases of juveniles or transplantation of broodstock are likely to add value to other forms of management. Where communities are persuaded that planting seed clams is a cost-effective way of maintaining recreational fisheries, studies will need to be done at each location to identify the optimum stocking levels to avoid density-dependent effects on growth and survival.

For the oyster industry on the east coast of the United States, an important need is to identify populations with greater resistance to the diseases MSX and Dermo and to use them in selective breeding programmes to produce disease-resistant individuals. Provided such selection does not affect other important adaptations, it may be possible to re-establish self-sustaining fisheries throughout much of the former distribution of this species. Another important issue is to pursue the research outlined in Section 3.2.5 to identify how and where to re-establish oyster reefs to re-create the hard substrata necessary for prolific settlement of spat.

3.3. ABALONE

3.3.1. Background and rationale for stock enhancement

Abalone (*Haliotis* species) are the most prized of the marine gastropod molluscs used for food. They are highly valued for the delicately flavoured meat of their foot muscle (Figure 3.10), and about half of the extant 56 species (Geiger, 2000) are harvested commercially, primarily for markets in east Asian countries. Abalone occur in greatest numbers along temperate rocky coastlines, where the macro-algae on which they feed are abundant, although a few of the smaller species are harvested only in tropical waters. They occur mostly at depths <20 m, but some species live at depths >50 m. Abalone are partially protected from fishing by the dense macro-algal cover and complex boulder habitat in which they live. The cryptic behaviour of the adult stage of many species also affords some protection, although other species are conspicuous as adults after they emerge from the sub-boulder habitat at onset of sexual maturity (Kojima, 1981; Prince et al., 1988a; Nash, 1992). Nevertheless, all species can be caught fairly easily during the day by divers using compressed air or by free diving.

The high market value of abalone and their vulnerability to capture have led to intense pressure on stocks worldwide; the general trend has been an increase in production to a maximum level followed by a decline, in some cases to trivial catches. In many countries, stringent management measures have been introduced to reverse such trends. In North America, for example, there is currently a moratorium on commercial, recreational and traditional fishing for the northern abalone *Haliotis kamtschatkana* in British Columbia

(Campbell, 2000a), a moratorium on fishing for all species of abalone in southern California, and only recreational fishing is allowed in northern California (Tegner, 2000). In 2001, the Californian white abalone (*H. sorenseni*) became the first marine invertebrate to be listed as an endangered species in the Unites States (Hobday *et al.*, 2001), and there is concern that Californian black abalone *H. cracherodii* populations may have declined to dangerously low levels (Altstatt *et al.*, 1996). In British Columbia, *H. kamtschatkana* is vulnerable to over-fishing, because recruitment is low and sporadic (Campbell, 2000b); in fact, there is evidence that populations of *H. kamtschatkana* there were declining before the fishery developed (Breen, 1986). Causes of abalone population declines in California include over-fishing (Davis, 2000; Tegner, 2000), climate change (Tegner, 2000) and disease (Haaker *et al.*, 1992; Bower, 2000). The southward spread of the sea otter on the west coast of the Unites States has also increased the mortality rate of abalone populations (Tegner, 2000).

Stocks of abalone have also declined in Japan, where trends in the catch have been reported variously as decreasing from 3,000 t yr^{-1} to 1,000 t yr^{-1} (Seki and Taniguchi, 2000), and declining to less than half (from ~6,200 t to ~2,300 t) between 1970 and 1995 (Masuda and Tsukamoto, 1998). The causes of these decreases include: intense fishing pressure (Masuda and Tsukamoto, 1998); size limits that are too low relative to size at onset of sexual maturity (Roberts *et al.*, 1999); greatly increased rates of infestation by boring polychaetes (Yanai *et al.*, 1995) and increased prevalence of disease (Nakatsugawa, 1993).

The abalone fishery in South Africa was relatively stable for many years until the upheavals in society in the 1980s and 1990s led to increased illegal fishing (Tarr, 2000). The problem has been exacerbated further in some areas by a recent influx of the rock lobster *Jasus lalandii*, which are reported to reduce survival of juvenile abalone by preying on the sea urchins under which they shelter (Tarr *et al.*, 1996).

Australia, the largest producer of abalone, has experienced some localised stock collapses (Shepherd *et al.*, 2001). In Tasmania, which supplies ~25% of the global wild abalone market, reductions in catch quotas (by 30% in 1989) has been used to lower harvests to sustainable levels (Nash, 1992). Even so, there are signs of stock declines (Tarbath *et al.*, 2003) and recruitment failure in some areas.

Figure 3.10 (A) A juvenile greenlip abalone *Haliotis laevigata* from southern Australia (photo: R. Kuiter). (B) The muscular foot of an abalone that enables these gastropods to grip the substratum firmly and is a delicacy in east Asia (Photo: R. Kuiter). (C) Typical cobblestone habitat for juvenile abalone *Haliotis rubra*, Tasmania, Australia (photo: S. Talbot). (D) Adult abalone *Haliotis iris* in New Zealand and a starfish predator, *Astrostole scabra* (photo: R. Naylor).

In New Zealand, there are seven important fisheries for the paua *Haliotis iris*. Three of these have been assessed, and commercial catches have been reduced because of concerns about sustainability (Annala *et al.*, 2002).

Several aspects of the biology of abalone, and the nature of abalone fisheries, have led managers to consider the use of cultured juveniles to restore severely depleted stocks or enhance production. For example, populations of abalone can be limited by the supply of juveniles, and stocks of some species in some countries suffer from recruitment over-fishing.

The earliest releases of hatchery-reared juveniles were in Japan in the 1960s in response to decreases in landings of abalone and other marine species (Kitada *et al.*, 1992). The purpose of the early Japanese stock enhancement programmes was to harvest the released animals once they reached marketable size but before they reached sexual maturity (Saito, 1984; Seki and Sano, 1998). As a result, little effort was made initially to assess the effect of hatchery releases on natural populations. However, when abalone landings continued to decline despite release of large numbers of juveniles (Figure 3.11), the objective of the releases changed; cultured abalone were allowed to contribute to recruitment to the population (Seki and Sano, 1998; Seki and Taniguchi, 2000) (i.e., the emphasis changed toward restocking).

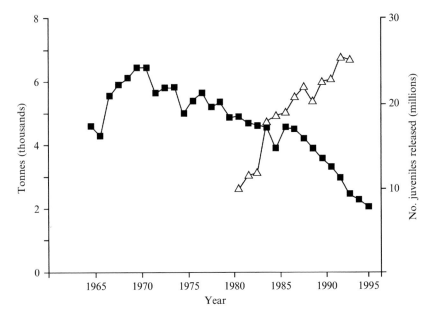

Figure 3.11   Total catch of abalone (■) and number of juveniles released (△) in Japan between 1964 and 1994 (redrawn from Masuda and Tsukamoto, 1998).

In southern California, releases of both larval and juvenile abalone have also been made with a view to rehabilitating depleted stocks (Tegner and Butler, 1989). In New Zealand and Australia, releases are still at the experimental scale and are being used to determine optimal release conditions and to assess whether post-release survival rates are great enough to be economically or practically feasible (Schiel, 1993; Preece et al., 1997; Shepherd et al., 2000; Heasman, 2002; Heasman et al., 2003).

Although the objectives of releasing hatchery-reared abalone have varied between stock enhancement and restocking, both among and within countries, we have placed our analysis of the progress made for this group under 'stock enhancement'. This was the original objective in several cases, including Japan, where the earliest and largest release programmes were initiated.

3.3.2. Supply of juveniles

Although it is possible to collect abalone larvae settling on natural substrata (McShane and Smith, 1991) and artificial collectors placed in the sea (Keesing et al., 1995; Nash et al., 1995; Rodda et al., 1997), and the density of settling juveniles can sometimes be high, this is not a practical way of obtaining sufficient postlarvae or juveniles for ongrowing and release. The main problems are that rates of settlement are too variable (McShane and Smith, 1991; McShane, 1995), and the removal of postlarvae is relatively difficult. Therefore, except for occasional transplantation of wild-caught juveniles among sites (Saito, 1979), larval and juvenile abalone for release in the wild have been produced in hatcheries. Hatchery and land-based nursery methods have been described extensively (e.g., Hahn, 1989). Swimming larvae are settled onto stacked plates coated with a 'diatom film' on which they feed. As the juveniles grow, the diet changes to macro-algae and, increasingly, artificial foods (Fleming, 1995; Jackson et al., 2000). Indeed, juvenile production methods are becoming more and more efficient with early weaning onto artificial diets at a size of 1–2 mm (Heasman, 2002; Heasman et al., 2003). The weaning is done before the juveniles lose condition from over-grazing the diatom film and results in greater rates of survival. Such improvements are very much needed to provide the large numbers of juveniles required for effective restocking and stock enhancement at relatively low cost.

3.3.3. Defects caused by rearing

Atypical (emergent, non-shelter-seeking) behaviour of hatchery-reared juvenile abalone has been reported in tank trials (Schiel and Welden, 1987) and in the sea (Ozumi, 1998). In comparisons of wild and hatchery-reared

Haliotis rufescens, Schiel and Welden (1987) found that cultivated abalone were slower to seek hiding places during the first several hours after release onto rocky substrata in aquaria but did exhibit appropriate behaviour relatively quickly.

Several methods have been proposed for pre-conditioning cultured juvenile abalone before release so that they behave cryptically when placed in the sea. These include: introducing predators to the rearing tanks; crushing some individuals in the tanks to release abalone fluids to induce crypsis in conspecifics, and providing complex substrata and wave action to promote shelter-seeking behaviour and fitness (Tegner, 2000). The effectiveness of these methods is unknown.

3.3.4. Release strategies

Mortality rates of juvenile abalone are often highest immediately after release (Momma, 1972; Tegner and Butler, 1985; Schiel, 1992, 1993; Hooker, 1998, cited in Roberts *et al.*, 1999; Ozumi, 1998; Seki and Sano, 1998; Heasman *et al.*, 2003). Not surprisingly, release strategies for cultured abalone have focused on reducing this mortality, although they have also been aimed at overcoming the relatively high cost of producing 'seed' by releasing juveniles at a smaller size (De Waal and Cook, 2001a; Heasman *et al.*, 2003). This imperative has even resulted in developing ways to release larvae (Tegner *et al.*, 1986; Tong *et al.*, 1987; Schiel, 1992, 1993; Preece *et al.*, 1997; Shepherd *et al.*, 2000; Heasman *et al.*, 2003).

Strategies to reduce the mortality of cultured abalone in the wild have included: releasing juveniles at sizes of up to ~40 mm shell length; placing the young abalone into suitable protective habitat; releasing them into temporary artificial shelter so that they acclimate and adopt the appropriate cryptic behaviour before emerging to seek refuge in the surrounding reef, and removing predators.

Suitable release sites have been identified as areas of shallow cobblestone habitat (Figure 3.10), which provide ample crevices of various sizes for shelter (Kojima, 1974, cited in Seki and Taniguchi, 2000; Saito, 1981; Roberts et al., 1999). The structure of the reef is likely to be more important than protection provided by associated biota. For example, De Waal and Cook (2001b) found that, contrary to previous belief, the primary factor determining survival of juvenile *Haliotis midae* in South Africa was substratum type (stacked boulders <50 cm diameter), rather than the shelter provided beneath the spines of the sea urchin *Parechinus angulosus* (Day and Branch, 2000).

Several types of artificial shelters or 'seeding devices' have been developed to allow newly released juveniles to acclimate and emerge safely into the natural habitat (Ebert and Ebert, 1988; McCormick *et al.*, 1994; Davis, 1995; Sweijd *et al.*, 1998; Cook and Sweijd, 1999a). The development of a release strategy using customised shelter overcomes the need for the pre-release behavioural conditioning mentioned in Section 3.3.3. By removing the effects of predators, artificial structures also provide the additional benefit of allowing researchers to measure natural rates of recruitment (Davis, 1995). Removal of predators has been attempted to reduce mortality of cultured juvenile abalone placed in the wild. For example, octopus were removed before release of two species of abalone in southern California (Tegner and Butler, 1985; Tegner, 1989); however, continual removal of predators was necessary to prevent high mortalities.

The release of larvae requires a different strategy than the release of juveniles but, here too, use of artificial structures has been important. Schiel (1992), Preece *et al.* (1997) and Heasman *et al.* (2003) released abalone larvae onto selected rocky habitats and retained them within experimental sites with tent-like enclosures of fine mesh until they settled. The inference was that, in experiments without tents, many larvae were advected elsewhere by currents. The rather unusual approach of releasing larvae, when the various historical attempts at stock enhancement for a range of marine species eventually led to abandonment of such practices (Blankenship and Leber, 1995; Nicosia and Lavalli, 1999 and references therein), is only made possible by the short duration of the larval stage of abalone (Prince *et al.*, 1987). In this regard, abalone are similar to topshell (Chapter 2, Section 2.2), and this aspect of their biology means that translocation of wild adult broodstock to form high-density aggregations can be considered as an alternative way of supplying increased numbers of larvae for rehabilitating depleted stocks (Tegner and Butler, 1985). To evaluate the biological and economic feasibility of this measure, Emmett and Jamieson (1989) transplanted large numbers of adult *Haliotis kamtschatkana* from a site where growth was poor to one where it was rapid. They found significant increase in growth and concluded that the economic feasibility of such transplants depends on recovery rates of the transplanted adults and the costs of carrying out the operation. Although this study did not consider the potential of transplanted abalone to enhance reproduction and recruitment, it confirms that relocated broodstock survive and grow well. More importantly, large increases in abundance of 0–3-yr-old juveniles were observed 3 yr after transplantation of tagged broodstock at a site in California whereas there was no increase at a reference site (Tegner, 1992). This indicates that translocations of adults can be an effective way of regenerating depleted stocks. Regrettably, abundance of the broodstock declined greatly in this study because of poaching (Tegner, 2000), underscoring the need for firm management of restocking programmes (see Chapter 6).

3.3.5. Use of artificial habitat

In addition to providing shelter for cultured larvae and juveniles at the time of release, artificial structures have also been used to enhance adult abalone habitat and influence natural settlement of larvae (Hayashi and Yamakawa, 1988; Jensen, 1999; Jensen *et al.*, 1999; Sahoo and Ohno, 2000). Abalone densities on artificial reefs may be higher than on surrounding natural reef (Kumaki *et al.*, 2001), but removal of other attached organisms may be needed to maintain suitable habitat (Kumaki *et al.*, 2001), suggesting that this may be impractical other than for experimental or small-scale operations. Similarly, continual re-arrangement of artificial reef structures was necessary to prevent occlusion by sand of interstitial spaces inhabited by abalone (Hayashi and Yamakawa, 1988). Culture of juvenile abalone on artificial rocky reefs covered with net in the intertidal zone may reduce predation and allow easier access for maintenance and husbandry (Guo *et al.*, 1996, 1998).

Development of concrete block structures in high-current areas with low-relief substrata in Japan has promoted larval settlement in the lee of these structures. Like other physical features (e.g., headlands, small islands and topographically complex substrata), these artificial habitats can potentially play a role in concentrating larvae to levels that are many times the 'background' concentration typical of surrounding areas (Davis, 1986; Tanaka *et al.*, 1986; Shepherd *et al.*, 1992a).

A related issue is that some of the plant and animal species associated with abalone habitats have been manipulated to increase the number of abalone. Uki (1981; cited in McCormick *et al.*, 1994) showed that production of abalone increased from 0–3 t yr^{-1} after experimental 'afforestation' of substratum dominated by crustose coralline algae. Conversely, Kim and Chang (1992) found that increases in structural complexity of the habitat by cultivation of seaweed only resulted in a temporary increase in abundance of juvenile abalone, because predatory fish (including wrasses) also increased greatly within the seaweed as it developed. Also, in areas from which the sea urchin *Centrostephanus rodgersii* was removed, there was a 10-fold increase in abalone abundance relative to control areas (Andrew *et al.*, 1998). The reasons for this are not clear, however, and could have been a direct effect of urchin removal or the consequent changes in habitat and increases in shelter (Andrew *et al.*, 1998).

3.3.6. Minimising genetic impacts

Cook and Sweijd (1999b) report significantly reduced genetic diversity and skewed haplotype frequencies in *Haliotis midae* in the abalone farms of South Africa relative to their source populations. Whether this was due to low numbers of broodstock or to selective breeding among broodstock is uncertain. At least one other study shows that selective breeding can be a problem. Using microsatellite analysis, Selvamani *et al.* (2001) found that nearly all juvenile *H. asinina* reared under controlled conditions had the same male parent, despite efforts to provide sperm from several males. They concluded that highly controlled breeding practices may be required to ensure that the genetic diversity of progeny from a captive breeding abalone population is maintained at the level of diversity of the original broodstock. Mgaya *et al.* (1995) reported loss of rare alleles of F_3 'stock' compared with wild populations of *H. tuberculata*, but little reduction in mean number of alleles per locus or levels of heterozygosity, and no evidence of inbreeding. Smith and Conroy (1992) also reported reduction in genetic variation and loss of rare alleles in hatchery-reared larvae of *H. iris* compared with the wild population. However, until genetic analyses of wild larvae (or single cohorts of juveniles) are made, it is not clear if this result is atypical of wild populations (Hedgecock, 1994a; Chapman *et al.*, 1999). Release of sterile, triploid abalone for 'put-and-take' sea ranching programmes would eliminate these genetic risks to natural populations. Research is in progress to assess the usefulness of this method (Heasman, 2002).

It is necessary to understand the spatial structure of abalone populations if further serial declines or collapses of abalone populations are to be averted (Shepherd and Rodda, 2001; Shepherd *et al.*, 2001). Abalone population structure has been inferred from information on larval behaviour, larval dispersal in relation to local hydrodynamics and coastal topography and genetic differences among populations. Genetic studies reveal that the spatial scale of connectivity between populations varies widely among species of abalone. Using neighbourhood size, or region of complete genetic mixing (random mating), as the measure of genetic connectedness of populations, neighbourhood sizes of near zero to ~500 km have been reported for abalone. For example, Jiang *et al.* (1995) found fixed genetic differences (suggesting a complete lack of genetic exchange) between neighbouring populations of *Haliotis diversicolor* ~35 km apart. High levels of genetic differentiation were recorded among populations of *H. discus hannai* (Hara and Fujio, 1992; Hara and Kikuchi, 1992) and *H. midae* (Sweijd, 1999, cited in Conod *et al.*, 2002) over a similar spatial scale. Neighbourhood sizes of near zero were reported for *H. laevigata* (Brown and Murray, 1992) and *H. corrugata* (Del Río-Portilla and González-Aviles, 2000), but sampling was done at intervals of 35–50 km, and so finer-scale sampling may have shown greater genetic similarity, with consequent increases in estimated neighbourhood size to above zero but less than 35–50 km. In contrast, some other species of abalone have much larger estimated neighbourhood sizes (e.g., ~100 km for *H. fulgens* [Zuñiga *et al.*, 2000] and 500 km for *H. rubra* [Brown, 1991; Conod *et al.*, 2002]).

It is difficult to reconcile the genetic and larval dispersal lines of evidence of population structure for some abalone species, especially *Haliotis rubra*. Localised larval dispersal of only a few tens of metres was inferred by Prince *et al*. (1987, 1988b) at sites where densities of sexually mature adult *H. rubra* were manipulated. Similarly, Andrew *et al*. (1998) found higher recruitment rates of *H. rubra* to areas from which sea urchins were removed and adult abalone were added than at control areas where urchins were removed but abalone were not added. These studies suggest very localised dispersal (and therefore small neighbourhood sizes) for this species, whereas the genetic studies suggest neighbourhood sizes of hundreds of kilometres (Brown, 1991; Brown and Murray, 1992; Conod *et al*., 2002). It seems likely that the high variability in population structure found among abalone species also exists among populations within species, with dispersal distances of larvae from largely self-recruiting populations being determined by local hydrodynamics, depth, bottom complexity and coastline configuration.

The most parsimonious explanation of the genetic and dispersal data for abalone is that there is infrequent, long-distance dissemination of larvae on a background of mostly localized dispersal, with frequency of long-distance dispersal being sufficient to prevent fine-scale genetic differentiation of populations (Conod *et al*., 2002; Elliott *et al*., 2002).

It is important to note that genetic affinities among neighbouring populations may be maintained by far lower rates of larval dispersal than those required to prevent depletion of populations by excessive fishing (or to assist their recovery). Large neighbourhood size therefore does not necessarily imply high resilience to fishing pressure or high rates of recovery of depleted populations by long-distance larval dispersal.

The variable spatial scales of larval dispersal and genetic population structure described previously may explain observed differences in resilience to fishing among species and among populations within species. For species with multiple, largely self-replenishing populations at small spatial scales, traditional coarse-scale management measures may not provide effective protection against localised over-fishing and depletion (Prince and Hilborn, 1998; Shepherd and Rodda, 2001; Shepherd *et al.*, 2001; see also Withler, 2000). Localised management by abalone fishing co-operatives or communities may be needed to ensure locally appropriate rates of fishing pressure to minimise the risk of depletion (Prince *et al.*, 1998).

It should not be difficult to obtain distinctive multi-locus genotypes of individual abalone broodstock using mtDNA and microsatellite loci (Kirby *et al.*, 1998; Selvamani *et al.*, 2001; Li *et al.*, 2003) to produce genetically tagged juveniles. Although this should facilitate identification of hatchery-reared juveniles in the wild (see Section 3.3.8), such genetic manipulations might seem to conflict with the desire to release juveniles that are as genetically similar as possible to the wild stock. However, if markers are hypervariable enough, all multi-locus genotypes will be individually rare, and there is no need to choose particular genotypes. Microsatellite loci are likely to fulfil these needs, making it possible to produce genetically tagged juveniles without compromising their genetic diversity or fitness greatly (Burton and Tegner, 2000; Bravington and Ward, 2004).

3.3.7. Diseases

Bower (2000) provides a detailed review of abalone diseases and parasites and the risks associated with transplantation of animals. The risks of transmitting these diseases and parasites among hatcheries, from hatcheries to wild populations, and between locations or even countries are high, and the consequences are severe (Kuris and Culver, 1999). Here, we present a small selection of the many cases described by Bower (2000) to illustrate the virulence of abalone diseases and the risks of spreading them.

Withering syndrome (WS) is a bacterial pathogen of several abalone species (especially *Haliotis cracherodii*) in California (Richards and Davis, 1993; Friedman *et al.*, 2000). WS has caused mortalities of up to 95–100% in some abalone populations since its discovery in 1986 (Haaker *et al.*, 1992; Richards, 2000). Given the virulence of this pathogen, the importance of strict protocols for transfers of Californian abalone cannot be overstated. The population impact of WS was one of the factors that led to closure of abalone fisheries in California (Davis, 2000). Discovery of WS in two isolated locations in the colder waters of northern California has been attributed to transplantation from areas or abalone farms now known to be infected with this organism (Finley *et al.*, 2001). A disease of cultured *H. discus hannai* with similar symptoms, and of unknown aetiology, has been reported on the northern coast of China (Guo *et al.*, 1999).

Labyrinthuloides haliotidis is a pathogen of small (<15 mm) *Haliotis kamtschatkana* in British Columbia that has caused high mortality in hatcheries and contributed to the closure of the first and, until recently, only abalone hatchery there (Bower, 2000). This organism has yet to be located in the wild, and so it is not known what role (if any) it has played in the decline of native populations of *H. kamtschatkana*. In Brittany and the Channel Islands, 60–80% of the European abalone *H. tuberculata* died from *Vibrio* infection in 1997–1999 (Nicolas *et al.*, 2002).

Introduced parasites have also caused great damage to stocks of abalone. The problems that have beset California and Japan provide salutary examples. Abalone farms in California became infected by a non-native sabellid polychaete (probably introduced from South Africa with a shipment of cultured abalone that were not quarantined) in the 1990s (McBride, 1998). Tubicolous polychaetes infect the shell of abalone, interfering with growth, and may cause severe shell deformities and even death. Infestation of some wild *Haliotis discus hannai* populations in Japan by the polychaete *Polydora* sp. have been as high as 100% (Kojima and Imajima, 1982; Yanai *et al.*, 1995). Increasing infestation rates of abalone by *Polydora* are believed to have contributed to the declining abalone catch in Japan. However, at this stage it is not known whether hatchery releases of abalone were implicated in the spread of *Polydora*.

3.3.8. Systems for marking juveniles to assess effectiveness of releases

There are at least two ways of marking hatchery-reared juvenile abalone *en masse* as an aid to assessing contribution of hatchery releases to increased catches: distinctive shell colour and genetic tags (see Section 3.3.6). Abalone reared in artificial systems generally have brightly coloured shells caused by diet (Leighton, 1961; Sakai, 1962; Olsen, 1968a,b; Horiguchi *et al.*, 1994). This is a relatively cheap and simple option for batch-marking cultured animals; however, identification becomes more difficult with time as overgrowth of coralline algae and other epi-biota increases.

Polymorphic microsatellite loci in various species of *Haliotis* (Kirby *et al.*, 1998; Evans *et al.*, 2000; Selvamani *et al.*, 2001; Conod *et al.*, 2002) provide a more complex and expensive way to identify juveniles, reared from parents of known genotype, released in the wild. Although the costs may be prohibitive for stock enhancement programmes, this technology will be necessary for assessing the contribution of progeny of hatchery-reared abalone to replenishment of populations in restocking programmes.

3.3.9. Status of stock enhancement initiatives

3.3.9.1. Releases of larvae

Evaluating the success of larval releases is difficult and time-consuming. In particular, the small size of the larvae, and the complexity of the substratum where they are released, makes it hard to estimate survival accurately (Schiel, 1993). Estimates of survival usually depend on waiting until the abalone grow to a size where they are more visible. However, such estimates are confounded by two countervailing sources of bias: continued emigration of juveniles from the release area with time and increasing likelihood of sighting as they grow larger and begin to emerge from the sub-boulder habitat (Shepherd *et al.*, 2000). Estimates are made easier when the release area is an isolated patch reef (artificial or natural), because absence of juveniles can be attributed to mortality.

Although early trials with releases of larvae indicated some potential for success (Tong et al., 1987), with survival rates of ~0.36%, subsequent attempts have resulted in lower survival and failed to demonstrate the cost-effectiveness of this approach. Studies involving *Haliotis iris* in New Zealand and *H. laevigata* and *H. rubra* in Australia are cases in point. Schiel (1992) released larvae of *H. iris* at a density of 20,000 m^{-2} using mesh tents for 24 h to retain larvae over coralline rocks and concluded that larval releases were not cost-effective because survival to a shell length of 5 mm was ~0.05%. McShane and Naylor (1995) reported survival of *H. iris* post-larvae that had been settled onto coralline-coated boulders in the hatchery at a density of 4,000 m^{-2} as 0.5–5% two days after settlement and <0.2% after 16 weeks. Experimental releases of larvae of *H. laevigata* and *H. rubra* into coralline-covered boulder habitat at densities of 2,000–120,000 m^{-2} resulted in variable survival of 0.02–7.8% at 6–9 d after settlement (Preece et al., 1997; Shepherd et al., 2000). For *H. laevigata* larvae released at 2,000 m^{-2}, ~5% survived to 6 d after settlement, 0.1% to 9 d and 0.03% to 11 mo. The lowest densities were most cost-effective, and survival may be greater at even lower densities (Preece et al., 1997; Shepherd et al., 2000). The variability in survival reflects the great patchiness of suitable microhabitat and predators, and advection from release sites by currents characteristic of topographically complex substrata such as boulders (Guichard and Bouchet, 1998; Shepherd et al., 2000).

In addition to being highly variable, survival of released postlarvae has often been strongly density-dependent (McShane, 1991; Schiel, 1993; Sasaki and Shepherd, 1995; Shepherd et al., 2000; Heasman, 2002; Heasman et al., 2003), although the strong relationship between release density and survival found by Schiel (1993) is difficult to interpret, because the main cause of mortality was burial by sand. Nevertheless, the consensus is that density-dependent mortality can occur because high densities of juvenile abalone are more likely to attract roving predators or promote competition for living space and food (Roberts et al., 1999; Shepherd et al., 2000; Sasaki and Shepherd, 2001).

Heasman et al. (2003) have shown that the carrying capacity of substrata with crustose coralline algae for postlarval abalone is a determinant of post-release survival. To avoid the high, density-dependent mortality that occurs when larvae and postlarvae are added to natural substrata (Preece et al., 1997; Shepherd et al., 2000; Heasman et al., 2003), methods need to be developed for large-scale release at low densities. Using SCUBA or com-pressed-air "hookah" equipment to do this would be impractical and prohibitively costly.

Although many of the releases of larvae have been at an experimental scale, larger long-term initiatives have been undertaken. One of the earliest operations, dating back to the 1960s in Mexico (Ortiz-Quintanilla, 1980), is still being maintained by some fishing co-operatives (Mazón-Sastegui *et al.*, 1996; Aviles, 2000), although recent attempts to assess its effectiveness have not been conclusive (S. Shepherd, personal communication 2003). In Japan, Seki and Taniguchi (2000) found that, despite the far lower unit cost of producing larvae, their rates of survival were too low to be economically competitive. They concluded that the greatest potential for restocking and stock enhancement of abalone was by releasing juveniles, not larvae. Experiments with *Haliotis rubra* in New South Wales (Australia) have also demonstrated that releasing juveniles at a mean shell length of 7–12 mm is more cost-effective than releasing larvae (Heasman, 2002). This concurs with Seki and Sano's (1998) results with *H. discus hannai.*

3.3.9.2. Releases of juveniles

Most releases of juvenile abalone have been in Japan, where the scale of operations over the past four decades has been large compared with many other groups of invertebrates (Saito, 1984; Uki, 1989). Quoting government statistics, Masuda and Tsukamoto (1998) reported the number of juvenile abalone released into fishing grounds in Japan increased from ~10 to ~30 million yr^{-1} between 1980 and 1992 (Figure 3.11). The accuracy of these figures is uncertain because they are at variance with other information, for example, release of about 150 million yr^{-1} since the early 1980s (Seki and Taniguchi, 2000). Releases in other countries have been much lower or experimental (Tegner and Butler, 1985; Schiel, 1993; McCormick *et al.*, 1994; Shepherd *et al.*, 2000; Heasman, 2002).

The success of releasing cultured abalone has generally been estimated from recovery of individuals 'tagged' with distinctively coloured shells (Section 3.3.8) (Tegner and Butler, 1985; Seki and Taniguchi, 2000), although physical tags have been used in some experiments (C. Dixon, personal communication, 2004). As with released larvae, estimation of survival is prone to bias by dispersal (Saito, 1984; Tegner, 2000; De Waal *et al.*, 2003) and crypsis, with probability of capture increasing with size (Shepherd *et al.*, 2000; Catchpole *et al.*, 2001). This is illustrated by a 3-yr study to assess the effectiveness of releasing *Haliotis discus hannai* in Korea, where the numbers of juveniles recaptured every 3 mo remained fairly constant, yielding a mean annual survival from the final recapture survey of 16% (calculated from Table 3 of Chang *et al.*, 1985). In this case, the diminishing number of juveniles was apparently counterbalanced by the increased probability of locating larger individuals.

Recaptures of released juvenile abalone are reported either as the 'spot' recapture rate at any one time during post-release surveys or as the 'cumulative recapture rate' of commercial-size individuals over several years. Kojima (1995) reported cumulative recapture rates of 12–51%, with rates from most stock enhancement efforts of 10–30% (Kojima, 1994, cited in Masuda and Tsukamoto, 1998; Kemuyama et al., 1997). The cumulative recapture rates estimated by Zhao et al. (1991a) were also 10–30%. Seki and Taniguchi (2000) estimated, without evidence, that their 'spot' recapture rates were ~50%. In contrast, Zhao et al. (1991b) presented quantitative evidence that 'spot' recapture rates were much lower (1.6–9.4%), with little indication of any size dependence in recapture rate in the ranges of 6–10 and 26–32 mm. When juveniles are released into cages or enclosures, dispersal can be prevented, and cumulative recapture rates can then be high (Guo et al., 1998).

Despite moderately high post-release survival rates of cultured juveniles in some areas of Japan (Table 3.2), and good contributions of hatchery-reared abalone to annual landings there, overall the abalone stock enhancement programmes in Japan have failed to augment, or even maintain, levels of catch since releases began (Figure 3.11). Kafuku and Ikenoue (1984) suggest that, despite the large scale of releases in Japan, the numbers of cultured juveniles have been too low to affect total catch. This is supported by the analysis of Seki and Sano (1998), who showed that 2–10 million 6-mo-old juvenile Haliotis discus hannai would need to be released to produce 12 t of adult stock, depending on mortality rate. When this is considered with reference to the numbers of juvenile abalone released annually in Japan (Figure 3.11), it is clear that the scale of abalone stock enhancement is small relative to the natural abalone population and annual harvest. This has led to a reappraisal of management of abalone stocks in Japan (Uki, 1989; Ikenoue and Kafuku, 1992; Masuda and Tsukamoto, 1998; Seki and Sano, 1998; Kiyomoto and Yamasaki, 1999; Seki and Taniguchi, 2000). Other measures, such as limits on total catch (Kiyomoto and Yamasaki, 1999), higher minimum size limits (Tanaka, 1988) and seasonal closures (Masuda and Tsukamoto, 1998), have been recommended. There has also been an increasing emphasis on using releases of juveniles to restock rather than enhance abalone resources (Seki and Taniguchi, 2000).

A similar message is emanating from the smaller-scale stock enhancement initiatives in California, where releases of juvenile abalone have not been effective. There has also been concern that release programmes have redirected funds from traditional fishery management. This has resulted in postponement for two decades of the necessary, but often politically difficult, decisions to reduce fishing pressure (Burton and Tegner, 2000; Davis, 2000; Tegner, 2000).

Table 3.2 Summary of juvenile abalone release experiments, modified from McCormick et al. (1994).

Country	Species	Initial length (mm)	Duration	Mean survival (%)	Comments	Reference
Japan	Haliotis discus discus	8–18	1 wk	0–52	Site-specific mortality	Kojima (1981)
		15–50	2 yr	13	Artificial reef	Sakamoto et al. (1984)
		15–40	2–6 yr	12–51	Results collected from fisheries captures (1983–92)	Kojima (1995)
	H. discus hannai	>40	1 yr	>60	Size relationship for survival	Inoue (1976)
		14.4	9 mo	49	Importance of site selection. Protection from predators	Takeichi et al. (1978)
		20–45	2 yr	51–65		Honma and Iioka (1980)
		15–21	1 yr	17–33	An artificial reef (concrete cribs)	Momma et al. (1980)
		20–50	2 yr	25–50		Takeichi (1981)
		<22	154 d	<10		Miyamoto et al. (1982)
		>22	154 d	23–31		Saito (1984)
		10–52	16 mo	0–30		Takeichi (1988);
		30		25–30		Inoguchi (1993)
		16–35	4 yr	5–10	Conditioning of seed in cage for 7 d. Use of oyster shell for releasing. Importance of habitat selection	Seki and Taniguchi (2000)
		24.5	28 mo	27		
		16.5	11 mo	37		
		16.5	5 mo	11–24		

					Notes	Reference
	H. d. hannai (wild)	6–10	18 mo	2–3	Survival rates as a function of size calculated from models	Zhao et al. (1991b)
		11–15	18 mo	4–6		
		16–20	18 mo	4–6		
		21–25	18 mo	4–9		
		26–32	18 mo	4–8		
		40–60	24 yr	2–31		Saito (1979)
	H. gigantea	>40	1 yr	>60		Inoue (1976)
	H. sieboldi	>40	1 yr	>60		Inoue (1976)
California	H. fulgens	25	1–35 d	92–99	PVC releasing module. Important transport effect	McCormick et al. (1994)
	H. rufescens	20–35	1 yr	<1	44% of dead shells recovered. Predator control.	Tegner and Butler (1985); Tegner (1989)
		40–80	1 yr	<1	Importance of emigration	
		15–90	100 wk	18–30	Artificial reef (assemblage of concrete blocks)	Davis (1995)
		8.2	2 yr	1	Releasing unit. Sea-urchins improve survival	Rogers-Bennett and Pearse (1998)
New Zealand	H. iris	3–13	2 yr	54	Good record of habitat characteristics. Record of density before and after seeding. Seeding by hand. Minimum survival rates recorded	Schiel (1993)
		3–22	1 yr	10		
		5–22	1.5 yr	7–24		
		9–49	10 mo	19–46		
		6–23	7 mo	8		

(Continued)

Table 3.2 (Continued)

Country	Species	Initial length (mm)	Duration	Mean survival (%)	Comments	Reference
		19	3 yr	24	Mortality caused by sediments, handling and bad seed quality	Schiel (1992)
		26 & 45	5.5 wk	65	Comparison of wild/hatchery, large/small, predators/no predators	
South Africa	*H. midae*	14.8 / 8.2	6 mo / 6 mo	27–39 / 26–28	Use of PVC releasing module. Minimum survival rates	Sweijd *et al.* (1998)
		24–28	2 mo	59		De Waal and Cook (2001b)
South Australia	*H. laevigata* (wild)	12–16 / 60–95	2 mo / 1 yr	24 / 44	Dependent on substratum	Shepherd (1986)
		25–31	9 mo	0–55	Importance of habitat: survival rate higher in 2-layer boulders than 1-layer	C. Dixon, personal communication (2004)
	H. rubra	12.1		8–16	Use of 3 release methods	Shepherd *et al.* (2000)
France	*H. tuberculata*	20	2.5 yr	40–50	Artificial habitat preventing migration and predation	Flassch and Aveline (1984)

For artificial releases of juvenile abalone to augment natural populations significantly, new methods are needed. In particular, it will be necessary to avoid forming aggregations of juveniles that attract predators and to release the juveniles at the sparse densities (1–2 individuals m^{-2}) that match the inherently low carrying capacity of crustose coralline algal habitats (Heasman *et al.*, 2003).

It is still unclear whether use of artificial habitats to enhance post-release survival of juvenile abalone may reverse these conclusions. In California, the longer-term benefits of higher rates of survival within artificial habitats (Davis, 1995) seem to be minimal because of high mortality once the abalone emerge and migrate to surrounding substrata (Rogers-Bennett and Pearse, 1998; Tegner, 2000). By contrast, relatively high survival of released juveniles in South Africa (Sweijd *et al.*, 1998), and absence of dead shells adjacent to release chambers once all abalone have emerged, suggest that protection during release is effective. High post-release survival rates have also been acheived for juvenile *Haliotis laevigata* (mean size, 28 mm shell length) placed on artificial reefs of small boulders two layers thick (C. Dixon, personal communication, 2004). In addition, high survival rates (90% over 18 mo) and annual return on investment >30% have been achieved from releases of 14–30 mm juvenile *H. discus hannai* placed on artificial reefs covered with nets in the intertidal zone (Guo *et al.*, 1998). Nevertheless, protective release chambers cannot be expected to improve survival unless they are located adjacent to suitable natural shelter. Otherwise, major mortality is likely to occur when abalone leave the structures and become exposed to predators.

The variable results from different studies outlined previously highlight the difficulty in extrapolating results from one time and place to another because of spatial and temporal patchiness in food, predators and shelter (Shepherd *et al.*, 2000). For each location, the challenge is to design cheap release devices that allow juveniles to be placed at low density on the large scales needed to contribute significantly to natural populations. However, the cost and aesthetic impact of the large numbers of artificial shelters that will be needed for releases of significant numbers of juveniles needs to be considered carefully before embarking on large-scale stock enhancement of abalone. The small, biodegradable shelters that each house 10–20 juveniles, now being developed for *Haliotis rubra* in Australia, promise to address these concerns (M. Heasman, personal communication 2004).

The best size at which to release juveniles still needs to be determined for most species of abalone. Although post-release survival rates generally increase with size at release (Table 3.3; Saito, 1984; Seki and Taniguchi, 2000), release operations are usually optimized when abalone are released at smaller sizes. Using post-release survival rates and hatchery production costs from four sites in Japan, Zhao *et al.* (1991b) showed that optimal release size

Table 3.3 Relationship between size and estimated survival rate of juvenile abalone *Haliotis discus* after release into the sea at several prefectures in Japan (note that duration of releases varied among studies).

Prefecture	Abalone length (mm)	Survival rate (per unit time) (%)	Reference
Fukuoka	34	18–21	Uchiba and Futajima (1980)
Fukushima	<10	<1	Saito (1984)
	16–20	<10	
	20–24	<20	
	24–28	<25	
Iwate	20	27.5	Tsuchida *et al.* (1971)
	30	64.6	Takeichi (1981)
	20	25.0	
	23	39.0	
	30	32.7	
	36	45.8	
	40	49.5	
	50	34.5	
Kanagawa	20	<10	Inoue (1976)
	25	25–30	
	30	30–60	
	40	70–80	
Kyoto	21.6	15.0	Nishimura and Tsuji (1980)
	30.8	32.1	
Yamagata	20	51.0	Honma and Iioka (1980)
	25	58.9	
	30	56.8	
	35	55.2	
	40	64.5	
	45	54.6	

was ~20 mm (range, 11–28 mm) but suggested that local determinants of growth and survival would modify this conclusion.

There have been no studies to determine whether released juvenile abalone augment or replace wild populations. This is an important issue, especially in Japan, where large-scale releases have been made, yet catches have declined steadily. At present, it is not possible to determine whether released juveniles slowed this decline or contributed to it by promoting local density-dependent interactions, introducing diseases or degrading the gene pool. Density-dependent interactions are a distinct possibility because abalone are usually released in relatively high-density aggregations in Japan (e.g., Seki and Taniguchi, 2000). This would affect survival if juveniles cannot disperse before resources such as shelter and food become limiting and where roving predators are common.

The density-dependence question is made more interesting because translocation of wild or cultured juvenile abalone outside their natural ranges has resulted in establishment of fisheries. Between 1950 and 1974, a total of 4.7 million wild-caught juvenile *Haliotis discus hannai* were transplanted into Funka Bay, Japan, where the species does not occur naturally (Saito, 1979). Total catch over the same period was 288 t. Capture rates of the released abalone ranged from 20–82% at several different locations within the bay and, although an economic analysis showed that the cost of juveniles equalled or exceeded their value when they entered the fishery, profits were gained from the capture of their progeny. The South African abalone *H. midae* has also been transplanted to a site on the west coast 400 km north of its natural range (Sweijd *et al.*, 1998). Two batches of juveniles (mean size, 8.2 mm and 14.8 mm) were released into modules attached to the substratum and allowed to emerge at will. Minimum survival estimates after 6 mo were 25–39%. A simple economic analysis using these survival rates suggested that commercial-scale releases in the area might be profitable (Sweijd *et al.*, 1998). For both these examples, it would be interesting to know whether the lack of resident conspecifics offset any adverse effects on growth and survival resulting from climatic differences at the margins of the species' ranges.

3.3.10. Management

An interesting problem for managers is that fisheries for some species of abalone are more resilient to fishing pressure than others. This is true for largely self-replenishing populations within a species as well (Shepherd *et al.*, 2001). A typical pattern in the early days of an abalone fishery is that many areas are fished but, over time, the number of productive fishing areas decreases. Some fishing areas, known by commercial abalone divers as 'non-recovery bottom', disappear, often quite quickly. At the other extreme, some 'hot spot' areas, known as 'nashiro' in Japan (Masuda and Tsukamoto, 1998), become the mainstay of the fishery. Reasons for these differences in resilience to fishing have not been determined scientifically, but almost certainly relate to differences in recruitment rate, which, in turn, is determined by hydrodynamic conditions, coastline configuration and complexity of the substratum. The implication is that highly productive populations of abalone at 'hot spots' are unlikely to need restocking if over-fished. Rather, they are probably best re-established by either removing fishing pressure for a period and/or by transplanting adult broodstock and allowing the naturally high rates of recruitment to replenish the population.

Abalone fisheries are managed by a variety of measures (Shepherd *et al.*, 1992b), but universal among these is a minimum legal size limit. Other measures include limits on total allowable catch (TAC) and number of licensed commercial fishermen, and seasonal or area closures. In general, the size limit is set with reference to the size at which sexual maturity is attained to ensure that, on average, abalone have had at least one, preferably two years to reproduce before becoming vulnerable to fishing (Nash, 1989, 1992). In principle, the size limit may be set at a level that provides sufficient opportunity for good egg production by the population to protect against 'recruitment over-fishing' (Nash, 1992). However, some populations of abalone (e.g., in California) have declined despite adequate opportunities for egg production conferred by an appropriate size limit (Tegner *et al.*, 1989) or even after fishing has been reduced or stopped (Tegner, 2000). By contrast, the rapid recovery of *Haliotis rubra* stocks in Tasmania (Australia) after a 30% reduction in TAC in 1988 (Nash, 1996) and a reduction of the Tasmanian east coast TAC in 2001 (D. Tarbath, personal communication 2003), were based on the expectation that recruitment would not fail. Clearly, the causes of the decline in California abalone stocks extended beyond over-fishing to include ocean climate change (Tegner and Dayton, 1987), which affected kelp populations and most probably the reproduction and recruitment of abalone (Tegner, 1993; Tegner *et al.*, 1996).

These two contrasting examples illustrate that it is not easy to determine whether restocking or stock enhancement are appropriate tools for managing abalone fisheries unless the reasons for the decline in abundance are well understood. In the California example, where environmental factors were implicated in the population decline, the existing (though low) natural recruitment rates may have been limited by reduced carrying capacity (e.g., not enough food for juveniles or adults in the prevailing nutrient-poor conditions). Attempts to boost recruitment or replenish populations by restocking under these adverse circumstances are unlikely to be an effective replenishment measure and may even exacerbate the problem. In the Tasmanian example, high levels of recruitment and rapid recovery of the population(s) after reduction in fishing pressure imply that here, too, release of juveniles would have been an inappropriate management action.

3.3.11. Problems to overcome

A limiting factor for restocking and stock enhancement programmes for abalone may be the inability of hatcheries to produce the required number of juveniles. Limits to hatchery production can be expected because of the relatively high costs of providing the surface area needed for attachment and feeding of juveniles, and the time required to rear them to a size at which their chances of survival are acceptable. Even in Japan, where the scale of releases is large, the numbers of juveniles released have proved insufficient to make a noticeable difference to the catch (Kafuku and Ikenoue, 1984), except at a local scale. Advances in hatchery production of abalone will be needed to help overcome this obstacle. Recent developments in intensive culture methods for rearing juveniles to a size of ~7 mm suggest that this may be possible. For example, year-round spawning of captive broodstock, and greatly intensified nursery production and early weaning of juveniles onto artificial diets, have resulted in a 10-fold increase in production efficiency (Heasman, 2002).

Although production costs can be reduced substantially by releasing larvae, the need to contain and protect them using specialized structures until they settle successfully will also limit the number of places that can receive larvae at reasonable cost. Alternatively, larvae may be released when they are competent to settle (Tegner, 2000), or as week-old postlarvae (Heasman et al., 2003), overcoming the need for such structures. Nevertheless, the very low survival rates achieved so far suggest that releases of larvae are not a viable option for restocking or enhancing populations of abalone. The difficulty in producing genetically diverse larvae and juveniles, which is in part related to the limited opportunities for paternity among broodstock, also needs to be overcome.

Even with the promise of increased efficiency in production of juveniles, it is still unclear whether hatchery and release methods can be developed to reduce the high variability in survival. Certainly, the current limitations on the overall number of animals that can be established successfully indicate strongly that the past emphases on stock enhancement of abalone should be re-orientated to restocking, as is now happening in Japan (Seki and Sano, 1998; Seki and Taniguchi, 2000). Provided there is a clear indication that releases will add value to other forms of management, and regulations protect the released animals, such investment in hatchery production will result in continued harvests rather than unprofitable 'put and take' operations.

Another potential impediment to such restocking programmes, however, is the prevalence of virulent diseases and parasites (see Section 3.3.7). In some places (e.g., British Columbia) pathogens have prevented production of juveniles in hatcheries. In others (e.g., California) imposition of stringent quarantine conditions in hatcheries to reduce the risks of transmitting severe diseases that occur during production of juveniles has led to the suggestion that the use of larvae may be the only practical option for restocking programmes (Tegner, 2000).

Inadequate consideration of possible density-dependence relationships could also negate the release of juvenile abalone. Although recruitment rates of postlarval and juvenile abalone are highly variable (Shepherd, 1990; McShane and Smith, 1991; McShane, 1995), and influenced by adult density (Prince et al., 1987, 1988b; McShane and Smith, 1988), habitat type (McShane, 1991) and environmental conditions (McShane and Smith, 1988), it is evident from catch curve analysis of age structure (Nash, 1992;

Nash *et al.*, 1994; unpublished data) that rates of recruitment to adult size can exhibit far less inter-annual variation than would be expected for species with such high fecundity (Koslow, 1992). This implies that mortality rates for the early life stages of abalone are density-dependent. Density-dependent mortality has also been observed in postlarval and small juvenile *Haliotis rubra* by Preece *et al.* (1997), Shepherd *et al.* (2000) and Heasman *et al.* (2003), but the size or density at which such mortality diminishes has not been established. Heasman *et al.* (2003) provided strong evidence of density-dependent survival after release of postlarvae into juvenile habitat: release densities varied by a factor of eight but were almost identical 8 wk later (Figure 3.12). Unless the conditions under which density-dependent mortality occurs are established, and releases are made at levels lower than those that normally induce density-dependent mortality, or at sizes above which density-dependent mortality is diminished, the benefits of adding cultured juveniles to the population will be negligible. To reiterate, there will be no justification for releasing juveniles into situations where the natural supply of juveniles is already great enough to cause density-dependent mortality.

3.3.12. Future research

In addition to finding ways to scale up production of disease-free, genetically diverse juveniles cost-effectively, viable restocking and stock enhancement of abalone will depend on development of improved release strategies for

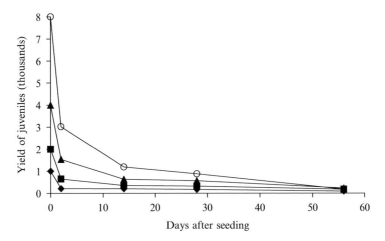

Figure 3.12 Effect of initial stocking density of postlarvae on mean yield of juvenile *Haliotis rubra* released onto cobblestone habitat at densities of 1,000 (◆), 2,000 (■), 4,000 (▲) and 8,000 (○) individuals per litre of mixed-grade rock (redrawn from Heasman *et al.*, 2003).

cultured juveniles and larvae. This research must centre around developing systems to release larger numbers of individuals so that they survive at rates approaching those of wild juveniles. To be effective, these investigations should be based on a thorough understanding of environmental carrying capacity and density-dependent mortality among juvenile abalone and must identify the size and density at release required to avoid triggering predation or food limitation. The alternative challenge is to design low-cost devices for large-scale, low-density releases of small (5–10 mm) juveniles that promote high survival by mitigating predation and providing sufficient resources for all individuals (Heasman *et al.*, 2003). Provided such devices work well, they should remove the need for pre-conditioning abalone to avoid predators after release.

Accurate estimates of post-release mortality are currently prone to bias because young abalone are cryptic and disperse from release areas (De Waal *et al.*, 2003). Sound experimental designs are needed to obtain good survival estimates. Multiple mark-recapture analyses can determine whether survival is dependent on time of release or size at release (Lebreton *et al.*, 1992; Schwarz *et al.*, 1993; Catchpole *et al.*, 2001). Displacements of individually tagged juveniles can also be measured to provide estimates of emigration from the release site. Survival estimates, corrected for emigration, can then be obtained (Arnason, 1972, 1973; Schwarz and Arnason 1990; Schwarz *et al.*, 1993). Nakai *et al.* (1993) have applied this approach to *Haliotis discus hannai*.

The high cost of operating hatcheries is an incentive to investigate whether replenishment of severely over-fished stocks of abalone can be achieved more effectively by translocation and clustering of broodstock (Rogers-Bennett and Pearse, 1998) or by other management measures. Such alternative interventions might include: reducing the total allowable catch; implementing and enforcing legal minimum size limits and/or area closures to protect the necessary levels of egg production or reducing competition with sea urchins (Andrew *et al.*, 1998; Andrew and O'Neill, 2000).

3.4. QUEEN CONCH

3.4.1. Background and rationale for stock enhancement

The queen conch *Strombus gigas* has been exploited in the Caribbean for centuries (Stoner, 1997b) and forms the second most valuable fishery resource in the region, after spiny lobsters. Production reached record levels during the 1990s (FAO, 2002; Table 3.4), with several countries reporting substantially increased landings since the 1980s (Figure 3.13). However, some of these

Table 3.4 Landings (tonnes live weight) of queen conch *Strombus gigas* from the Caribbean between 1950 and 1999. Note that improved reporting accounts for much of the increases since the 1970s. Blank spaces in the table indicate that no data are available; +++ = substantial landings from Mexico that include many species of strombids; rf = recreational fishing only. Data from FAO (2002) and sources indicated below.

Country	1950–1954	1955–1959	1960–1964	1965–1969	1970–1974	1975–1979	1980–1984	1985–1989	1990–1994	1995–1999	Source
Anguilla					5	25	50	82	51	45	
Antigua and Barbuda								370	624	211	
Bahamas				600	857	1,292	2,274	2,247	2,322	2,873	
Belize	700	1,000	1,050	2,000	4,986	2,918	1,202	779	1,085	934	
British Virgin Islands									32	122	
Colombia					239	51	318	2,522	543	866	Mora (1994)
Cuba					1,864	5,273	2,077	1,523	312	3,301	Ferrer and Alcolada (1994)
Dominican Republic					556	1,257	5910	11,251	19,340	9,633	
Grenada				500		24	53	91	35	39	
Guadeloupe		300	500	1,100	1,000	1,250	1,460	1,910	2,440	2,450	
Haiti							200	1,350	1,930	1,780	

Years

											Reference
Honduras	+++		500	800	244	403	342	913	2,600	4,431	
Jamaica		+++						1,300	7,600	9,870	
Mexico			+++	+++	+++	+++	+++	+++	+++	+++	
Netherlands Antilles								70	70	48	Rodriguez Gil (1994)
Nicaragua										371	
Puerto Rico										675	
Saint Kitts and Nevis									21	233	
Saint Lucia								5	41	108	
Saint Vincent/Grenadines									32	93	
Turks and Caicos Is.	300	500	800	1,000	302	2,846	2,508	3,753	2,820	3,570	
U.S.A. Virgin Islands							70	60	140	140	
U.S.A.—Florida				100	rf	rf	rf	Closed	Closed	Closed	Glazer and Berg (1994)
Venezuela				50	605	42	135	349	62	Closed	Rodriguez and Posada (1994)
Total	1,000	1,800	2,850	6,150	10,658	15,381	16,599	28,575	42,100	41,793	

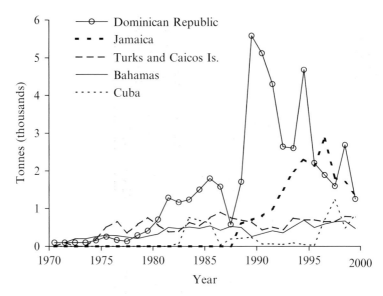

Figure 3.13 Landings (live weight) of queen conch *Strombus gigas* from selected countries in the Caribbean between 1970 and 1999. The decline in production in Jamaica reflects development and subsequent over-exploitation of the offshore fishery at Pedro Bank. No explanation is available for the increase and subsequent decline of the fishery in the Dominican Republic. Data derived from FAO (2002).

apparent increases are probably due to improved reporting, the inclusion of other species, the lack of a historical record or the expansion of local fisheries to deeper waters on oceanic banks, (e.g., Pedro Bank, southwest of Jamaica). Mexico is probably one of the largest producers of queen conch, but uncertainty arises because the landing statistics include other species of strombids. In most countries, however, easily accessible nearshore stocks have declined drastically (Aiken *et al.*, 1997), and it is clear that conservation and management of queen conch require urgent attention to ensure that this important species continues to deliver benefits to the Small Island Developing States of the region (Chakalall and Cochrane, 1997).

Release of cultured juveniles as a means of obtaining greater catches was first suggested by Berg (1976) and pursued further by Brownell *et al.* (1977). Since then, much work has been done to describe the ecology of queen conch to determine how to produce and release juveniles and to evaluate the effectiveness of stock enhancement as a means of increasing catches (see reviews by Acosta, 1994; Appeldoorn and Rodriguez, 1994; Creswell, 1994; Dalton, 1994; Davis, 1994; Stoner, 1997a; Iversen and Jory, 1997). This large body of research has provided a clear picture of the reasons why over-fishing has reduced yields, the biology of the species and the various factors that need to be considered to evaluate the potential and cost–benefit of stock enhancement as a tool for managing queen conch fisheries.

3.4.2. Supply of juveniles

D'Asaro (1965) was the first to rear conch in the laboratory. Production systems for juveniles were then developed by Ballantine and Appeldoorn (1983) and subsequently modified to provide greater output and increased reliability (Davis et al., 1987, 1992; Heyman et al., 1989; Davis and Dalton, 1991; Glazer et al., 1997b; Shawl et al., 2003).

To be assured of an adequate supply of eggs, hatcheries establish an 'egg farm'—an enclosed area of seabed stocked with adult conch before the onset of the spawning season. A typical egg-mass contains about 400,000 eggs. These are cleaned and incubated until the eggs hatch. Veligers are best stocked in 1,000 l parabolic tanks at a density of 150 l^{-1}. About 75% of veligers normally survive metamorphosis, and 16% survive to postlarvae of 12–25 mm shell length (SL) (Davis, 1994).

Settlement can be induced in competent larvae at 16–25 d by adding algal extracts (*Laurencia poitei*), potassium chloride or epiphytes on seagrass blades (Siddall, 1983; Davis et al., 1990; Davis, 1994). However, hydrogen peroxide is superior to these agents (Boettcher et al., 1997) and now used routinely to induce settlement (M. Davis, personal communication 2002). Larvae that do not settle die. Growth, development and survival of settled larvae are satisfactory at a wide range of temperatures (24–32°C) and salinities (30–40 ppt) (Davis, 2000). Attempts to raise the young outside these limits result in arrested development and high mortality. Juveniles reach 50–60 mm SL in 6 mo in hatcheries, and, at this size, they spend almost all their time buried under sand. Above 60 mm SL, they are buried part of the time (Stoner, 1994).

There are few published estimates of the cost of producing juvenile conch. Davis et al. (1987) described the operations of a conch farm in the Turks and Caicos Islands, which is still operational. In 1987, it was estimated that seed conch of 24–45 mm SL could be produced for US$0.15 each. Glazer et al. (1997b) estimated that a single person operating a highly automated state-run hatchery in Florida could produce 15,000 postlarval conch per month. However, no estimate of costs is provided.

3.4.3. Defects caused by rearing

Stoner and Davis (1994) found that mortality of tethered, hatchery-reared conch was approximately 55% over 90 d, compared with <40% for wild conch. The inferior survival of hatchery-reared conch occurred because they did not burrow as often and because they had thinner shells. Growth rates of hatchery-reared conch were also initially slower than for wild individuals held in the same enclosures, although shell spination, weight and growth rates converged quite rapidly (Stoner, 1999). These deficits are not generally considered to be impediments to release programmes, however. For example, Marshall and Lipcius (1999) state that the survival rates of hatchery-reared juvenile conch relative to those of wild juveniles were not significant compared with the influence of release habitat on survival. They concluded that hatchery-reared conch could be used for replenishment of wild stocks if they were released in suitable habitats. Also, Delgado *et al.* (2000) demonstrated that some of the problems of cultured queen conch could be overcome. They reported that juvenile hatchery-reared conch exposed to predatory spiny lobsters burrowed more frequently than controls and had thicker shells, although they also grew more slowly. These behaviours and attributes would presumably give increased survival in the wild.

3.4.4. Release strategies

Much work has gone into identifying the best habitats for the release of juvenile queen conch and to reducing predation. Iversen (1982) was the first to recognize that juveniles occur only near cays in areas with strong currents. This was confirmed by Stoner *et al.* (1996a), who found that important conch nurseries were always located in tidal channels that were flushed regularly with clear oceanic water. These nurseries were permanent, although the exact locations of concentrations of juveniles varied from year to year (Jones and Stoner, 1997). Shell middens were located along shores close to these concentrations, attesting to their permanence (Stoner, 1997b).

Initially, low abundance of small (12–25 mm SL) conch were observed on the surface of the substratum in dense seagrass beds in such locations (Sandt and Stoner, 1993). Subsequently, however, high concentrations of juveniles <10 mm SL were found in seagrass beds using a suction dredge (Stoner *et al.*, 1994, 1998). The larger juveniles, 35–54 mm SL, occur in coarse sand closer to shore, where they bury to a depth of 3–4 cm by day and emerge shortly before sunset (Sandt and Stoner, 1993).

It is now clear that juvenile queen conch only grow and survive well in known juvenile conch habitat because 1-yr-old conch transplanted to apparently suitable areas not inhabited by juveniles had high mortality rates and slow growth (Stoner and Sandt, 1991, 1992). In general, the carrying capacity is ~ 2 individuals m^{-2}. However, in recognized nursery areas, juvenile abundances are correlated with seagrass density up to 608–864 shoots m^{-2} and decline at higher densities (Stoner and Waite, 1990;

Stoner *et al.*, 1999). Thus, queen conch have extremely specific nursery habitat requirements at small sizes, and any departure from these conditions leads to high rates of mortality (Stoner and Glazer, 1998). Release programmes should, therefore, identify the location of functional nursery areas and release cultured juveniles only in such places.

Studies on predation of queen conch show that newly settled individuals of 1 mm SL are consumed by a large array of small predators, including polychaetes, crustaceans, molluscs and fishes. With increasing size, crabs and spiny lobsters progressively gain in importance as predators, whereas predatory gastropods become the principal cause of mortality once queen conch reach sizes >90 mm SL (Table 3.5). All these predators can occur at high densities and exert significant influences on abundances of later stages of queen conch when they emerge to feed (Iversen *et al.*, 1986, 1987, 1990; Ray-Culp *et al.*, 1997, 1999). However, mortality and predation rates vary greatly. In the Bahamas and Florida Keys, for example, the instantaneous natural mortality rate (M) of immature conch (200 mm SL) varied inter-annually from <2.0–12.0 yr^{-1}, and mortality was greatest in summer. Mortality rates of queen conch of 100 mm SL were similar and varied with sites, seasons, densities and hatchery-related deficits in behaviour and shell form (Stoner and Glazer, 1998).

The potential for predation to have a dramatic effect on survival of released queen conch has been demonstrated by tethering experiments. For example, Ray *et al.* (1994) attached hatchery-reared individuals of 20–40 mm SL in an established aggregation of juveniles and recorded 100% mortality in 54 d. Another group of tethered conch of 22 mm SL had 50–90% mortality in 11 d (Ray and Stoner, 1995). Not surprisingly, mortality rates during initial experimental releases of hatchery-reared queen conch were extremely high (Appeldoorn *et al.*, 1983; Jory and Iversen, 1983; Laughlin and Weil, 1983; Siddall, 1983; Appeldoorn, 1985), and annual survival rates of unpenned juveniles were ~2% (Iversen *et al.*, 1986), although they may have been higher (but tagged animals were difficult to relocate). Survival was subsequently improved greatly (28–73%) by placing released juveniles in protective pens.

A minimum size at release of 55 mm SL was recommended by Jory and Iversen (1983), although smaller juveniles can be released effectively into shallow, fenced enclosures that provide partial protection from predators (Davis *et al.*, 1992; Dalton, 1994). Ray *et al.* (1994) recommended a larger minimum release size of 75–90 mm SL but recognized that size at release for restocking is a tradeoff between survival rates and costs. Stocking density has also been investigated. There is a strong negative correlation between mortality of conch of 60–80 mm SL and density (Ray and Stoner, 1994).

Table 3.5 Taxa that prey on queen conch *Strombus gigas* of different sizes. Note also that younger stages of predators that attack large conch are also likely to eat smaller animals.

Taxon	<5	5–10	10–30	30–50	50–70	70–90	>90	Source
Polychaeta								
Glyceridae	?							1
Nereidae	? A							1
Sigalionidae	?							1
Spionidae	?							1
Syllidae	?							1
Crustacea								
Diogenidae	C			CP	CP	CP	CP	1, 2, 4
Calappidae	C	C	C	C	C			1, 2, 4
Majiidae	C							1
Menippidae				C	C	C	C	4
Portunidae	AC	AC	AC		C			1
Alpheidae	AC							1
Palinuridae	C				C	C	C	1, 2
Gastropoda								
Muricidae							PB	4
Fasciolaria sp.							P	4
Pleuropoca sp.			P	P	P	P	P	4
Pisces								
Daysatidae							C	2
Carangidae							C	3
Monacanthidae	F							1
Labridae								4
Tetraodontidae	F							1
Chelonia								
Caretta caretta							C	2

Abbreviations indicate the cause of mortality: A, aperture broken; B, shell bored; C, shell cut or crushed; F, foot or eye stalks eaten; P, aperture penetrated; ?, unknown. Sources: (1) Ray-Culp *et al.* (1997); (2) Iversen *et al.* (1986); (3) Randall (1964); (4) Jory and Iversen (1983).

From these observations, Stoner and Glazer (1998) recommended that: (1) queen conch should be released at sizes >70 mm SL; (2) every effort should be made to increase seed quality; (3) historically important nursery sites should be stocked and (4) releases should be made during autumn to avoid the high predation rates that occur in summer. They also recommended that stocking densities should be at least 1 conch m^{-2}, and that several high-quality sites should be stocked over more than 1 yr to overcome the observed high interannual variation in survival rates.

3.4.5. Minimising genetic impacts

The limited genetic studies done to date suggest how the release of cultured queen conch may be done in a responsible way. Berg *et al.* (1984) examined 10 polymorphic loci from diverse stocks and concluded that Bermuda has an isolated self-sustaining population of conch, whereas populations in Belize and the Turks and Caicos Islands were similar but differentiated from those in the Grenadines. Subsequent studies covering a large number of sites in the Caribbean, the Bahamas and Florida demonstrated that there was a high degree of genetic similarity throughout the region, sustained by larval drift and gene flow (Mitton *et al.*, 1989; Campton *et al.*, 1992). Local patchiness in genetic characteristics probably resulted from 'presettlement stochastic events and processes'. These studies suggest that juveniles can be released at a substantial distance from the source of the egg masses without negative impacts on local gene frequencies, provided the released conch have been derived from a large number of egg masses. It will be important to determine whether the more sensitive techniques now available for assessing differences in population genetics support the results of these earlier analyses.

3.4.6. Disease risks

No disease problems, or impacts on other species, have been reported after the release of cultured conch in the wild.

3.4.7. Systems for marking juveniles to assess effectiveness of releases

Queen conch have been marked by numbered streamer tags glued to the spire of the shell or by spaghetti tags tied around the spire during small-scale field experiments. However, relocation of tagged animals proved difficult, because queen conch spend most daylight hours buried and because such tags become overgrown with algae and other encrusting organisms (Appeldoorn and Ballantine, 1983; Stoner and Davis, 1994). These problems were overcome by Glazer *et al.* (1997a), who used metal detectors to trace aluminum tags attached to the spires of queen conch shells and were able to relocate buried individuals of 75 mm SL with a high degree of reliability. Although it is effective for small-scale experiments, the system developed by Glazer *et al.* (1997a) would not be suitable for monitoring the outcome of mass-releases of juveniles in a stock enhancement programme. An effective method for tagging large numbers of juveniles remains to be developed.

3.4.8. Status of stock enhancement initiatives

In the early 1980s, the potential for extensive mariculture of queen conch was considered to be low because of high hatchery costs, lack of mass-rearing techniques and high predation on juveniles (Iversen, 1982). Since then, hatchery techniques have been refined progressively and costs reduced. However, although all steps necessary for responsible release of cultured queen conch have been developed, the accumulated evidence still indicates that stock enhancement is unlikely to be cost-effective because the expense involved in producing juveniles, coupled with very high mortality rates after release, would exceed of the value of the surviving adults.

Another major problem limiting the potential for stock enhancement of queen conch is the limited availability of nursery habitats. The following brief description of the larval ecology of queen conch explains the problem. Adult conch usually migrate to coarse sand and rubble in slightly deeper waters (6–10 m) to lay their egg masses (Davis *et al.*, 1984; Stoner *et al.*, 1992). The resulting veligers are distributed widely by currents. Those that are transported into deep water perish (Davis, 2001), but others that are competent metamorphose and settle anywhere in the shallows where epiphytic algae occur on seagrass blades or where the red alga *Laurenci poeti* lives. Both types of algae are known to release substances that trigger metamorphosis of conch larvae (Davis *et al.*, 1990; Davis and Stoner, 1994; Stoner *et al.*, 1996b). However, juvenile conch are found only in well-defined, persistent nursery areas in seagrass beds, places flushed regularly by tidal flow of water from offshore (see Section 3.4.4). Despite extensive studies by Stoner and colleagues, no reasons have been found why particular nursery areas persist on a historical scale, whereas apparently similar adjacent areas are devoid of significant numbers of juvenile conch. There are, therefore, a limited number of areas in which juveniles can be released successfully.

The large body of research on culture and release methods has delivered two major benefits. First, it paved the way for farming the species where market conditions are appropriate. For example, queen conch are grown to 60–90 mm SL in on-shore nurseries and large shallow-water enclosures in Providenciales, Turks and Caicos Islands and marketed at this size. Second, availability of cultured juveniles may be useful to managers who wish to use restocking programmes to re-establish spawning biomass at sites that are largely self-recruiting but have been heavily over-fished.

3.4.9. Management

A wide range of management measures has been used for harvesting queen conch in the Caribbean. In most places, these include prohibition of taking immature animals (i.e., individuals that have not developed a flared lip to the shell) (Figure 3.14), production quotas, closed seasons and creation of fishery reserves. However, enforcement of these regulations is a problem in most countries (Chakalall and Cochrane, 1997).

A concern for management is that reduced population densities might affect formation of mating aggregations (Appeldoorn, 1988). This has been substantiated by Stoner and Ray-Culp (2000), who reported an 'Allee effect' in populations of queen conch in the Bahamas (i.e., a population density threshold below which reproduction does not occur). They found that mating virtually ceased at densities <50 conch ha^{-1} and that increased densities resulted in an asymptote in egg production at around 200 conch ha^{-1}. This finding has important implications for the region. Of the 14 areas where abundance of conch has been assessed and compared, eight had <25 individuals ha^{-1}, five had $<200\,ha^{-1}$ and Los Roques National Park in Venezuela had 886 conch ha^{-1} (Aiken et al., 1997). The 'Allee effect' would explain the non-recovery of stocks from Florida and Bermuda, despite protection for many years. It also provides a guide to the numbers of large cultured juveniles that need to be established in programmes designed to restore spawning biomass.

Inadequate knowledge of population structure is another problem for management. Stoner (1997a) has demonstrated that meta-populations are likely to exist for queen conch in parts of their distribution. Unless the structure of the stock is known with respect to 'sink' and 'source' areas, it is not possible for managers to gauge whether the local fishery will benefit from efforts to increase spawning biomass or whether this will depend mainly on the number of spawners further up-current in a meta-population.

In some countries, protected areas may assist recovery of conch abundance, although studies to demonstrate this unequivocally are still lacking. For example, densities of queen conch on algal plains in a lobster and conch reserve in the Turks and Caicos Islands were much greater (2,162 ha^{-1}) than in adjacent exploited areas (687 ha^{-1}) (Tewfik and Bene, 2000). Also, Stoner and Ray (1996) found large differences in stock densities in a protected area and an unprotected area in the Bahamas, and densities of larvae in the protected area were orders of magnitude greater. However, because of prevailing currents, they postulated that much of the larval production would be transported elsewhere in the northern Bahamas.

3.4.10. Future research

Although most aspects of the biology and ecology of queen conch have been studied, and hatchery and juvenile rearing techniques have advanced to the point where commercial cultivation of juveniles is possible, no attempts have been made to re-establish spawning populations (>200 adults ha^{-1}) in protected areas by restocking with large juveniles. There are many areas of the Caribbean where restocking may be required to restore populations of queen

Figure 3.14 (A) Queen conch *Strombus gigas* the basis of subsistence, artisanal and export-driven fisheries in the Caribbean. The flared outer lip of the shell only develops after maturity (photo: C. Berg). (B) Juvenile queen conch grazing in shallow seagrass (photo: A. Stoner). (C) Although some queen conch shells are used for ornamental purposes, most are discarded at landing sites where they form imposing 'middens', some of which have accumulated over centuries (photo: C. Hesse).

conch in a reasonable time. The practicalities of doing so and the spawning success of released cultured animals need to be determined so that managers can decide if restocking programmes are viable tools. Estimating the spawning success of released queen conch should be relatively easy because they lay egg masses that can be collected and monitored. Another important area of research to gauge the potential benefit of restocking programmes is to determine patterns of larval dispersal because, unless a large proportion of larvae are retained locally, benefits are unlikely to accrue to groups making the investments.

3.5. SHRIMP

3.5.1. Background and rationale for stock enhancement

Shrimp of the family Penaeidae, also known as prawns in some countries, are distributed mainly in tropical and subtropical latitudes, although some species extend into warm temperate regions where sea water temperatures are generally above 15° C (Dall *et al.*, 1990). In developed countries, shrimp are usually caught by trawling, whereas in developing nations they are also captured by a variety of other methods, such as trammel-nets, gill nets, bag-nets and push-nets. Significant fisheries for shrimp are found in southeast and central Asia, the southern United States and Mexico, Central and South America, China and Australia. Most fisheries for shrimp are fully or over-exploited and characterized by high fishing mortality (Garcia, 1983; Gulland and Rothschild, 1984; Ye, 2000).

Four main life-cycle categories have been described for penaeid shrimp, three of which involve an estuarine and coastal phase (Dall *et al.*, 1990). Most penaeids that grow to a large size spawn in offshore waters and the planktonic larvae are transported to shallow coastal areas where they settle as postlarvae in different nursery habitats 3–4 wk after the eggs are released. The nursery habitats for juvenile stages vary among species. For example, saltmarsh provides critical nursery habitat for juvenile brown shrimp *Penaeus aztecus* (Zimmerman *et al.*, 1984); the tiger prawns *P. esculentus* and *P. semisulcatus* require seagrass and algae (Loneragan *et al.*, 1998); juvenile banana prawns *P. merguiensis* are found along mangrove-lined mudbanks (Staples *et al.*, 1985) and western king prawns *P. latisulcatus* are associated with sandy substratum (Potter *et al.*, 1991)[1].

---

[1] Note that subgenera of *Penaeus* were elevated to genera by Perez-Farfante and Kensley (1997). There is some controversy over this revised nomenclature, and the older names are used in this section (following Baldwin *et al.*, 1998 and Lavery *et al.*, 2004).

Until relatively recently, it was thought that the large interannual variations in catches for many shrimp fisheries were caused mainly by environmental variation. However, over the past 15–20 yr, clear examples of the
effects of over-fishing on spawning stocks (recruitment over-fishing) have
been demonstrated for several species (e.g., *Penaeus esculentus* in Australia
[Penn and Caputi, 1986; Penn *et al.*, 1995; Wang and Die, 1996] and
P. setiferus in Mexico [Gracia, 1991, 1996] [see also Penn *et al.*, 1997])
(Figure 3.15). The decline of annual catches of *P. esculentus* in Exmouth
Gulf in Australia from the mid–1970s until the early 1980s was arrested only
after the fishery was closed for a year; catches then recovered (Figure 3.15).
There are other cases where large declines in catches of shrimp have been
recorded but the cause of the decline has not been identified clearly. For
example, catches of kuruma shrimp *P. japonicus* in Japan decreased in the
1960s (Figure 3.15) until large-scale stock enhancement was initiated in the
1980s (Fushimi, 1999; Imamura, 1999). The decline of *P. japonicus* was
attributed to a loss of coastal habitat and degradation of coastal water
quality during the period of rapid economic growth in Japan (Fushimi,
1999). Landings of all shrimp from Campeche Banks, Mexico, comprising
>80% *P. duorarum*, declined from >20,000 t in the 1960s and 1970s to
~5,000 t in the 1990s (Figure 3.15), apparently as a result of over-fishing
and deterioration in habitat quality. A change in the ecosystem may also
have occurred (Ramirez-Rodriguez *et al.*, 2000). A recent meta-analysis of
published data from a range of shrimp populations found that recruitment
was related to abundance of spawners and concluded that shrimp populations should be managed to maintain sufficient mature adults to provide an
abundant supply of juveniles (Ye, 2000).

The great economic value of shrimp fisheries has led to very high exploitation and prompted several countries to assess the potential for stock
enhancement programmes. Indeed, large-scale stock enhancement of shrimp
fisheries has been underway since the 1960s in Japan and the 1980s in China
(Table 3.6). In Japan, enhancement was initiated not only in response to the
decline of wild fisheries but also because of loss of coastal habitats. Large-
scale stock enhancement programmes in Japan have been developed through
the establishment of national and regional (prefecture) facilities dedicated to
rebuilding and enhancing wild stocks (Imamura, 1999). However, Japan has
also had the capacity to produce excess postlarvae for use in shrimp stock
enhancement programmes from hatchery facilities designed to supply the
aquaculture industry. For example, in 1981, 513 million postlarvae were
produced, and 60% were released to the sea because of the scarcity of land
to farm shrimp, limiting aquaculture production to about 1,800 t at the time
(Uno, 1985). The total Japanese production of shrimp postlarvae in 1996
was approximately 458 million, with 275 million released for stock enhancement (Imamura, 1999). In China, implementation of stock enhancement has

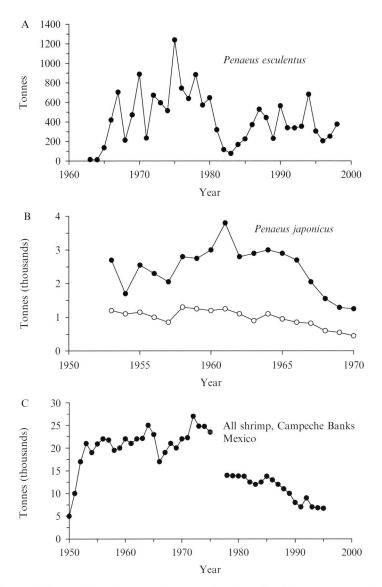

Figure 3.15 Variation in annual catch of shrimp for (A) *Penaeus esculentus* in Exmouth Gulf, Australia (data provided by Department of Fisheries, Western Australia), (B) *Penaeus japonicus* for the whole of Japan (●) and the Seto Inland Sea (○) before large-scale stock enhancement (redrawn from Fushimi, 1999) and (C) the Campeche Banks, Mexico (redrawn from Ramirez Rodriguez *et al.*, 2000).

Table 3.6 Summary of shrimp stock enhancement initiatives.

Species (country)	Years	Location	Source of animals	Scale of releases (M = million, B = billion)	Tagging method	Funding	Reference
Penaeus japonicus (Japan)	1964–present	All Japan	Prefecture hatcheries	200–300 M	Length composition, tail-fin clip, coded microwire tags	Japanese Government to prefectures	Kurata (1981); Kitada (1999); Fushimi (1999)
		Seto Inland Sea		100–150 M			Miyajima and Toyota (2002); Tanida *et al.* (2002); K. Hamasaki, personal communication (2004)
		Hamana Lake		2–11 M			
P. chinensis (China)	1980s–present	Bohai Sea	Double stocked prawn ponds	1 B (max = 2.2 B)	Some streamer tags	Chinese government to aquaculture for production	Liu (1990); Xu *et al.* (1997); Q. Wang personal communication (2002)
		Northern: Yellow Sea		800 M juveniles (max = 2.2 B) 300 M postlarvae (max = 520 M)			

P. aztecus, P. setiferus and P. duorarum (U.S.A.)	1971–1974	Southern: Shandong Florida	Hatcheries	800 M juveniles, now 300 M 16–52 M 3–137 M	None	Marifarms Inc.	Kittaka (1981)
P. semisulcatus, P. japonicus and Metapenaeus affinis (Kuwait)	1972–1978	Kuwait	Hatcheries, small tanks	4–25 M	Dies, tags	Kuwait Institute for Scientific Research	Farmer (1981)
P. monodon[a] (Taiwan)	1985–1992	Taiwan	Hatcheries and ponds	3M	Normal release season, some tags		Liao (1999); Su and Liao (1999)
(Sri Lanka)	1995–1997	Rekawa Lagoon	Commercial	55,000–70,000	Length composition		Davenport et al. (1999)
P. esculentus (Australia)	2001–2002	Exmouth Gulf	Raceways	250,000	Micro-satellite DNA	Industry (M.G. Kailis)	Loneragan et al. (2003, 2004)

[a]Between 1985 and 1992, these releases also included P. japonicus and P. semisulcatus.

also been an additional benefit from the large shrimp aquaculture industry that produced up to 200,000 t yr^{-1}. Excess postlarvae, mainly *Penaeus chinensis* (= *P. orientalis*), were grown to a larger size and released in the north of the country. Interest in shrimp stock enhancement in most other countries (e.g., the United States, Kuwait and Taiwan) (Table 3.6) has also developed because of the ability to produce postlarvae for aquaculture. In Kuwait, Japanese fishery experts at the Kuwait Institute for Scientific Research were involved in investigations on large-scale stock enhancement of *P. semisulcatus* to increase recruitment to the fishery (Farmer, 1981).

In contrast, stock enhancement of brown tiger prawns *Penaeus esculentus* in Exmouth Gulf, Australia, has been investigated as a means of increasing the long-term average catch by $\sim$100 t yr^{-1} and reducing inter-annual variation in catches (Loneragan *et al.*, 2001a, 2003, 2004). There was no existing shrimp aquaculture industry in the region, and facilities were developed to produce shrimp for stock enhancement alone (Crocos *et al.*, 2003). In addition to the potential benefits of increased, less variable catches, stock enhancement of *P. esculentus* in Exmouth Gulf was expected to provide a more continuous supply of shrimp to markets and an increase in efficiency of the fleet and factory processing operations. Data on survival, growth and migration could be obtained from augmentation of the shrimp population in Exmouth Gulf with marked individuals. This information would increase the ability of managers to obtain maximum benefits from the resource in two ways: by providing improved parameter estimates for stock assessment models and by identifying better harvesting strategies. The goal of increased, less variable catches of *P. esculentus* from stock enhancement was based on two assumptions: first, that the carrying capacity of Exmouth Gulf for this species is not often reached, and second, that the major factors affecting mortality act earlier in the life history (i.e., at the mysis, protozooal or postlarval stages), rather than at the size of the released juvenile shrimp (0.5–1.0 g).

3.5.2. Supply of juveniles

All shrimp postlarvae used to provide juveniles for stock enhancement programmes are produced in hatcheries using standard techniques. Mated females are typically collected from the wild and induced to spawn, although domesticated broodstock are being developed for some species (e.g., *Penaeus vannamei*, *P. japonicus* and *P. monodon*). In Japan, postlarvae of *P. japonicus* are produced by facilities in the prefectures funded by the Japanese government (Imamura, 1999). In 1997, 17 national and 57 prefectural hatcheries were active in producing shrimp for release in the wild. The national facilities develop the techniques for the culture of different species, whereas the prefectures complete the large-scale production of juveniles for stock enhancement (Imamura, 1999). Since the early 1980s, about 300–500 million juvenile *P. japonicus* have been produced and released in the wild in Japan each year (Kitada, 1999; Table 3.6). The massive stock enhancement programmes for *P. chinensis* in China, with releases of up to several billion shrimp each year, have relied on the large aquaculture industry with double stocking of ponds and release of excess juveniles at a size of ~30 mm total length (about 8 mm carapace length) (Wang *et al.*, 2002; Q. Wang, personal communication 2002).

In most other shrimp stock enhancement programmes, larvae have been produced in hatcheries and grown to release size in tank systems or released at small sizes and provided with some protection from predators (Table 3.7). In Exmouth Gulf, Australia, larvae were transferred from the hatchery to concrete raceways about 20 d after reaching the postlarval stage (PL 20) and grown to release size (0.5–1.0 g, 7–12 mm carapace length) (Crocos *et al.*, 2003). Four raceways produced 250,000 juvenile shrimp for release. About 180 raceways, costing AUD$4.5 million (US$2.9 million), would be needed to rear the estimated 24 million juvenile shrimp of 1 g to achieve a potential increase in catch of 100 t (Ye *et al.*, 2003, 2005).

3.5.3. Defects caused by rearing

Great care is needed in producing shrimp for stock enhancement pro-grammes, because they are highly vulnerable to predation (see Section 3.5.4) and must have appropriate behaviour for survival. The appropriate behaviour patterns vary among species and with ontogeny. Small postlarvae of *Penaeus semisulcatus*, a species that associates predominantly with sea-grass and algae as juveniles (Loneragan *et al.*, 1998), do not select seagrass in preference to bare sand until they reach a size of about 1.8 mm carapace length (CL), whereas larger postlarvae and juveniles (>2 mm CL) show strong selection for seagrass (Liu and Loneragan, 1997). *P. esculentus,* another species associated with seagrass and algal habitats as juveniles (Figure 3.16), later burrow in the substratum at a size of ~12 mm CL, when the leaves of most seagrass species are too small to provide effective camouflage (Kenyon *et al.*, 1995). *P. japonicus* change progressively from planktonic to benthic behaviour between 10–30 mm total length (2–8 mm CL) and are predominantly benthic at a size of 30 mm total length (Fushimi, 1999). In addition, juveniles of species with a preference for sandy substra-tum (e.g., *P. latisulcatus*) are likely to bury at smaller sizes than *P. esculentus,* whereas some species found in turbid waters (e.g., *P. merguiensis*) rarely bury in the substratum (Dall *et al.*, 1990).

Despite this sound knowledge of the ecology of juvenile shrimp in the wild, little information is available on the condition and behaviour of

Figure 3.16 (A) Juvenile of the shrimp *Penaeus esculentus* associated with sea-grass *Zostera capricorni*, a critical nursery habitat for this species (photo: CSIRO Marine Research). (B) The features of cultured *Penaeus esculentus* that are checked to assess fitness for release in stock enhancement trials: 1 = rostral spines, 2 = antennae, 3 = antennal scales, 4 = carapace, 5 = pereiopods, 6 = telson, 7 = pleopods and 8 = uropods (photo: P. Crocos, CSIRO Marine Research).

cultured juveniles before and immediately after release. In particular, it is still unclear whether they have the appropriate species-specific behaviour to ensure high survival rates immediately after release. Reports indicate, however, that cultured shrimp show some morphological defects related to hatchery rearing and high-density juvenile production. For example, culture of *Penaeus japonicus* in concrete tanks to sizes >30 mm total length (about ~8 mm CL) can lead to loss of limbs and damage to walking legs and swimmerets (Kurata, 1981). Similarly, all released 1-g juveniles of *P. esculentus* (12 mm CL) grown at high densities in raceways (2,000–3,000 postlarvae m^{-3}) had lost antennae, and most were also missing walking legs (Crocos *et al.*, 2003). Further laboratory studies of *P. esculentus* grown at densities of 75, 2860, 5720 and 11430 individuals m^{-3} found that those held at densities of ≥2,860 m^{-3} had significantly greater damage than those reared at 75 m^{-3} (Figure 3.16). However, the shrimp recovered from damage within 7 d of transfer to a density of 75 m^{-3}, probably because they were growing quickly and moulting every 3–4 d (Heales *et al.*, 1996, Crocos *et al.*, 2003). Laboratory experiments also show that burrowing behaviour is affected by the feeding regime and that shrimp fed just before transfer to low-density culture burrow more than those that are not fed (Crocos *et al.*, 2003).

The only observations of the fitness of cultured *Penaeus esculentus* released in the wild indicate that juveniles grown at high densities in raceways and released into Exmouth Gulf exhibited the natural escape response, a sudden 'flick backwards' (Kenyon *et al.*, 1995) when approached by fish predators (Crocos *et al.*, 2003). However, some individuals released high in the water column did not swim to the bottom immediately (Crocos *et al.*, 2003).

3.5.4. Release strategies

Strategies for releasing shrimp in stock enhancement programmes have seldom been evaluated systematically, although several factors affecting the success of releases have been reported, particularly size at release, release habitat, release season, stocking density and predation. These findings are summarised in the following.

3.5.4.1. Size of released shrimp

Most stock enhancement programmes for shrimp, including those in Japan, China, the United States, Kuwait and Sri Lanka, started releasing shrimp at early life-history stages and correspondingly small sizes (Table 3.7). In Japan and China, this practice soon changed to releasing larger juveniles of ∼30 mm total length (∼8 mm CL) (Liu, 1990; Tanida *et al.*, 2002). This size is similar to the optimal size at release estimated from a bio-economic model for stock enhancement of *Penaeus esculentus* in Exmouth Gulf, Australia (0.5–1.0 g, or 8–12 mm CL) (Ye *et al.*, 2003; 2005). Simulations from this model indicated that releases at a size of 1 g were likely to be more profitable than those at 0.5 g. The lower economic benefits predicted for the 0.5-g shrimp are due to greater exposure time to mortality and density-dependent effects that typically affect

Table 3.7 Strategies used to release penaeid shrimp in stock enhancement programmes and trials.

Species	Release habitat/location	Size/stage at release	Release season	Recovery rates (% of releases)	Reference
Penaeus japonicus	Saltmarsh	Postlarvae	June–July (no natural recruitment)	8.4[a]	Kurata (1981)
	Shallow coastal water	10–30 mm TL (2–8 mm CL)[b]	August– September	22–27.6[a]	Miyajima and Toyota (2002)
				18[c]	Tanida *et al.* (2002)
P. chinensis	From aquaculture ponds	Postlarvae 30 mm TL (7–10 mm CL)	July–August	35.6 (13.7–62.9) 4.5–10	Liu (1990) Q. Wang personal communication (2002)
P. aztecus	Pens	Postlarvae[d]	Spring	11.7	Kittaka (1981)
P. setiferus	Netted nursery area	0.3 g	Early summer	12.5	Kittaka (1981)

Species	Habitat/site	Size released	Timing	Recovery	Reference
P. duorarum	Netted growing area	1.1 g	July (mid-summer)	Negligible recovery in bay (migration)	Kittaka (1981)
P. semisulcatus	Selection of suitable nursery sites (e.g., lack of predators, suitable substratum, suitable depth)	Postlarvae stage (PL) 20 direct, then PL 20 to 'naturalization' pond for several months	Not reported	Not reported	Farmer (1981)
P. monodon	Nursery cages, sites in lagoon	PL 22 (13 mm TL) Later released at 21 mm TL	Outside normal recruitment period	2.8, 3.5	Davenport *et al.* (1999)
P. esculentus	Seagrass	7–12 mm CL (0.5–1.0 g)	December	Not monitored	Loneragan *et al.* (2003, 2004)

[a] Recovery rates estimated from cohort analysis.
[b] TL = total length; CL = carapace length; approximate conversion from TL to CL is TL = 4 CL.
[c] Recovery rates estimated from uropod (tail-fin) clipping.
[d] Released at 20 mg.

smaller shrimp in the wild that are not balanced by the lower production costs (Ye *et al.*, 2003; 2005). To enhance the fishery for *P. esculentus* by 100 t per year, the bio-economic model estimated that 30 million shrimp of 0.5 g would be needed compared with 24 million shrimp of 1 g.

During the initial phase of shrimp stock enhancement, when postlarvae were released, attempts were made to provide protection from predation. Alternatively, releases were made where densities of predators were low. In Japan, postlarval *Penaeus japonicus* were released into artificial tidelands with restricted access for predators (Kurata, 1981; Fushimi, 1999), whereas in the United States (*P. aztecus*, *P. duorarum* and *P. setiferus*) and Sri Lanka (*P. monodon*), postlarvae were placed in netted enclosures or cages (Kittaka, 1981; Davenport *et al.*, 1999). In Sri Lanka, juvenile *P. monodon* were subsequently released at a larger size from the enclosures into a natural lagoon. In the United States, however, after shrimp reached a size of ~0.3 g, the pen was opened to allow them to move to larger, netted areas (Kittaka, 1981). The shrimp were retained in this area for 1 or 2 mo until they reached a size of ~1.1 g. The net was then removed, allowing the shrimp to move into growing areas further from shore where efforts had been made to remove predators by trawling and application of piscicides (Kittaka, 1981). The recovery rates from releases of postlarvae have sometimes been low. For example, only 2.8% and 3.5% of *P. monodon* postlarvae were caught subsequently during 2 yr of stock enhancement (Davenport *et al.*, 1999). This was despite an estimated survival of 70–80% while the postlarvae were protected in nursery cages for 2 wk. The release of postlarvae in Japan and China was apparently abandoned because of poor rates of recovery.

In recent years, the size at release of *Penaeus chinensis* in China has been reduced from 8 mm CL, as used in the 1980s and early 1990s, to ~4 mm CL (Q. Wang, personal communication 2002). The reason for the change in release size is not clear. It may be related to the large outbreaks of disease among cultured shrimp recorded in China and the need to reduce stocking density in ponds earlier in the production cycle to minimise risk of disease (see also Section 3.5.6). In addition, smaller release sizes may have been used to reduce production costs. This has become an increasingly important consideration for stock enhancement in China since 1993, when the country embraced a free-market economic policy (Xu *et al.*, 1997; Wang *et al.*, 2002; Bartley *et al.*, 2004).

3.5.4.2. Release habitat

Hatchery-reared shrimp have usually been released in intertidal and shallow subtidal waters, where postlarvae of most species settle and juveniles reside for several months (Loneragan *et al.*, 1994, Rothlisberg *et al.*, 1996; Tanida *et al.*, 2002). Little reference has been made to other attributes of release habitat for shrimp in stock enhancement programmes, except that predators are likely to affect survival after release. For example, no *Penaeus chinensis* were recovered from releases in clear waters near the mouths of estuaries, probably a result of predation (Liu, 1990; Xu *et al.*, 1997). High rates of predation on postlarval *P. japonicus* by gobies have also been reported (Kurata, 1981; Fushimi, 1999), and many species of fish prey on shrimp at a range of sizes (Robertson, 1988; Minello *et al.*, 1989; Haywood *et al.*, 1998). Even relatively low rates of consumption by a particular species can be significant if the predator is abundant. For example, on intertidal seagrass beds, juvenile *P. semisulcatus* comprised 76% of the diet of barramundi *Lates calcarifer* compared with only 26% of the diet of queenfish *Scomberoides commersonianus*, but the impact of queenfish on the population of *P. semi-sulcatus* was much greater than that of barramundi because of their greater abundance (Haywood *et al.*, 1998). Furthermore, greater numbers of shrimp have been found in the diets of fish as shrimp densities increased (Salini *et al.*, 1990).

For species that associate closely with aquatic vegetation as juveniles (e.g., *Penaeus esculentus*), releasing juveniles in areas of high plant biomass is likely to minimise losses to predation and highlights the importance of choosing the best release habitats for shrimp. This has been demonstrated by laboratory experiments in which predation by a visual fish predator on juvenile *P. esculentus* was three times lower among the tall, broad-leaved seagrass *Cymodocea serrulata* than over bare sand and 50% lower among *C. serrulata* than the short, thin-leaved seagrass *Halodule uninervis* (Kenyon *et al.*, 1995). Furthermore, areas with high seagrass biomass ($100\ \mathrm{g\,m^{-2}}$) support higher densities of juvenile *P. esculentus* than those with lower biomass (Loneragan *et al.*, 1998). The growth of *P. semisulcatus*, a species with similar juvenile habitat requirements to *P. esculentus*, is also faster, and survival is greater, when seagrass biomass is higher (Kenyon *et al.*, 1995; Loneragan *et al.*, 2001b). The initial releases of *P. esculentus* in Exmouth Gulf were therefore made in shallow seagrass beds.

3.5.4.3. Release season

Few studies have been done on the best time of year to release juvenile shrimp. It is interesting to note, however, that releases have not always occurred during the peak times for natural recruitment. For example, cultured juveniles of *Penaeus japonicus* in Japan (Kurata, 1981; Fushimi, 1999), *P. chinensis* in China (Liu, 1990) and *P. monodon* in Sri Lanka (Davenport *et al.*, 1999) were released outside the normal recruitment period. This was done so that released shrimp could be distinguished from wild

recruits on the bases of length–frequency distributions to provide an indication of the success of stock enhancement (see Section 3.5.8). In both Japan
and China, releases were shifted by only 1 or 2 mo but, in Sri Lanka,
P. monodon were released 6 mo outside the natural time of recruitment.
The effects on survival of releasing juveniles at times outside the natural peak
recruitment periods are unknown.

3.5.4.4. *Stocking density*

Investigations of the effects of stocking density on growth and survival of
released shrimp have been limited to small-scale studies on postlarval *Penaeus
japonicus*, and on juvenile *P. semisulcatus*. Survival of postlarval *P. japonicus*
declined at stocking densities >100 m^{-2}, whereas growth started to slow at
a density of 200 to 300 m^{-2}. Consequently, 200 to 300 m^{-2} was suggested as
the optimal density for release into acclimation enclosures in the field
(Fushimi, 1999). The growth rate of juvenile *P. semisulcatus* in a high biomass
seagrass bed (70 g m^{-2}) declined to lower than average field growth rates at
stocking densities of >10 m^{-2} (Loneragan *et al.*, 2001b). However, the effect of
stocking density on growth and survival is likely to be modified by the release
habitat. For example, growth of *P. semisulcatus* in a seagrass bed of low
biomass (7 g m^{-2}) was slower and declined at a lower density compared
with growth rates in the high-biomass seagrass bed described previously
(Loneragan *et al.*, 2001b).

3.5.5. Minimising genetic impacts

Comparison of the genetic diversity of *Penaeus chinensis* in China, where
there is a large shrimp aquaculture industry and a history of large-scale stock
enhancement programmes (see Section 3.5.8), with the diversity of the same
species along the Korean coast, provides a sobering message about the
potential effects on wild stocks of releasing hatchery-reared juveniles. The
genetic diversity of *P. chinensis* (from randomly amplified polymorphic
DNA) measured by both polymorphic loci and the heterozygosity of the
broodstock, was lower in China (25.5% and 0.0198, respectively) than in
Korea (32.4% and 0.0307, respectively) (Q. Wang, personal communication
2002). This suggests that the lower genetic diversity in China was a result of
cultured shrimp released during stock enhancement programmes, or es-
caping from farms, breeding with wild stock and reducing heterozygosity
among progeny. These findings highlight the need to examine the genetic
diversity of *P. japonicus* in Japan where large-scale stock enhancement has

been practiced for a longer period than in China, albeit on a somewhat smaller scale (Table 3.6). These studies are even more urgent considering the bottleneck effects and reduction in heterozygosity reported during the experimental production of *P. japonicus* (Sbordoni *et al.*, 1986).

Protocols to minimise the genetic risks to shrimp in the wild have been developed for *P. esculentus* in Exmouth Gulf, Australia (Loneragan *et al.*, 2001a). These include: (1) determining the genetic structure of the wild population before stock enhancement; (2) using broodstock from the target population selected for enhancement (i.e., no domestication of broodstock); (3) randomly collecting broodstock (to avoid family groups); (4) tracing individual families during rearing using genetic markers; (5) releasing the same number of individuals from each of the captive-bred families and (6) monitoring the effects of the release using molecular methods (e.g., microsatellite DNA markers).

Microsatellite DNA studies indicate that there is only one population of *Penaeus esculentus* in Exmouth Gulf (Lehnert *et al.*, 2003; Meadows *et al.*, 2003). However, in the northern prawn fishery of tropical Australia, which occurs along 6000 km of coastline, it is likely that there are several populations of *P. esculentus* and *P. semisulcatus* (Condie *et al.*, 1999).

3.5.6. Disease risks

Disease has been a widespread problem in many countries producing shrimp by aquaculture, particularly in Asia and the Americas (Baldock *et al.*, 1999; Subasinghe *et al.*, 2001). The production of shrimp from aquaculture in Taiwan collapsed from about 80,000 t in 1988 to 9,000 t in 1990, mainly because of disease. An epidemic of white spot syndrome virus (WSSV) in Sri Lanka in 1996 was estimated to have caused an 85% loss in the total area of farms producing shrimp. The annual aquaculture production of *Penaeus chinensis* in China declined from 140,000 t in 1992 to 30,000 t in 1993 because of an outbreak of WSSV. This disease also affected stock enhancement, because it greatly reduced the number of shrimp released, and it lowered recovery rates (Q. Wang, personal communication 2002). Outbreaks of WSSV have also caused major losses in shrimp production from aquaculture in Thailand and the Americas (Subasinghe *et al.*, 2001). Although infections of WSSV have been reported in wild populations of a number of shrimp species (*P. monodon, P. japonicus, P. semisulcatus, P. penicillatus, P. duorarum* and *Metapenaeus ensis*), the levels of infection are generally much lower than in farmed shrimp (Baldock *et al.*, 1999). However, there is strong circumstantial evidence that the incidence of disease among wild shrimp is greater where the culture of shrimp in coastal ponds is common. For example, a much higher proportion of the wild broodstock of *P. monodon* tested positive for WSSV from the Gulf of Thailand (15%) than from the Andaman Sea (3–5%) (Baldock *et al.,* 1999). These differences in infection rates could be caused by escapes from aquaculture in the Gulf of Thailand (Baldock *et al.*, 1999). The major collapses of shrimp production, and the suggestion of an impact from aquaculture and stock enhancement on the disease status of wild populations, highlight the importance of developing a rigorous protocol to assess and reduce the risk of diseases, including establishing the disease status of wild populations, for any shrimp stock enhancement programme.

Protocols for monitoring disease have been developed as part of the feasibility study to assess the potential for stock enhancement of *Penaeus esculentus* in Exmouth Gulf, Australia (Loneragan *et al.*, 2001a). In this programme, health profiles are completed on representative samples from individual batches of postlarvae. From each batch of postlarvae/juveniles, the histopathology is examined for 150 shrimp selected at random. Adopting this approach gives 95% confidence of detecting disease at a prevalence rate of 2%. Polymerase chain reaction (PCR) primers have been developed for shrimp viruses, and their use will be incorporated in the disease protocol to further reduce the risk of releasing diseased cultured shrimp during stock enhancement. If high mortalities are detected in a batch, further screening for disease is carried out, and batches are not released if there is: (1) a virus associated with a lesion; (2) a protozoon associated with an inflammatory or degenerative lesion; (3) a metazoan parasite associated with lesions that are known or suspected to be pathogenic to other species (e.g., pearl oysters); (4) a fungal infection that causes lesions; (5) bacteria or rickettsiales associated with lesions or inflammation; (6) unexplained lesions and (7) inexplicable mortality in the batch that the certifying pathologist considers unacceptable. In addition to these tests, the disease status of the wild populations needs to be established and monitored during the stock enhancement programme.

3.5.7. Systems for marking juveniles to assess effectiveness of releases

Rothlisberg *et al.* (1999) summarised the attributes of an effective tag for marking shrimp to assess the effectiveness of a stock enhancement programme. Ideally, such a tag should be: (1) small enough to mark early life history stages; (2) detectable at all subsequent life history stages; (3) unique to the local population and able to identify individuals or cohorts; (4) inexpensive and quick to apply and detect; (5) either transmissible or non-transmissible to subsequent generations and (6) harmless to both the shrimp and consumer.

Some physical marking systems (e.g., eye-stalk rings and coded-wire tags) (Su and Liao, 1999; NMT, 2003) and fluorescent elastomer tags (Godin et al., 1996), have been applied successfully to smaller shrimp and may provide a suitable mark for experiments designed to evaluate the effectiveness of different release strategies. Another biological tag, uropod or tail fin clipping, has provided an effective mark for estimating the recapture rates from stock enhancement of Penaeus japonicus (Miyajima and Toyoto, 2002; Tanida et al., 2002).

Stable isotopes and genetic tags provide other options for marking shrimp. Research on naturally occurring stable isotopes ($\delta^{13}C$ and $\delta^{15}N$) and enriched ^{15}N (Preston et al., 1996) has shown that they have potential to provide a short-term chemical tag to monitor the fate of small shrimp during the first few weeks after release (Crocos et al., 2003).

Rothlisberg et al. (1999) also described some of the options for genetic tags of penaeid shrimp, including three different types of DNA markers: mitochondrial (Benzie et al., 1993), microsatellites (Meadows et al., 2003) and amplified fragment length polymorphisms (AFLPs) (Moore et al., 1999). In Penaeus japonicus, 12 microsatellite loci displayed between 4 and 24 alleles and heterozygosities of 50–90% in wild individuals. AFLPs exhibited high levels of variation, with more than 570 polymorphic loci being defined using 40 different combinations of primers. The high level of variation and the relative ease of isolation and characterization of AFLPs make them the markers of choice for screening and monitoring genetic diversity in P. japonicus stock enhancement programmes (Rothlisberg et al., 1999).

Eight microsatellite DNA loci have been identified, developed and deployed to establish the genetic stock structure of Penaeus esculentus and to evaluate their potential for discriminating between released and wild shrimp (Lehnert et al., 2003; Meadows et al., 2003). The statistical methods developed to assess the effectiveness of these microsatellites reveal that if the genotypes of both parents are known, then eight loci would be sufficient to monitor the success of a stock enhancement programme (Bravington and Ward, 2004). However, in most commercial shrimp production systems, females of species with a closed thyleca (like P. esculentus) mate in the wild before they are collected and spawned in hatcheries, so it is not possible to assess the genotype of the fathers. It may be possible to deduce the paternal genotype by sampling larvae, or a mixed larval homogenate, from mothers of known genotype but, if this is not feasible, several (3–6) new loci would need to be developed and deployed to reliably discriminate between the released and wild shrimp.

3.5.8. Status of stock enhancement initiatives

Despite the number of species and countries in which stock enhancement of shrimp has been attempted (Table 3.6), only the programmes for *Penaeus japonicus* in Japan and *P. chinensis* in China have been maintained over many years. Releases of *P. japonicus* in Japan started with production of 1.6 million mysis-stage larvae in 1963, increasing to 10 million juveniles in 1965. Since the early 1980s, the number of juvenile *P. japonicus* released throughout Japan has ranged from about 300–500 million per year and has been relatively stable at average annual releases of approximately 300 million juveniles (Figure 3.17; Fushimi, 1999; Kitada, 1999). The major release area in Japan is the Seto Inland Sea, where about half of the total number of cultured shrimp are released into the wild (Fushimi, 1999).

In China, from 1985 until 1993, an annual average of about 1 billion juvenile *Penaeus chinensis*, of 25 mm total length, were produced and released in the Bohai Sea (maximum release = 2.2 billion) (Table 3.6). In addition, ~300 million juveniles were released annually in the Haiyangdao Fishing Ground of the northern Yellow Sea and 800 million in the Qinghai Fishing Ground, southern Shandong (maximum release = 1.25 billion) (Q. Wang, personal communication 2002, Table 3.6). Juvenile *P. chinensis* have also been produced and released in Xianshan Bay and Dongwuyang Bay, areas to the south of where this species is found naturally, and these releases have been referred to as transplantations (Q. Wang, personal communication 2002). *P. monodon* and *P. japonicus* have also been transplanted in China. However, since 1993, the scale of shrimp stock enhancement programmes in China has been reduced, and shrimp have been released at smaller sizes (Q. Wang, personal communication 2002; see also Section 3.5.4). For example, in the Qinghai Fishing Ground, the average annual releases have been reduced from approximately 800 million to 300 million, and some of the production laboratories for stock enhancement are no longer carrying out transplantations (Xu *et al.*, 1997; Q. Wang, personal communication, 2002).

It is hard to find information on the performance and obstacles faced by the Chinese and Japanese stock enhancement programmes for shrimp. Evaluation is also difficult, because there is little documentation on the rationale for the scale of releases, selection of optimal release sites or strategies and management regimes. This is also true for other continuing shrimp stock enhancement programmes and for countries where releases of shrimp have ceased (Table 3.6). Research and development for stock enhancement of *Penaeus esculentus* in Australia has been suspended because the revised bio-economic model indicated lower levels of potential profit than previously predicted, and the industry partner (the M. G. Kailis Group of Companies) currently has higher priorities for investment in mariculture (Loneragan *et al.*, 2003, 2004).

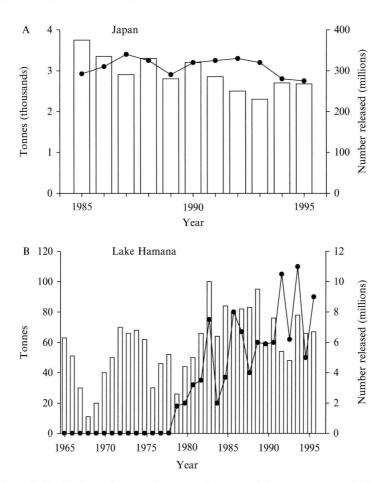

Figure 3.17 Variation in annual commercial catch of *Penaeus japonicus* (columns) and number of juveniles released (●) for (A) the whole of Japan, and (B) Lake Hamana in Shizuoka Prefecture (redrawn from Fushimi, 1999).

The limited information on success of stock enhancement programmes for shrimp, including reports of recovery rates of released juveniles, and the effects of density-dependence on survival, is described in the following.

3.5.8.1. Recovery rates

Contributions of released shrimp to commercial catches have been estimated by length–frequency (or cohort) analysis and by marking a proportion of the released shrimp, either with tags or by uropod clipping (see Sections 3.5.4 and 3.5.7). Recapture rates have been variable. The most encouraging results come from the Seto Inland Sea of Japan, where 22 and 27.6% of released *Penaeus japonicus* at Akoho City, Hyogo Prefecture, were caught in 2 consecutive years (Tanida *et al.*, 2002, Table 3.7). The shrimp were recovered in the waters adjacent to six neighbouring prefectures (Tanida *et al.*, 2002). Recovery rates of up to 18% have also been estimated for releases of *P. japonicus* made at night in Kyota Prefecture (Miyajima and Toyota, 2002). These recapture rates are considerably higher than the 6.8–10.2% reported for *P. japonicus* transplantations in China (Q. Wang, personal communication, 2002). Encouraging predictions also come from the bio-economic model for *P. esculentus* stock enhancement in Australia, which estimates a recovery rate of approximately 18.8%, assuming an average size at capture of 25 g (Ye *et al.*, 2003, 2005). In general, recapture rates for *P. japonicus* are higher than those of 5–12.5% reported for *P. chinensis* (Xu *et al.*, 1997; Q. Wang, personal communication, 2002; Table 3.7). (Note that we have given more credence to the lower survival rates reported by Xu *et al.*, 1997 and Q. Wang, personal communication, 2002, than those by Liu, 1990 for postlarvae, see Table 3.7, because of the apparently large confidence limits of the latter and the fact that results were not representative of other areas in China). Xu *et al.* (1997) estimate that a release of 500,000 juvenile *P. chinensis* is required to produce 1 t of catch, compared with about 200,000 juvenile *P. japonicus*, corresponding to recapture rates of 10% and 25% respectively, assuming that the average size at capture is 20 g. The lower rates of recovery of *P. chinensis* relative to *P. japonicus* may be because *P. chinensis* completes a long over-wintering migration before returning to coastal spawning and nursery grounds (Liu, 1990), whereas *P. japonicus* does not migrate on this scale. In the United States, recovery rates of *P. duorarum* were lower than those of *P. aztecus* and *P. setiferus* (Table 3.7), also possibly because of greater movement by this species (Kittaka, 1981).

Some other attempts at shrimp stock enhancement have been less encouraging. For example, of 55,000–70,000 postlarvae of *Penaeus monodon* released in a lagoon in Sri Lanka, only 2.8–3.5% were recaptured (Davenport *et al.*, 1999). This may have been because the shrimp were released out of season and/or experienced high mortality immediately after release from the postlarval holding cages. However, these results should not be used to evaluate the potential of hatchery releases to increase production, because the juvenile habitat requirements of *P. monodon* were not assessed, and the survival of juveniles was not monitored. Despite the low overall recovery rates of *P. monodon* in Sri Lanka, the total shrimp catch (including *P. indicus*) was worth 33% more during stock enhancement than in earlier years, and the annual catch of *P. monodon* increased by 1400% after release of juveniles (Davenport *et al.*, 1999). In general, there is a widespread need among shrimp stock enhancement initiatives for the rigorous evaluation of optimal release strategies proposed by Leber (1999, 2002).

3.5.8.2. Density dependence

The available data on releases of cultured juveniles for both *Penaeus japonicus* in Japan and *P. chinensis* in China suggest that a density-dependent effect on populations of shrimp may become significant as increased numbers of juveniles are released (Figures 3.18 and 3.19).The commercial catches of *P. japonicus* from the whole of Japan became more variable as the number of shrimp released for stock enhancement increased from 300–350 million (catch = 2200–3400 t) compared with releases just below 300 million (catch = 2800–3600 t, Figure 3.18). This is also evident at the local scale. For example, in Lake Hamana, Shizuoka Prefecture, commercial catches of about 80 t were obtained with releases of 4 million juvenile *P. japonicus* but did not improve consistently as the number of released juveniles increased up to 11 million (Figure 3.18). However, average catches during stock enhancement in Lake Hamana increased from about 47 t before releases to 68 t afterwards, and the variability in catches decreased (CV before = 41% vs. 19% afterwards, Figure 3.17; Fushimi, 1999). Thus, the stated objective of stock enhancement programmes for *P. japonicus*, to compensate for the effects of loss of coastal habitats through land reclamation on postlarval settlement and early juvenile growth, are being achieved to some extent. An important question, however, concerns the cost/benefit of restocking. Do the benefits exceed the costs? Early estimates of shrimp stock enhancement in Japan estimated a return of Yen 2.74 per postlarva and a cost/benefit ratio (including costs for producing shrimp and depreciation of facilities) of 1:1.8 (Kurata, 1981).

Data on stock enhancement of *Penaeus chinensis* in Jiaozhou Bay on the Yellow Sea coast of the Shandong Province, China, between 1981 and 1987 are summarised by Liu (1990). Shrimp releases started in 1984 and continued until 1986 (Figure 3.19). No shrimp were available for release in 1987, possibly because of disease problems. The population size of *P. chinensis* in Jiaozhou Bay was estimated from trawl surveys. From these limited data based on 3 yr of stock enhancement, the population size of *P. chinensis* declined from 40 million to 30 million as the number of shrimp released increased from 120 million to 200 million (Liu, 1990; Figure 3.19). These data further highlight the need to take density-dependent effects into account when assessing the potential benefits in stock enhancement of shrimp. Information on the cost-effectiveness of shrimp stock enhancement in China is limited, although a cost/benefit ratio of 1:5.2 has been estimated for the transplantation of *P. chinensis* in Hangzhou Bay (Xu *et al.*, 1997). It is interesting to note, however, that many of the large-scale stock enhancement efforts in China were not continued by fishermen's groups when support from the government ceased (see Section 3.5.9).

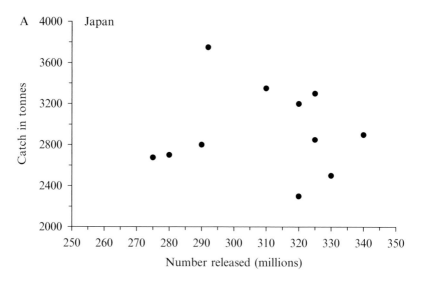

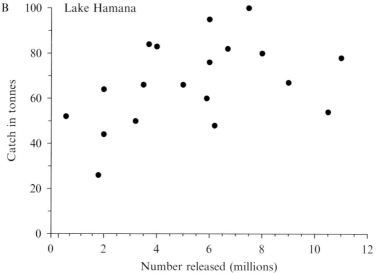

Figure 3.18 Relationships between number of *Penaeus japonicus* released and annual commercial shrimp catch in tonnes for (A) the whole of Japan, and (B) Lake Hamana in Shizuoka Prefecture (redrawn from Fushimi, 1999).

Density-dependent effects were included in the revised bio-economic model for stock enhancement of *Penaeus esculentus* in Exmouth Gulf, Australia (Ye *et al.*, 2003; 2005). In this study, the historical stock-recruitment relationship for the fishery was examined (Caputi *et al.*, 1998) to

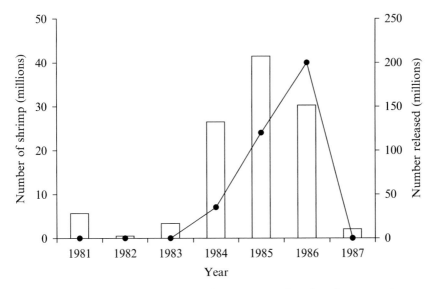

Figure 3.19 Relationship between estimated population size of *Penaeus chinensis* (columns) and number of juveniles released (●) in Jiaozhou Bay, Shandong Province, China between 1981 and 1987 (redrawn from Liu, 1990).

evaluate the possible effects of density-dependence resulting from releases of juveniles (Loneragan *et al.*, 2003; Ye *et al.*, 2003, 2005). This analysis suggested that density-dependent mortality could range from 8–23% for the 7 wk between release and emigration from the nursery ground. However, because of a lack of more accurate information on the values for density-dependent mortality, a uniform distribution was used in the simulation modelling to assess the bio-economics of the proposed stock enhancement. Incorporating parameters for post-release and density-dependent mortality increased the estimate of the numbers of juveniles required for a simulated 100 t increase in catch from 14 million to 24 million (Loneragan *et al.*, 2004; Ye *et al.*, 2003, 2005). This increase in the number of juveniles required to compensate for density dependence and post-release mortality highlights the need for experimental studies to evaluate optimal stocking densities.

3.5.9. Management

In many developing countries, managing shrimp fisheries is difficult because of the large number of people engaged in fishing and the diversity of fishing operations, which vary from subsistence to large trawlers. In most of the developed countries where shrimp are caught, management has proved to be easier. In general, shrimp fisheries there are managed with regulations on fishing practices (i.e., input controls), where attempts are made to limit the amount and location of fishing effort rather than output controls (i.e., quota systems). The management measures used for shrimp fisheries in developed countries include regulating the number of boats, limiting the number of days fished, limiting the head-rope length of the trawl net, and controlling where and when fishing takes place (e.g., through seasonal and/or spatial closures to protect small shrimp). After the decline in the Campeche Banks shrimp fishery in Mexico, closed areas (particularly within 15 miles of the coast), a closed season and mesh size restrictions were introduced (Ramirez-Rodriguez *et al.*, 2000).

The management regime for the *Penaeus esculentus* trawl fishery in Exmouth Gulf, Australia, combines a variety of measures, including seasonal closures, in an effort to make it sustainable. These measures are designed to allow a constant and sufficient escapement of shrimp each year to provide an optimal spawning stock, irrespective of the strength of annual recruitment (Penn *et al.*, 1997). The overall management regime of the fishery in Exmouth Gulf also uses area closures and a variety of seasonal closures to protect juvenile shrimp. Arrangements for the possible incorporation of stock enhancement as an additional tool for managing the shrimp fishery in Exmouth Gulf have been relatively straightforward, because all but one of the vessels in this fishery belong to one company. Hence, the beneficiaries of stock enhancement are clearly identified, and the costs can be allocated accordingly. Strong management regimes and incentives for investment were some of the main reasons for choosing to evaluate the potential of shrimp stock enhancement in Exmouth Gulf (Loneragan *et al.*, 2003, 2004).

In Japan, there is a move to reduce and eventually withdraw national funding for stock enhancement of *Penaeus japonicus* and to hand over this responsibility to the fishing industry (Imamura, 1999). In many prefectures, in addition to funds available from prefectural and municipal governments, financial resources for stock enhancement are also collected from the fishing co-operatives (Imamura, 1999). For example, this system has been intro-duced for stock enhancement of flounder in Fukushima and Aomori Pre-fectures, where individual commercial fishermen pay 4–5% of the value of landings toward the costs of producing juveniles (Kitada, 1999). These funding arrangements between different levels of government and the fishing co-operatives indicate that it may be possible to introduce a user-pay system for shrimp enhancement, provided it is possible to apply the principle that those groups that pay for the production of juvenile shrimp will be the ones entitled to harvest them (Kitada, 1999). The present challenge for *P. japonicus* fisheries in Japan, which include both gill net fishers and traw-lers, is to ensure the proper management of the wild stock and to use stock enhancement only as one of several management options to help sustain production (Kitada, 1999). The scientific evaluation of stocking effectiveness and the responsible use of hatcheries to produce juveniles are key areas for future stock enhancement of *P. japonicus* in Japan (Kitada, 1999).

The enhancement and transplantation programmes for shrimp in China have been affected by the introduction in 1993 of the new economic policy based on the market economy (Xu *et al.*, 1997; Q. Wang, personal communication, 2002). Releasing shrimp into waters outside their normal range in China (i.e., transplantation into Dongwuyang Bay and Xian Shan Bay) had ceased by 1996 because of a lack of funds (Q. Wang, personal communication 2002). In Xiang Shan Bay, where transplantation stopped in 1994, there was no practical way of recovering the costs of releases from fishermen (Xu *et al.*, 1997). Many of the fishermen in Xiang Shan Bay are organized into co-operatives, and it is possible that they might re-initiate hatchery releases in the region. However, this would require co-operation with the provincial government to control access to trawling and gill netting in the bay. A second reason for stopping hatchery releases was the failure to establish a self-replenishing population of shrimp in the area (Xu *et al.*, 1997). This is perhaps not surprising, because the shrimp were released into areas where they had not previously been found, and because *P. chinensis* undergoes an over-wintering, offshore migration of ~1,000 km into warmer waters before returning to coastal spawning grounds (Liu, 1990).

3.5.10. Marketing

Like scallops, shrimp are a luxury food and are marketed in many ways, including boiled, dried, frozen, fresh and live. However, unlike scallops, shrimp stock enhancement has not yet led to massive increases in shrimp production. Rather, in Japan and possibly China, it has been used to stabilize production after declines in catch. Considering the massive size of the shrimp aquaculture industry, any decrease in the value of shrimp because of over supply is unlikely to be driven by successful stock enhancement. However, any reduction in market value through increased supply is likely to be less severe for stock enhancement programmes than for aquaculture, because markets place a premium value on wild shrimp.

3.5.11. Problems to overcome

There are few impediments to the production of shrimp for stock enhancement in countries with large aquaculture industries, such as China and Japan. However, in nations with more limited infrastructure for shrimp aquaculture, such as Australia, production of the large numbers of shrimp needed for effective stock enhancement requires significant capital investment in both hatchery and juvenile production facilities. For example, a release of 24 million juvenile *P. esculentus* of 1 g in Exmouth Gulf would require AUD $3.8 million to expand hatchery facilities, and a further AUD $4.5 million to build the raceways for juvenile production (Loneragan *et al.*, 2003, Ye *et al.*, 2003, 2005).

Another major impediment to further development of stock enhancement programmes for shrimp is lack of rigorous evaluation of current initiatives in the context of the management regime and dynamics of the fishery. Unless stakeholders are sure that the additional revenue will exceed the costs, and that production cannot be increased by other less expensive methods, there is little incentive to invest in stock enhancement. In short, stock enhancement is only one management option, and other strategies, such as habitat improvement, reductions in fishing effort and introduction of spatial and seasonal closures, may lead to greater production than release of cultured juveniles (Hilborn, 1998; Leber, 1999, 2002; see Section 3.6.3). The development of the strategy for stock enhancement of *Penaeus esculentus* in Exmouth Gulf illustrates the process needed to design and evaluate stock enhancement programmes for shrimp. A steering committee with representation from managers, industry and scientists from all disciplines (aquaculture, ecology, genetics, statistics, fishery modeling), met regularly to guide development and assessment of the initiative (Loneragan *et al.*, 2003, 2004). The bio-economic model constructed during this process represented all components of the stock enhancement system, evaluated the potential economic returns from releases and helped researchers adopt a holistic approach to stock enhancement (*sensu* Leber, 2002).

The lack of a clear mechanism to identify who pays for the cost of producing and releasing juveniles, and who benefits from such investments, is inhibiting shrimp stock enhancement in some countries. In particular, the previously large programme in China is undergoing difficulties because neither the state-run enterprises nor other agencies are keen to invest in stock enhancement because of the difficulty in identifying who benefits from the harvest of released shrimp (Q. Wang, personal communication, 2002). As described in Section 3.5.9, this resulted in a reduction, and finally cessation, of hatchery releases into the Bohai Sea, and a much lower level of stock enhancement in the northern Yellow Sea in recent years (Q. Wang, personal communication, 2002).

3.5.12. Future research

Large-scale annual releases of >100 million juvenile shrimp have been made for many years in Japan (Kurata, 1981; Fushimi, 1999) and China (Liu, 1990; Xu *et al.*, 1997; Wang *et al.*, 2002). The success and impacts of these programmes need urgent evaluation, particularly the questions of whether stock enhancement adds value to other forms of management; whether the infrastructure required to produce the millions of juveniles needed can be made available cost-effectively, and whether the beneficiaries of releases are in a position to pay the costs. In addition, two important pieces of biological research are needed to identify the full potential of stock enhancement for shrimp. The first is development of optimal release strategies, including the effects of release size, season, and site on survival, and the effects of stocking density on survival and growth in the nursery grounds. Such studies have been recommended as the next stage of research and development for stock enhancement of *Penaeus esculentus* in Exmouth Gulf, Australia, which would then be followed by a pilot release of 1–3 million juveniles (Loneragan *et al.*, 2003, 2004). Identification of optimal release strategies should greatly reduce the current uncertainty of estimates of mortality and density dependence during the months after release. The potential benefit of stock enhancement programmes for shrimp would also be improved by research on ways to maximise the health and fitness of released shrimp before release.

The second area of biological research for shrimp stock enhancement concerns the use of a marking system to follow the fate of hatchery-reared juveniles after release. This will depend on the species of interest and the numbers that need to be marked to determine whether stock enhancement has been effective or whether it has simply displaced wild shrimp. Genetic tags hold much promise but will require considerable work. This is highlighted by research to date on *Penaeus esculentus*, which demonstrates that it is still not possible to discriminate unambiguously between released and wild shrimp without knowledge of the paternal genotype. Testing the feasibility of mass-genotyping larvae to deduce paternal genotype is now required. If this does not prove to be successful, an additional 3–6 microsatellite loci would need to be identified, developed and deployed over and above the eight already available (Lehnert *et al.*, 2003; Loneragan *et al.*, 2003, 2004).

3.6. SPINY LOBSTERS

3.6.1. Background and rationale for stock enhancement

Spiny lobsters (Palinuridae), also known as rock lobsters, crayfish and langoustes in some countries, differ substantially from homarid lobsters (Section 3.7), both in biology and zoogeography. The Palinuridae lack large claws (chelae) and, whereas homarids are restricted to temperate waters, palinurids are found worldwide from the tropics to cool temperate, even sub-Antarctic, regions (Holthuis, 1991). Species of Palinuridae can be extremely widespread (e.g., *Panulirus ornatus, P. homarus, P. argus*) or endemic to small portions of coastline (e.g., *P. cygnus, P. interruptus*) or islands (e.g., *Jasus frontalis, J. paulensis*). The most striking difference between the spiny lobsters and homarid lobsters, however, is the larval phase. The Palinuridae and their close relatives the Scyllaridae have long-lived, leaf-like larval stages called phyllosomas. This extraordinary larval phase can last a few months in some scyllarids (e.g., *Thenus orientalis* [Mikami and Greenwood, 1997a]), and up to 2 yr in some palinurids (e.g., *J. edwardsii* [Lesser, 1978; Booth, 1979, 1994; Kittaka, 1994, 2000]). The complex life cycle of palinurids is not restricted to the long pelagic phase; for several species such as *P. argus* (Butler and Herrnkind, 1997), there is also usually a shallow coastal nursery phase followed by migration offshore into deeper water (Figure 3.20).

Considerable variation occurs in the natural abundance of spiny lobsters as a result of the vagaries of the larval and nursery phases. This variation in abundance, the vulnerability of many species to capture and their high market value (84,273 t valued at ca. US$927 million in 2000)[2] have led to concerted efforts to: (1) understand the dynamics of spiny lobster stocks; (2) develop ways to maximise catches from wild fisheries on a sustainable basis (Phillips *et al.*, 1994; Phillips and Kittaka, 2000); (3) manage shared spiny lobster resources co-operatively (Cochrane *et al.*, 2004) and (4) increase production further through aquaculture and stock enhancement (Lee and Wickens, 1992; Wickens and Lee, 2002). Until recently, most of the effort directed toward increasing and sustaining production has centred on management of wild stocks, although interest is now growing rapidly in evaluating the scope for aquaculture and stock enhancement. Researchers have recognised the difficulties in the assessment of population structure, size and 'health' (trends in abundance) (Addison, 1997; Hilborn, 1997). Catch per unit effort (CPUE) and catch size distributions have proved to be good short-term estimators of the effects of fishing on spiny lobsters, but poorer in the long term. Also, selectivity and catchability of various types of gear, and migrations of spiny lobsters, affect estimates of size/age in the population. Indeed, it is has often been difficult to detect changes in size distribution of the catch in response to fishing at high levels of exploitation. Estimation of annual recruitment is much easier, because annual catch is equivalent to annual recruitment when fishing is intense.

Other tools that have proved to be particularly useful for managers of the large fishery for *Panulirus cygnus* in Western Australia are abundance

---

[2] The total palinurid catch was derived from FAO's FishStat Plus database, and value was estimated using an approximate value of US$11,000 t^{-1}, based on consultation with FAO commodity statisticians.

Open Ocean	Reef	Islands-Shallow Coastal Nursery

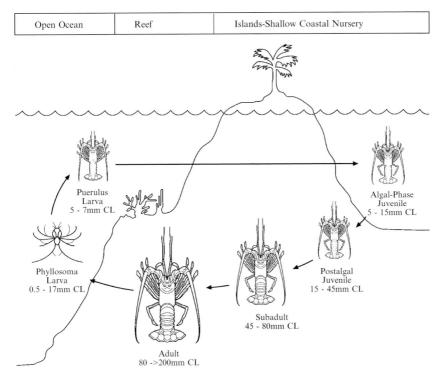

Figure 3.20   The life history of the Caribbean spiny lobster, *Panulirus argus* (redrawn from Butler and Herrnkind, 1997). CL, carapace length.

indices for puerulus larvae and juveniles as predictors of subsequent catches of legal-sized animals (Phillips, 1986; Caputi *et al.*, 1995a; Phillips *et al.*, 2000). This approach is also being implemented, or is under development, for fisheries for *P. argus* in Cuba (Cruz *et al.*, 1995), *Jasus edwardsii* in New Zealand (Booth, 1994) and *P. japonicus* in Japan (Nonaka *et al.*, 2000).

Estimating the relationship between spawning stock size and subsequent recruitment is recognised as a way to assess whether spawning is declining to a level that would jeopardise recruitment. This approach was applied in the fishery for *Panulirus cygnus*, where early analysis showed evidence of recruitment over-fishing (Morgan *et al.*, 1982). Later, Caputi and Brown (1993) and Caputi *et al.* (1995b) warned that more work was needed on estimating the influence of environmental factors and fishing effort to deliver an unambiguous estimate of spawning stock size.

A related issue is the difficulty in assessing the relative contributions of the various components of spawning biomass to recruitment, given the long larval life spans and the potential for dispersal. Settling juveniles can be produced either by sub-legal sized lobsters in shallow water, the population actively subjected to fishing, or from the less vulnerable individuals associated with offshore refugia. More work is needed on genetic stock structure, adult migration, and larval dispersal to define the spawning biomass in areas that are fished and not fished (Lipcius *et al.*, 1997; Griffin *et al.*, 2001) (for examples of related research see Rothlisberg *et al.*, 1996; Condie *et al.*, 1999). It is also necessary to understand how environmental factors might affect adult and sub-adult growth and mortality, and larval dispersal at local and regional scales in relation to management areas.

The preceding issues make it difficult to identify clear reference points for recruitment over-fishing in many spiny lobster fisheries. However, some indicators have been identified. For example, *Panulirus argus* in the Gulf of Mexico and off the southeastern United States is considered over-fished when: (1) the eggs per recruit ratio of the exploited population to the unexploited population is reduced below 5%; and (2) recruitment to the fishery declines for 3 consecutive fishing years (Addison, 1997).

Despite the difficulty in aquiring data for the various stock assessment approaches for spiny lobster fisheries (Hilborn, 1997), the two major fisheries for spiny lobsters (Figure 3.21) are managed better than stocks of many

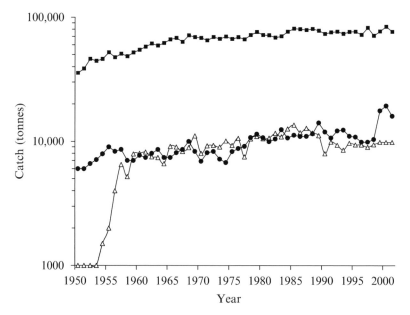

Figure 3.21 Variation in total catch of spiny lobsters *Panulirus cygnus* from Western Australia (●), *P. argus* from Cuba (△) and all exploited species of Palinuridae worldwide (■) from 1950–2001 (data from FAO).

other species of fish and invertebrates. In these two fisheries, much of the variation in catch is related to changes in the natural supply of juveniles. Similar patterns are also evident for many of the smaller fisheries and, overall, total landings of spiny lobsters worldwide have been fairly constant since 1970 (Figure 3.21).

Two main types of interventions are underway to increase production of spiny lobsters further. The first is culture of juveniles. Rearing of spiny lobsters from egg to post-puerulus has been achieved at an experimental scale for a number of species in very small numbers (Illingworth et al., 1997; Kittaka, 1997; Kittaka and Booth, 1994, 2000; Nonaka et al., 2000) (Table 3.8), but there is general agreement that use of hatchery-reared juveniles for aquaculture or stock enhancement will not be feasible for some time. Instead, the capture and culture of puerulus larvae is being developed. Progress is still generally at the research level (Phillips and Evans, 1997; Booth and Kittaka, 2000) (Table 3.9), and no large-scale stock enhancement has yet been undertaken. However, aquaculture based on growing pueruli and wild-caught juveniles produced >2,000 t in 2000, worth more than US$75 million (FAO, 2003a; Williams et al., 2002). Most of this production came from *Panulirus ornatus* in Vietnam, with only minor production in the Philippines, India, Taiwan and Singapore.

The second type of intervention to enhance production of spiny lobster fisheries provides additional habitat for the vulnerable, settling puerulus larvae, which typically undergo severe mortality from predation during the first benthic year (Herrnkind and Butler, 1994; Butler et al., 1997; Cruz and Phillips, 2000).

As both of these interventions are under investigation and development for a number of species and in a variety of geographic areas, we have departed from the sequence of headings used to report the progress of stock enhancement initiatives for the other species. Instead, we have described the research on stock enhancement, and the status of the various initiatives, for spiny lobsters by country.

3.6.2. Stock enhancement-related research

3.6.2.1. Australia

Australia is currently the world's largest producer of rock (spiny) lobsters. The recent catches of 18,000–20,000 t are ~25% of total world landings (FAO, 2003a). The catch is dominated by *Panulirus cygnus* but also includes *P. ornatus* and *Jasus edwardsii*. Rock lobster research is centred in three states: Western Australia *(P. cygnus)*, Tasmania *(J. edwardsii)* and Queensland *(P. ornatus)*. Recently, the national research effort on enhancement of rock lobsters has been brought together into the Rock Lobster Enhancement

Table 3.8   Summary of attempts at rearing larvae of spiny lobsters (Palinuridae) and slipper lobsters (Scyllaridae) to the postlarval stage in research facilities.

Species	Duration in culture (days)	Reference
Spiny lobsters		
Jasus lalandii	306	Kittaka (1988)
J. verreauxi	189–359	Kittaka *et al.* (1997)
J. lalandii/edwardsii hybrid	319	Kittaka *et al.* (1988)
J. edwardsii	416	Illingworth *et al.* (1997)
	212–302	Kittaka *et al.* (unpublished), cited in Kittaka (2000)
Panulirus japonicus	307–391	Kittaka and Kimura (1989) Yamakawa *et al.* (1989)
	231–417 (mean 319)	Sekine *et al.* (2000)
	787 (at 20°C)	Matsuda and Yamakawa (1997)
	462 (at 22°C)	
	305 (at 24°C)	
	324 (at 26°C)	
P. longipes	281–294	Matsuda and Yamakawa (2000)
Palinurus elephas	132–148	Kittaka and Ikegami (1988)
	65–69	Kittaka *et al.* (2001)
Panulirus argus	Incomplete	Moe (1991)
P. interruptus	Incomplete	Dexter (1972)
P. homarus	Incomplete	Radhakrishnan and Vijayakumaran (1995)
P. penicillatus	Incomplete	Minagawa (1990)
P. cygnus	Incomplete	Ito (1990) cited in Ito (1995)
Slipper lobsters		
Thenus orientalis	29	Mikami and Greenwood (1997a)
	28–45	Mikami and Greenwood (1997b)
Thenus sp.	28	Mikami and Greenwood (1997b)
Ibacus ciliatus	76	Takahashi and Saisho (1978)
I. novemdentatus	65	Takahashi and Saisho (1978)
Scyllarus americanus	32–40	Robertson (1968)
S. demani	42–53	Ito and Lucas (1990)
Scyllarides latus	Incomplete	Bianchini *et al.* (1996)

and Aquaculture Subprogram under the aegis of the Fisheries Research and Development Corporation, which also includes collaboration with researchers in New Zealand. The research includes: propagation and larval development, nutrition, culture systems and enhancement-based ecology.

3.6.2.1.1. *Western Australia:* Panulirus cygnus

Western Australia has a long history of investigating temporal and spatial variation in recruitment of puerulus larvae of *Panulirus cygnus* to inshore reefs based on artificial collectors (see Figure 3.22 and later). Puerulus

Table 3.9 Summary of growth rates of newly settled spiny lobster juveniles (puerulus larvae) in research facilities: CL = carapace length; BL = body length.

Species	Size attained	Duration of trial	Reference
Jasus edwardsii			
	37–42 mm CL	ca. 1 yr	Mills *et al.* (2004)
	12–14 g (at 10°C)	1 yr	Hooker *et al.* (1997)
	31–34 g (at 18°C)	1 yr	
	180 g (ca. 18°C)	ca. 3.3 yr	
	36.9 g 42.1 mm CL	1 yr	Jeffs and James (2001)
Palinurus	21–66 mm CL	0.75–1.9 yr	Kittaka *et al.* (2001)
elephas	260 g	2.0 yr	Cited in Wickens and Lee (2002)
Panulirus	450 g	1 yr	Lellis (1991)
argus	1.4 kg (110 mm CL)	2 yr	Sharp *et al.* (2000)[a]
	42 mm CL	1 yr	
P. japonicus	76 mm BL	4 mo	Yamakawa *et al.* (1989)
	330 g	16 mo	Cited in Lee and Wickens (1992)
	35–40 mm CL	1 yr	Norman *et al.* (1994)
P. homarus	330–800 g	16–18 mo	Chen (1990)
	200 g	8–9 mo	Rahman and Srikashnadhas (1994)
P. cygnus	ca. 30 mm CL (at 20°C)	1 yr	Phillips *et al.* (1983)
	ca. 38 mm CL (at 25°C)		
	76 mm CL	2.1 yr	
P. ornatus	225 g	0.75 yr	Jones *et al.* (2001)
	61.8 mm CL		

[a]Based on tag recovery of juvenile lobsters at liberty—extrapolation from graph.

settlement indices, along with juvenile abundance estimates, have been used for >20 yr to predict future commercial catches 3–4 yr in advance (Phillips, 1986; Caputi *et al.*, 1995a; Phillips *et al.*, 2000). Based on this extensive body of knowledge, managers of the fishery are now considering the potential role of capturing and growing pueruli for farming operations or rearing them to a size where their survival should be much greater and then releasing them to enhance production. Initially, this research is focusing on two key issues: (1) developing commercial-scale puerulus collectors that are more efficient and cost-effective than the ones used for scientific monitoring of settlement rates and (2) ensuring that harvesting pueruli will not reduce the subsequent catch. The latter point is quite contentious with fishermen because, in the absence of evidence to the contrary, they are concerned that the tight relationship between puerulus settlement and subsequent catch indicates

that collection of puerulus for aquaculture or stock enhancement may pose a threat to their livelihoods. Furthermore, legislators and managers need a sound basis for making changes to the regulations for this well-managed and lucrative fishery.

3.6.2.1.1.1. Puerulus collectors

The original puerulus collectors were designed for monitoring recruitment dynamics and catch prediction (Phillips, 1972). Phillips and Booth (1994) reviewed the effectiveness of puerulus collectors worldwide and found they are quite species-specific and that the numbers of pueruli caught are too low for commercial-scale aquaculture or stock enhancement programmes. More recently, Phillips *et al.* (2001) compared several different designs to assess their efficacy for commercial collection of puerulus of *Panulirus cygnus*. They assessed catch rates for 11 different collector types and sizes, and for the effects of location (depth and proximity to the coastline), proximity to nearby collectors and the frequency of collection. They found that 'sandwich' collectors containing on artificial seaweed had the highest catch rates and that only collectors set inshore at the surface (at depths <5 m) caught pueruli. There seemed to be a linear relationship between size of collectors and catch (ca. 20 pueruli m^{-2} over 2 mo of collection). The ratio of catch on collectors serviced frequently (seven times a month) to those cleared once a month was 2.7:1. In addition, placing a collector next to a neighbouring one improved the catch. Overall, however, catch rates were quite low (tens of pueruli collector^{-1} mo^{-1}), leading to the conclusion that commercial operations based on larger collectors, serviced more frequently ($<$ monthly), may be viable but would need to be assessed using a cost/benefit analysis.

3.6.2.1.1.2. Biological neutrality

Phillips and Melville-Smith (2001) examined the impact of possible puerulus collection on future catches of *Panulirus cygnus* to determine what management measures might be required to maintain biological neutrality. They used regional subsets of existing data on puerulus settlement, juvenile densities and mortalities and recruitment rates to the fishery to assess differences in the contribution of particular year classes of pueruli to commercial catches using a non-linear model. They concluded that: (1) recruitment to the fishery is most dependent on settlement of pueruli 3 and 4 yr earlier; (2)

Figure 3.22 (A) Collector used to catch puerulus larvae of the spiny lobster *Panulirus cygnus* in Western Australia (photo: D. Dennis). (B) Newly settled (clear) puerulus larva of the spiny lobster *Panulirus versicolor* and individuals collected a couple of days after settlement with their full colouration (photo: C. Hair). (C) Adult rock (spiny) lobster *Jasus edwardsii* from southern Australia and New Zealand (photo: R. Kuiter).

rates of recruitment based on settlement 3 and 4 yr earlier varied by region; (3) regional and year-class factors have to be taken into account in harvesting strategies; (4) only 2–3% of pueruli survive the first year on nursery reefs (comparable to estimates for *P. argus* [Herrnkind and Butler, 1994; Butler *et al.*, 1997]), but mortality is both density dependent and declines in subsequent years (ca. 20% survival in the second year) and (5) removal of 20 million pueruli in an average settlement year (in the area from 29° S to 30° S when ca. 600 million pueruli settle) would result in only a 1% decrease in the number of legal-sized spiny lobsters caught, equivalent to an effort of ca. 23,000 pot lifts for the season.

Given the early phase of the investigation, with only two seasons of data over a limited spatial scale, Phillips and Melville-Smith (2001) recommended further research on habitat availability and use, and early mortality rates, to clarify the actual fishing effort reduction necessary to compensate the reproductive capacity of the population if pueruli were removed for aquaculture. This information is needed to permit fishermen to assess the costs and benefits of trading-off reductions in effort for the proceeds from sales of pueruli for aquaculture or stock enhancement operations.

3.6.2.1.2. *Tasmania:* Jasus edwardsii

Research in Tasmania on aquaculture and stock enhancement of *Jasus edwardsii* has increased greatly in recent years. It is focused on propagation (Phleger *et al.*, 2001; Ritar, 2001; Nelson *et al.*, 2002; Ritar *et al.*, 2002), juvenile collection and on-growing (Crear *et al.*, 2000, 2002; Thomas *et al.*, 2003) and post-release mortality (Mills *et al.*, 2004).

Phleger *et al.* (2001) have assessed the lipid and fatty acid composition of wild-caught larvae from Tasmania and New Zealand to identify their nutritional requirements and improve artificial feeds. They found that phyllosoma and nektonic puerulus larvae have low storage lipids (triacylglycerol) and that phospholipid was the major lipid class. More detailed analysis of the essential omega–3 fatty acids (DHA, EPA and AA) showed that the ratios between them changed during development. These findings have been incorporated into enriched *Artemia* diets for larvae in culture, which now approximate the composition of wild phyllosomas (Nelson *et al.*, 2002). Crear *et al.* (2000, 2002) have examined the use of formulated diets for growing juvenile (2–5 g) *Jasus edwardsii*. They found that diets for the shrimp *Penaeus monodon* were promising and could be a useful partial substitute, but growth rates were not as high as attained with fresh mussels *Mytilus edulis*. They also report that provision of shelter increased survival, but not growth rate.

Natural mortality of *Jasus edwardsii* 1 yr after settlement was ~97% (Mills *et al.*, 2004), which is in stark contrast to mortality rates of <10% for juveniles raised in tanks (Thomas *et al.*, 2003). Therefore, collection of postlarvae and subsequent release 1 yr later provides a potential means of stock enhancement. Mills *et al.* (2004) demonstrate that this potential is unlikely to be affected by the time the juveniles spend in culture. They monitored survival and behaviour of released *J. edwardsii* using acoustic tags to quantify the distance and direction of movements, predator avoidance and habitat choice. All behaviours were comparable between lobsters grown in captivity for 1 yr and wild-caught (local and translocated) individuals. There were slight differences in food selection between the treatment groups, which could be attributed to naiveté or local differences in food availability within the study site.

3.6.2.1.3. *Queensland:* Panulirus ornatus

Research on the tropical ornate rock lobster *Panulirus ornatus* indicates that it has several potential advantages for culture over other panulirids (Linton, 1998). These include: (1) a relatively short larval life of about 6 mo (Pitcher *et al.*, 1995); (2) high reproductive capacity with an ability to produce up to three broods in 6 mo (MacFarlane and Moore, 1986) and (3) rapid growth. Several studies confirm the latter attribute; for example, *P. ornatus* reached a size of 70–100 mm CL in 2 yr in the wild at rates up to 1.4 mm a week (Dennis *et al.*, 1997; Skewes *et al.*, 1997) and an average of 330 g in 16 mo when stocked in ponds in Taiwan at 25 g (Chen, 1990). The most encouraging information on growth rate, however, is from Jones *et al.* (2001), who measured the effect of density on growth and survival of wild-caught juvenile *P. ornatus* (mean weight 3.24 g, 13.8 mm CL) stocked in a flow-through raceway system at three densities (14, 29 and 43 m^{-2}). After 272 d, results from the density treatments did not differ significantly, and the mean size for all individuals was 225.3 g (61.8 mm CL). This represents a specific growth rate of 1.56% d^{-1} and equates to growth from 3 g to 1 kg within 18 mo. Mean survival was 52.5%.

A possible problem with *Panulirus ornatus* is that collection of puerulus larvae has proved to be difficult in Australia (but see results for Vietnam below). Pueruli are found sporadically in a number of natural habitats (e.g., coastal reefs) and on man-made structures (wharf pilings and fish farm netting) but have not settled on standard collectors in appreciable numbers (Linton, 1998). In addition, legislative barriers and perceived conflict with fishermen currently prevent large-scale collection of *P. ornatus* pueruli in Queensland for aquaculture or stock enhancement.

3.6.2.2. *New Zealand*

Experimental rearing of the spiny lobster *Jasus edwardsii* in land-based systems has been underway in New Zealand for >20 yr using wild pueruli (Booth and Kittaka, 2000; Jeffs and Hooker, 2000). Recently, the larval life cycle has been closed on an experimental scale by production of mature F1 individuals (Illingworth *et al.*, 1997; Kittaka *et al.* (unpublished) cited in Kittaka, 2000). In addition, workers in New Zealand are actively investigating the scope for commercial-scale collection and growing of juvenile *J. edwardsii* (see Section 3.6.3).

An economic study to assess the feasibility of land-based aquaculture of *Jasus edwardsii* showed that it was unlikely to be profitable (Jeffs and Hooker, 2000). This analysis was based on a number of experimental studies (e.g., growth rate of puerulus in captivity at different water temperatures and stocking densities) (Hooker *et al.*, 1997; James *et al.*, 2001), extrapolations from other types of commercial aquaculture in New Zealand (e.g., abalone), and the best-case scenario of seed costs (collected from the wild), food conversion efficiencies and growth rates over a 10-yr investment and production cycle. Infrastructure costs and operating expenses (feed and labour costs) were the areas where the most savings had to be made. To investigate ways of reducing infrastructure costs for aquaculture operations, Jeffs and James (2001) tested an experimental offshore cage culture system. They measured growth and survival at three sites on the north island of New Zealand. Not surprisingly, the highest growth rates were achieved at the lowest latitudes where the water was warmest, although mortality rates were the inverse. They concluded that sea-cage culture of *J. edwardsii* is feasible and compatible with other types of offshore aquaculture.

To facilitate development of the aquaculture of *Jasus edwardsii*, the New Zealand government has instituted a quota tradeoff, whereby for every tonne of quota withdrawn from the commercial fishery, 40,000 pueruli and early post-pueruli can be taken locally (Booth *et al.*, 1999). The initial attempts to catch puerulus larvae of *J. edwardsii* using collectors designed for recruitment studies were not encouraging. Peak settlement rate was only 0.9 puerulus collector^{-1} d^{-1} (Booth and Tarring, 1986), with mean rates of 0.5–0.75 puerulus collector^{-1} d^{-1} (Hayakawa *et al.*, 1990). However, the situation has now changed, and monthly catches in the thousands have been made using improved 'crevice' collectors during peak seasons in geographic 'hot spots' (Booth *et al.*, 1999; Andrew Jeffs, personal communication, 2003).

Although the developments in New Zealand are currently aimed at aquaculture, not stock enhancement, they are laying the foundation for the supply of cultured juveniles by means of effort reduction (quota retirement in exchange for pueruli). To determine whether the cultured juveniles will be suitable for stock enhancement, Oliver *et al.* (2001) investigated whether there were any behavioural deficiencies in the captive-reared animals. They found that when cultured individuals fed by day and not exposed to predators were released they adopted nocturnal emergence patterns and responded to predators in the same way as wild conspecifics.

3.6.2.3. Vietnam

The recent development of aquaculture of *Panulirus ornatus* in Vietnam provides a good example of increases in spiny lobster production based on collection of puerulus larvae (Williams, 2004). In addition to using a variety of collectors for settled juveniles (coral blocks and timber poles with drilled holes and bunches of net), puerulus larvae are caught by stationary tunnel nets that have red or white fluorescent lights at the mouth. The nets (15–20 m wide, 3 m deep and 15 m long) are set twice a night (at 20:00 and 01:00) for 4 h. Catches are variable (0–50 lift^{-1}), but typically yield 1 to 2 puerulus lift^{-1} over the season (peaks from November to April). This immense effort yields 3.5 million 'seed' annually. The low CPUE is compensated by very high prices, ~US$6.00 in 2004 for a single unpigmented puerulus, and up to US$11.00 for a juvenile of 15 mm CL grown for 2 mo.

The collected juveniles are reared in >30,000 simple net pens in shallow embayments. A group of growers rear the post-pueruli to 50 g and then sell them to other farmers for growing on to market size. The juveniles are first fed on shellfish and later on trash fish (by-catch and by-product). The lobsters are harvested 12–20 mo later at 0.5–1.0 kg, depending on the season and market opportunity, providing further evidence of the fast growth of *Panulirus ornatus* in culture. Production was ~2,000 t in 2002, (Williams *et al.*, 2002). Most of the production of *P. ornatus* is from two provinces (Khanh Hoa and Phu Yen) in central Vietnam (12–15° N). Tuan *et al.* (2000) estimate that survival during growing on is approximately 90%, which is greater than reported by Jones *et al.* (2001) for *P. ornatus* in land-based tanks and comparable to trials with *Jasus edwardsii* (Thomas *et al.*, 2003). Rates of juvenile survival during the first benthic year in the wild are unknown. Potential to increase production through the capture and culture of puerulus of *P. ornatus* has been identified by Hair *et al.* (2002), who put forward a promising scenario based on a hypothetical total natural supply of 100 million pueruli per annum and the assumptions of: (1) 5% survival after the first benthic year, in line with data for other species (e.g., *P. argus*) (Herrnkind and Butler, 1994; Butler *et al.*, 1997) and (2) 90% survival in culture. In their analysis, removal of 2 million pueruli for growing on would reduce recruitment to the fishery from 5 million to 4.9 million lobsters, but yield 1.8 million market-sized animals, a net gain of 1.7 million individuals.

For the capture and culture of puerulus to be considered a sustainable form of production in Vietnam, as opposed to short-term economic gain, the following questions need to be answered:

1. What is the location, population size and catch rate of the wild adults producing the puerulus larvae? Apparently, the fishery for *Panulirus ornatus* in the immediate vicinity of this captive industry has been

nonexistent since about 1995 and was one of the motivations for going into this style of production (Williams *et al.*, 2002; Williams, 2004).

2. What is the spatial and temporal variation in abundance of larvae? Given the local population decline, what is the 'upstream' source of the locally caught pueruli?
3. What is the actual survival of pueruli after settlement in Vietnam?
4. What proportions of pueruli are collected for growing on?
5. Are the initial high rates of survival in captivity vulnerable to diseases?
6. What is the number of adults reproducing in captivity and the fate of larvae hatched from the captive stock?
7. What is the impact of lobster farming on stocks of species used for food?
8. Can the cage environment and food conversion ratios be improved with formulated diets? Conversion ratios using trash fish can be as poor as 28:1 (Tuan *et al.*, 2000).

3.6.2.4. The Americas and the Caribbean

Unlike much of Asia-Pacific, where most research has been on larval and postlarval (puerulus) biology and ecology, investigations to improve production of spiny lobster fisheries in the Americas and the Caribbean have dealt largely with the role and suitability of artificial habitats. As general background for this approach, it is worth noting that use of artificial habitats to enhance commercial and recreational fisheries is widespread but controversial (Grove *et al.*, 1991; Polovina, 1991; Seaman and Sprague, 1991). Key questions concern whether artificial habitats: (1) enhance production by creating places that promote settlement and recruitment of juveniles, (2) provide shelter from predation and/or substratum for growth of food or prey or (3) simply concentrate existing stocks, and their predators, making them more vulnerable to exploitation and natural sources of mortality (Figure 3.23) (see Bohnsack, 1989; Bohnsack *et al.*, 1997; Lindberg, 1997 for general debate and reviews). The production/attraction issue, together with related questions about optimum designs for additional habitat, have been the focus of many studies in Florida, Cuba and Mexico seeking to assess the potential for artificial habitats to increase production of the key species in the region, *Panulirus argus*.

3.6.2.4.1. Assessing the habitat requirements of juveniles
The habitat requirements of juvenile *Panulirus argus* are now well known. At the end of the 9- to 12-mo planktonic phase, pueruli settle preferentially on hard substrata covered in macro-algae (e.g., *Laurencia* spp.) and seagrass

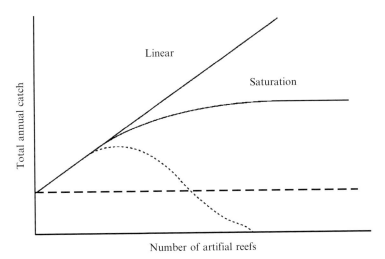

Figure 3.23 Predicted effects on catch of the 'attraction' and 'production' hypotheses for artificial reefs. The production hypothesis predicts increased catch as a function of the amount of artificial reef deployed (solid lines). The attraction hypothesis (dotted line) predicts an initial increase in catch followed by a decline and, in a worst case scenario, a decline to levels below catch without artificial reefs (dashed line) (redrawn from Bohnsack, 1989).

(e.g., *Thalassia testudinum*), where they metamorphose to 'algal-phase' juveniles (Figure 3.20) (Butler and Herrnkind, 1997; Herrnkind *et al.*, 1999). These settlement habitats are complex and offer both food and shelter (Kanciruk, 1980; Caddy, 1986). Settled juveniles remain solitary and associated with aquatic vegetation, until they reach a size of 15 mm CL. They then become gregarious, living with other juveniles in crevices that afford protection from predation (Childress and Herrnkind, 1994). Mortality during the first year of benthic life (algal and post-algal phases) is usually 96–99% (Herrnkind and Butler, 1994; Butler *et al.*, 1997). Given this high mortality and the potential limitations in settlement substrata (food and shelter), the early benthic phase of *P. argus* seems to represent a bottleneck to recruitment and, ultimately, catch (Caddy, 1986).

3.6.2.4.2. Artificial habitats for juveniles

To assess the potential for artificial habitats to improve settlement success of pueruli, Butler *et al.* (1997) examined both the 'supply side' hypothesis, in which recruitment success is determined by the numbers of postlarvae entering the nursery ground (Fogarty *et al.*, 1991), and the 'shelter bottleneck' hypothesis, in which recruitment is regulated by availability of settlement and post-settlement habitat (Caddy, 1986).

This 'shelter-bottleneck' hypothesis was tested in an experiment by adding both artificial shelters and additional juveniles to algal and hard bottom sites (Butler and Herrnkind, 1997). Numbers of small post-algal juveniles (<35 mm CL) increased significantly at all sites where shelter was enhanced, but not at unmanipulated sites. The addition of juveniles did not increase the numbers further. Thus, availability of crevice refuges limits the number of *Panulirus argus* in nursery grounds, given the pre-requisite conditions of sufficient pueruli and protective vegetated substrata during the settlement phase. Tag-recovery data also indicated that rates of survival corresponded to availability of shelter; recovery rates were six times higher in the high-shelter (artificial cement blocks) than in the low-shelter (natural) habitat.

During the preceding experiment, a massive cyanobacteria bloom reduced the number of marked sponge dens by 50% over hundreds of km^2. These dens housed 60% of all post-algal stage juveniles before the die-off and only 25% afterwards (Herrnkind *et al.*, 1997, 1999). To test the efficacy of large-scale shelter replacement under these circumstances, 1-ha arrays of cement block shelters (240 blocks ha^{-1}), approximating the density of sponge crevices in the region, were used. Within 6 mo, abundance of new crevice-dwelling spiny lobsters (<25 mm CL) increased two- to three-fold in shelter-supplemented and natural sponge sites, but not in the sponge die-off sites. Immigration of older juveniles (35–50 mm CL) and longer residency times were also noted.

In summary, small artificial shelters match the qualities of natural structures and can provide suitable additional habitat or ameliorate large-scale habitat loss for small juvenile *Panulirus argus*. Supply of both puerulus larvae and algal substrata varied in space and time. Neither alone predicted recruitment to subsequent life history stages. Herrnkind *et al.* (1999) proposed a conceptual model in which the carrying capacity of the environment (i.e., the number of crevices within algal cover) for post-settlement juveniles is the key determinant in establishing a relationship between abundance of immigrating puerulus larvae and subsequent successful recruitment of juveniles (Figure 3.24). They caution, however, that the full potential of using artificial shelters at a large scale to enhance recruitment of *P. argus* cannot be assessed without further research on the effect of competitors and predators, and an economic cost–benefit analysis.

3.6.2.4.3. Artificial habitat for sub-adults and adults

Far more research has been done on the use of larger artificial structures (*jaulas, sombras, pesqueros, casitas cubanas,* or simply *casitas*), and the habitation of larger *Panulirus argus* (Eggleston *et al.,* 1992; Herrnkind *et al.,* 1999, Briones-Fourzán *et al.,* 2000; Cruz and Phillips, 2000; Losada-Tosteson and Posada, 2001). *Casitas* are low shelters (Figure 3.25) placed in seagrass or algal meadows to provide daytime refuge for sub-adult and adult

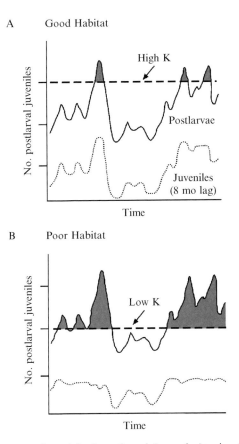

Figure 3.24 Conceptual model of postlarval (puerulus) spiny lobster settlement and subsequent juvenile abundance in (A) 'good' (many crevices) and (B) 'poor' (few crevices) nursery habitat. K = carrying capacity of sheltering habitat. Shaded areas under the curve represent mortality of puerulus larvae as a result of insufficient shelter (redrawn from Herrnkind *et al.*, 1999).

spiny lobsters that forage in these habitats at night. Throughout the Caribbean, fishermen use these structures to concentrate lobsters in areas that lack natural shelter. In Cuba, at least 250,000 *pesqueros* are used to attract 30–50% of the national catch (Cruz and Phillips, 2000). The structures vary in shape, size and construction but, because of shortages of natural materials (i.e., palm and mangrove trees) in coastal zones, they are now often made of fibre-cement, ferrocement or old tyres. The spiny lobsters are removed from the shelters by diving and spearing, dip netting, herding into a net down-current and/or surrounding the shelter with a seine or bag net (Cruz and Phillips, 2000).

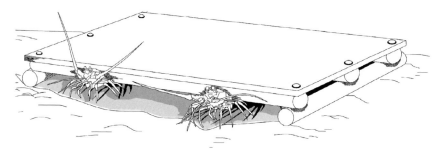

Figure 3.25   A *casita* used to attract and catch the spiny lobster *Panulirus argus* in the Caribbean (redrawn from Eggleston *et al.*, 1992).

Fishermen have refined the designs and placement of these artificial structures through trial and error. Important criteria are: season; size and habitat-specific behaviour of spiny lobsters, including degree of gregarious behaviour through olfaction; size of artificial structure; number and shape of openings, and composition, abundance and behaviour of the predator guild (Spanier and Zimmer-Faust, 1988; Spanier *et al.*, 1988; Lozano-Álvarez *et al.*, 1994; Spanier, 1994; Arce *et al.*, 1997; Sosa-Cordero *et al.*, 1998).

The efficiency of using *casitas* to concentrate spiny lobsters and facilitate harvest has been demonstrated by cost/benefit analysis (Seijo *et al.*, 1991) and their use is now widespread. However, their role in enhancing populations of *Panulirus argus* is equivocal, and perhaps even problematic. There is concern that the concentrated spiny lobsters are subject to both increased fishing mortality and predation intensity (Eggleston *et al.*, 1992). The relationships between predation and size of spiny lobster and interactions with the availability, location and dimensions of shelters were examined by Eggleston *et al.* (1992). They tethered animals at various distances from shelter and found that mortality of sub-adult spiny lobsters (46–65 mm CL) varied significantly as a function of body size and distance from refuge but not availability of shelter or location (proximity to coastline or offshore reef). These relationships emerged because spiny lobsters gain a size refuge as they grow, and predators have strong affinities to *casitas* and therefore limited foraging distances (<15–40 m, depending on species) away from these structures. Eggleston *et al.* (1992) also concluded that, for small sub-adult lobsters (46–55 mm CL), small *casitas* increase production by increasing survival in nursery grounds, but that survival of larger (56–65 mm CL) tethered sub-adults was lower in large *casitas*. Therefore, one size of *casita* does not fit all. Rather, the artificial shelters must be scaled according to body size of spiny lobsters if they are to afford increased survival across the size distribution of the population in the nursery grounds. To assess the efficacy of *casitas* as a stock enhancement tool, more research is needed on: size-specific survival and growth rates of spiny lobsters; local and regional population structure of *P. argus*; and benthic community structure (Eggleston *et al.*, 1992).

Briones-Fourzán *et al.* (2000) also cite three other cases where placement of *casitas* either: (1) increased the number of spiny lobsters in proportion to the number of shelters over time compared with control sites (Florida Bay in the United States, Lipcius and Eggleston, unpublished data); (2) increased density of juvenile spiny lobsters at sites where the shelters were placed (Puerto Morales Mexico, Lozano-Álvarez *et al.*, 1998 [cited in Briones-Fourzán *et al.*, 2000, unpublished data]) or (3) simply concentrated relatively large juvenile and sub-adult (40–75 mm CL) spiny lobsters migrating between nursery grounds and offshore reefs (Exuma Cays Bahamas, Eggleston and Lipcius, unpublished data). They conclude that the efficacy of *casitas* is determined largely by a combination of abundance of puerulus larvae, extent of available settlement substrata, and availability of natural shelters.

In the Mexican Bahía de la Ascension fishery, *casitas* have been used for >25 yr (Briones-Fourzán *et al.*, 2000). In 1985, there were approximately 20,000 *casitas* on the grounds, until hurricane Gilbert in 1988 reduced the number by 50%. By 1999, the number of *casitas* had increased to almost 17,000, but catches have remained low (mean catch pre-Gilbert ca. 50 t of tails; post-Gilbert ca. 30 t). In this case, increased numbers of *casitas* have not overcome recruitment failure. In the Cuban Gulf of Batabanó fishery, the number of *pesqueros* (*casitas*) climbed steadily from ~2,000 in the late 1970s to 5,000 in 1995, and fishing effort (inspections of *pesqueros*) rose from 4.0 to 12.5 × 10^5 from 1975 to 1987 (Cruz and Phillips, 2000). During this period, catch was relatively constant, but CPUE declined dramatically until 1990 when effort was restricted. These two examples suggest that *casitas* placed in the fishing grounds (as opposed to nursery areas) have not augmented the population but simply re-distributed the spiny lobsters among more refuges. This would support the 'attraction' hypothesis of artificial reefs, with the attendant risk of over-exploitation (dotted line in Figure 3.23) (Bohnsack, 1989).

3.6.2.5. Hawaii

Information on the fishery for *Panulirus marginatus* in Hawaii related to the potential for stock enhancement comes from comprehensive surveys of abundance (pre-exploitation and commercial CPUE data) and habitat undertaken to assess variation in landings among banks in the Northwestern Islands (Parrish and Polovina, 1994). The main findings of the surveys were:

(1) mean landings varied 300- to 400-fold over 6 yr across 13 reefs; (2) larval abundances were generally comparable across reefs and (3) the amount of various habitat types (gross topography and small-scale benthic shelter) varied significantly. Analysis of these data showed that there was a non-linear relationship between abundance of sub-legal spiny lobsters and habitat of intermediate relief (5–30 cm), which was made up largely of multi-species, macro-algal stands. Because the productive banks had much more of this intermediate relief habitat than the unproductive one, the results of surveys from Hawaii support the findings of Herrnkind and Butler (1986) and Herrnkind *et al.* (1999) that the amount of suitable habitat is a bottleneck to production of adult spiny lobsters. Another interesting outcome was that the pueruli recruited directly to the adult habitat because of the special conditions at the isolated islands (Parrish and Polovina, 1994).

3.6.2.6. Japan

Japanese scientists have been attempting to culture palinurid larvae for >90 yr (Nonaka *et al.*, 2000). Kittaka (1988) was the first to succeed in rearing larvae through to the puerulus stage. Since then, six other palinurid and two scyllarid species have been raised in Japan at an experimental scale (Table 3.8) (Kittaka, 2000). However, many biological and technical problems remain to be solved in establishing commercial-scale production (Kittaka, 1997; Nonaka *et al.,* 2000).

Attempts to collect puerulus larvae in Japan have been intermittent, small-scale and restricted to studies of recruitment dynamics. There seem to be relationships among abundance of pueruli, proximity of the central stream of the Kuroshio current, and lunar periodicity, but the numbers of larvae are low even at peak times on Phillips-style collectors (4–8 collector^{-1} day^{-1}) (Nonaka *et al.*, 2000). To date, there has been no attempt at large-scale systematic collection of pueruli for monitoring recruitment, growing on or stock enhancement. However, some of the basic ecological information required to adopt this approach is in place. Norman *et al.* (1994) studied the habitat selection of *Panulirus japonicus* pueruli in the wild and post-settlement growth rates in the wild and in the laboratory. They found that: (1) pueruli settled in a variety of holes (initially pholad holes) on shallow (2- to 3-m deep) reefs at densities of 0.2–0.3 individuals m^{-2}; (2) growth rates of both wild and laboratory-reared juveniles were relatively rapid, from 7.5 mm to 35 or 40 mm CL in 1 yr (Table 3.9) and (3) recruitment to the fishery occurred at 42 mm CL during the second winter after settlement.

Japanese investigators have also been examining use of artificial reefs to create or enhance spiny lobster fishing grounds since 1933 (Nonaka *et al.*,

2000). Oshima (1976, cited in Nonaka *et al.*, 2000) calculated that every 100 m² of artificial reef contributed 111–331 kg to the annual catch of *Panulirus japonicus*. There is concern, however, that the increased catch from poor fishing grounds is probably due to attraction rather than production (Nonaka *et al.*, 2000). For example, they describe development of artificial reefs, begun in 1976, at 10 locations in eight prefectures that concentrated the resource but did not provide evidence of improved conditions for recruitment of wild pueruli. The main problems were inadequate estimates of settlement and post-settlement dispersion and distribution.

Nonaka *et al.* (2000) conclude that: (1) stock enhancement of spiny lobsters in Japan is likely to depend on increasing the supply of pueruli through habitat enhancement, and (2) more research is needed to estimate the numbers settling and post-settlement survival, and to establish economic criteria to evaluate the cost-effectiveness of interventions.

3.6.3. Status of stock enhancement initiatives

Although there has been extensive research that could underpin the aquaculture of spiny lobsters, and eventually stock enhancement, large-scale release of juveniles in stock enhancement programmes is still some way off. The two countries that are in the best position to test this form of management on a meaningful scale are Australia and New Zealand.

3.6.3.1. Australia

In Tasmania, three operators currently hold permits to collect and grow pueruli of *Jasus edwardsii* (D. Mills, personal communication, 2003). Each permit allows a maximum of 50,000 pueruli to be collected annually. At this stage, all operators are working on an experimental scale and refining the design of collectors. Permit conditions require that after 1 yr (or when the juveniles reach a size of 35 mm CL, whichever comes first), the number to be released will be 5% of the pueruli harvested and an additional 20% of juveniles surviving at that time. The numbers to be released were arrived at through a process of negotiation between stakeholders (primarily the aquaculture industry and fishermen) rather than rigorous scientific investigation. Although the research to date suggests that cultured *P. edwardsii* will survive well when released, there has been no progress on estimating mortality of juveniles during the first benthic year beyond the short-term observations (11 d) by Mills *et al.* (2004). If growing on of pueruli proves to be profitable, and survival of naturally settling juveniles is ~5%, as found for other species, the model in place in Tasmania should result in stock enhancement through the release of the additional 20% of cultured animals.

3.6.3.2. New Zealand

A different approach is being taken in New Zealand. There, all spiny lobster stocks are managed under a Quota Management System, which is broken into regional Quota Management Areas. Within each of these areas, commercial fishermen own, and can trade, the right to catch a certain number of legal-sized animals (an Individual Transferable Quota). A trial harvesting of pueruli, initially for 3 yr, with a subsequent review, was initiated in 1996 by the national representative body of the Rock Lobster Quota Holders. This allowed the potential of rock lobster farming to be explored while protecting both the rights of existing fishermen and the industry (A. Jeffs, personal communication, 2003). The basis of the arrangement is that anyone can apply for a special permit to harvest pueruli provided they withdraw commercial quota at a rate of 1 t for access to 40,000 pueruli (or 3 kg live weight). The New Zealand government does not require a proportion of juveniles to be returned to the wild at a later date.

Three groups have gained permits over the past 6 yr and are using enhanced 'crevice' collectors to catch pueruli in settlement 'hot spots'. The first two groups focused on land-based farming, but only one is continuing on a small scale. The third group, which consists of a consortium of mussel and abalone farming companies, has commenced more recently and is having considerable success with developing sea-cage farming systems.

To assess the potential for using reared pueruli for stock enhancement, national scientists are monitoring survival of cultured juveniles released back to the wild. A national programme is also underway to test whether provision of small-scale artificial habitats in settlement and nursery areas will enhance production.

3.6.4. Minimising genetic impacts

The information on stock structure of palinurids is limited, although some significant studies have been made in the Caribbean, Australia and New Zealand. Even so, the identity of many stocks remains unclear. Analysis of *Panulirus argus* mitochondrial DNA samples, ranging from Panama and Venezuela in the south to Bermuda in the north, indicates high levels of gene flow throughout the Caribbean (Silberman *et al.*, 1994). However, the population in the Caribbean is distinct from the one in the Western Atlantic off Brazil (Sarver *et al.*, 1998).

Using allozymes, Smith *et al.* (1980) found no differences between samples of *Jasus edwardsii* from three locations in New Zealand but did find a slight difference between *J. edwardsii* in New Zealand and *J. novaehollandiae* in Tasmania, Australia. They suggested that the differences between samples from New Zealand and Australia were too small to separate the species but supported the argument for separate populations of a single species. Booth *et al.* (1990), using a larger genetic data set and additional biological information, confirmed that *J. edwardsii* and *J. novaehollandiae* were the same species (*J. edwardsii* has priority). They also confirmed that there was a single population in New Zealand but found three separate populations in Australia, one in Tasmania and two in the southern Australian mainland. Subsequently, however, studies by Ovenden *et al.* (1992) and Ovenden and Brasher (1994), using mitochondrial DNA analysis, found no evidence of population subdivision for *J. edwardsii* within or between New Zealand and Australia over a distance of 4600 km. Considering this conflicting genetic information, a relatively conservative approach is recommended toward movement of broodstock for hatchery-rearing trials and wild-caught pueruli for on-growing or stock enhancement. Further population analysis with more powerful genetic techniques (e.g., microsatellite DNA analysis, see Chapter 7, Section 7.2) is required. This conservative approach would restrict translocations between countries and confine releases to regions within countries where broodstock or pueruli were collected.

3.6.5. Disease risks

Comprehensive reviews by Sindermann (1990), Bower *et al.* (1994) and Evans *et al.* (2000) indicate that the principal diseases of spiny lobsters are bacterial (gaffkemia caused by *Aerococcus viridans*, shell disease, vibriosis); fungal (systemic and superficial) and parasitic (egg predators). No viral diseases have been described. The relatively small complement of diseases probably reflects the general lack of a commercial aquaculture industry that might bring about environmental stressors and more infectious conditions (e.g., crowding and handling) and a commercial imperative to find both the cause of the disease and a solution. This situation is now changing, however, with reports of mass mortality (20–30% per week) of *Panulirus ornatus* in some cage-culture operations in Khanh Hoa Province, Vietnam, where poor husbandry, poor water quality and limited water circulation have been implicated in promoting lethal infection. To date, it is not known whether the presence of the high numbers of *Vibrio* bacteria and *Fusarium* fungi associated with *P. ornatus* in Vietnam were the causes of the mortalities or secondary to the poor water quality (K. Williams, personal communication, 2003).

3.6.6. Future research

Three areas of research are needed to help make stock enhancement of spiny lobsters a reality through the release of on-grown postlarvae or juveniles or the use of additional habitat for settling juveniles. These three areas are: (1) larval and juvenile production; (2) optimised release strategies and (3) large-scale experiments using additional habitat in places where it is practicable to do so.

3.6.6.1. Larval and juvenile production

Meaningful levels of stock enhancement based on release of cultured juveniles depends on availability of large numbers (10^5–10^6) of 'seed', and so much effort still needs to go into identifying where and how to catch pueruli. The large-scale capture of puerulus larvae occurs only in a few places and is expensive. Jeffs and Hooker (2000) estimated the cost of wild-caught pueruli of *Jasus edwardsii* at between US$0.20 and 0.70 each. Reduction of costs will come by continuing research already underway to find those places where pueruli settle consistently in high numbers; by developing larger, efficient collectors or from alternative methods like the tunnel nets used in Vietnam.

   Clearly, production of large numbers of juveniles in hatcheries would remove the uncertainties of relying on wild pueruli. Research in this area should continue until it is clear whether the problems associated with the complicated larval phase of spiny lobsters, which currently limit cost-effective mass production, are likely to be overcome.

3.6.6.2. Optimised release strategies

The extensive research on habitat requirements, movement and mortality of juvenile spiny lobsters should now be applied in comprehensive field experiments to identify how and where to release reared juveniles, including into artificial habitats, and how to maximise survival. This has largely been done for *Panulirus argus* at a few sites but needs to be demonstrated for the other major species of spiny lobsters in Asia-Pacific. The resulting information can then be put into a bio-economic model to assess the optimal size (a tradeoff between production costs and mortality rates), location and time (to optimise survival) for release (see Loneragan *et al.*, 2003, for an example with penaeid shrimp).

3.6.6.3. Large-scale tests of artificial habitat for juveniles

Because survival of several species of spiny lobsters during the first benthic year is typically <5%, and two- to three-fold increases in survival of *Panulirus argus* occurred when additional juvenile habitat was provided over a 1-ha grid, there is substantial potential for enhancing production by reducing predation in this simple way. Where prevailing sea conditions make it practicable to install shelter with the required crevice sizes, three pieces of research are needed. To determine whether this intervention makes a significant contribution to the strength of settling year classes it will be necessary to: (1) identify nursery areas accurately; (2) test whether the number of individuals recruiting to nursery areas with large-scale installation of habitats contribute to the population of sub-adults and (3) monitor the sub-adults and animals of sub-legal size to ensure that any gains in abundance made during the first benthic year are not lost subsequently due to density dependence. Such gains will need to be well above the range in natural variation in the supply of pueruli and be cost-effective compared with other management options to rebuild or enhance stocks.

3.7. LOBSTERS

3.7.1. Background and rationale for stock enhancement

After more than a century of study, much is known about the biology and ecology of European and American lobsters (*Homarus gammarus* and *H. americanus*). The genus is confined to the Atlantic Ocean, from the Mediterranean to Norway, excluding the Baltic, and the east coast of North America, from North Carolina to Newfoundland. *Homarus americanus* ranges from the intertidal to depths of 700 m, whereas *H. gammarus* lives at depths <65 m.

Both species support important trap fisheries and stocks are exploited heavily. Increased prices have ensured unremitting fishing effort whenever catch rates have declined. In the case of *H. americanus*, catches have varied greatly during the past century; total annual landings in North America fell from >50,000 t yr^{-1} in the 1890s to <20,000 t in 1920. Canadian landings increased quite steadily after 1940, from around 5,000 t yr^{-1} to 15,000 t in 1975, whereas annual landings from the United States fluctuated around 20,000 t between 1920 and 1975. Since 1975, catches from both the United States and Canadian waters have increased considerably to reach a combined total of >80,000 t yr^{-1} (Nicosia and Lavalli, 1999; Browne *et al.*, 2001; FAO, 2002) (Figure 3.26).

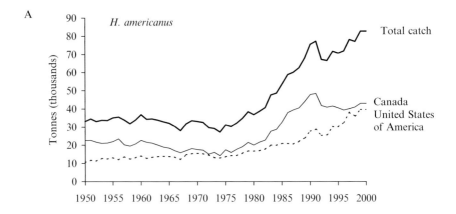

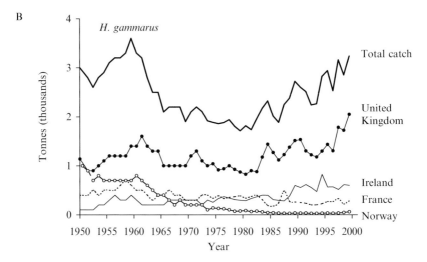

Figure 3.26   Landings of lobsters from 1950–1999; (A) *Homarus americanus* in
North America, and (B) *H. gammarus* in Europe (data from FAO, 2002).

Stocks of *Homarus gammarus* have never been as large as those of
H. americanus, and total landings reached 3,800 t in 1932, declined to about
1,800 t in 1980 and then rose again, somewhat erratically, to >3,000 t in
1999, largely as a result of increased catches in the United Kingdom
and Ireland (Addison and Bannister, 1994; FAO, 2002) (Figure 3.26). Stocks
declined to negligible levels in Norway, Sweden, Denmark, the Netherlands
and Germany by the mid–1980s (van der Meeren, 1991b; Browne *et al.*,
2001). Likewise, stocks in Italy, notably in the Adriatic, have collapsed

(Scovacricchi, 1999; Scovacricchi *et al.,* 1999), as have stocks in Turkey, the former Yugoslavia, Algeria, Spain and Portugal (Browne *et al.*, 2001).

The reasons for the resurgence of *Homarus americanus* remain obscure. Release from predation as a result of declines in stocks of Atlantic cod *Gadus morhua*, American plaice *Hippoglossoides platessoides* and other benthic-feeding fish are unlikely to account for the recovery, because these species rarely prey on lobsters (Hanson and Lanteigne, 2000). The fact that there was a widespread increase in catches in the absence of any significant change in management regimes suggests that the recovery may have resulted from higher recruitment rates related to a physical change in the environment or to a decline in abundance of some key predator on the larval or post-settlement stages (Nicosia and Lavalli, 1999). Oceanographic factors, such as seawater temperature, can influence lobster fecundity and the survival and behaviour of larvae (Waddy *et al.*, 1995; Palma *et al.*, 1999). The gradual warming of ocean waters in the Gulf of Maine and other areas along the east coast of the United States during the past two decades may be responsible for the decrease in the size at which female lobsters become sexually mature (Landers *et al.*, 2001). It may also account for the concomitant increase in the number of undersized, egg-producing females. The following brief description of the biology of lobsters reveals why it is difficult to understand patterns of mortality and abundance in these species.

After a relatively brief planktonic existence lobster larvae settle on the seabed at the fourth moult stage. The early benthic stages of both *Homarus americanus* and *H. gammarus* build burrows, usually under or near loose rocks or cobblestones, where they are protected to some degree from predators. The juvenile lobsters emerge briefly as they grow (Cobb and Wang, 1985), although they feed mostly on infauna such as small bivalves and crabs available in the burrow or on plankton wafted into the burrow by pleopod fanning (Barshaw, 1989; Lavalli, 1991; Wickins *et al.*, 1996; Sainte-Marie and Chabot, 2002).

Burrow-dwelling juveniles in cobblestone habitats have been sampled successfully in the United States using air-lift systems (Wahle and Steneck, 1991). Even so, estimates of settlement rates and survival during the first 2–3 yr of life are extremely difficult to obtain. Attempts to locate juvenile *Homarus gammarus* <3 yr old in Europe have generally been unsuccessful (Linnane *et al.*, 1999, 2001), either because densities are extremely low or because the juvenile habitat has yet to be identified. However, studies in a mesocosm showed that *H. gammarus* would colonize substrata of cobblestones or mussel shells in preference to sand or coralline algae (Linnane *et al.*, 2000a), and in this respect their behaviour is similar to *H. americanus*. Early juveniles of both *H. americanus* and *H. gammarus* have also been found under rocks in the lower intertidal zone (Cowan, 1999; Linnane *et al.*, 2000b), but this does not seem to be a nursery habitat of much significance.

Juveniles leave their burrows after 2.5–3.5 yr, at about 40 mm carapace length (CL), and become free-ranging sub-adults inhabiting 'dens' in crevices or holes, preferably in rocky habitat. However, lobsters will also excavate burrows if rocky habitat is unavailable. Lobsters cannot cope with strong currents, and where these occur their distribution is determined by availability of suitable shelter. Small lobsters are found where modest-sized rocks shield them, but large lobsters are forced to move elsewhere. Tidal currents also influence feeding, which is limited to times of slack water (Howard, 1988). Lobsters are relatively sedentary and will often inhabit the same den for extended periods. If captured and released fairly close to their den (<1 km), they are able to relocate it but are unable to do so if they are displaced by 5 km (van der Meeren, 1997). Unlike *Homarus gammarus*, some *H. americanus* undertake quite extensive offshore migrations, probably maintaining themselves in a temperature range of 8–14°C (Cobb and Wang, 1985). Migration rate in *H. americanus* is also related to size/age and origin, with larger and older animals from offshore populations traveling distances >90 km (Estrella and Morrisey, 1997).

Males of both species grow larger and faster than females (Campbell, 1983), and there are large variations in size at age (Bannister *et al.*, 1994). Studies on lipofuscin accumulations in eyestalk ganglia in *Homarus gammarus* that had been marked with microwire tags 4.5–8.5 yr previously have provided an age–lipofuscin correlation (Sheehy *et al.*, 1999). The oldest male studied was about 42 yr, and the oldest female was estimated to be 72 yr. If the estimates of Sheehy *et al.* (1999) are correct, the age of recruitment to the fishery will be highly protracted, and some slow-growing individuals may never reach the legal minimum size. Also, the fishery will actively select against fast growth.

Female *Homarus gammarus* attain maturity at 92.5–96 mm CL in Ireland (Tully *et al.*, 2001) but maturity in the range of 77–90 mm CL has been reported elsewhere (Cobb and Wang, 1985). Female *H. americanus* mature at 55–110 mm CL, although much depends on water temperature and latitude (Fogarty, 1995; Waddy *et al.*, 1995). It is not known when male lobsters become sexually active. Fecundity (F) is relatively low in *H. gammarus* (F = 0.0064 CL $(mm)^{3.1554}$), and egg size increases with body size (Tully *et al.*, 2001). The fecundity is greater in *H. americanus* (F = 0.00256 CL $(mm)^{3.409}$) (Campbell and Robinson, 1983), and there is substantial variation between populations and/or the results of various studies (Tully *et al.*, 2001).

Despite some important gaps in knowledge of the biology of the species, several attempts have been made to increase catches through release of hatchery-reared juveniles. The earliest work on cultivation of lobsters from egg to sub-adult stages was done in France in 1858, with the hope that this might 'regenerate the fishery' (Latrouite and Lorec, 1991). This was soon followed by artificial propagation work in the United States, Norway, Canada and the United Kingdom. The sole objective of these early hatcheries was to hatch lobster eggs artificially and release the first-stage larvae into local waters. These efforts had been abandoned by 1955 because of the costs and inability to demonstrate any benefits from releasing lobsters at such an early stage of development (Nicosia and Lavalli, 1999). The fact that lobsters can now be reared to larger sizes (Figure 3.27), albeit in relatively limited numbers, has re-awakened interest in releasing juveniles in the wild.

There have been concerns, however, about whether significant numbers of cultured juveniles would survive to exploitable size and whether the extent of juvenile habitat would limit the degree to which stocks could be augmented by hatchery-reared animals (Bannister and Addison, 1998). The latter concern was raised in relation to a study in the Gulf of St. Lawrence, Canada, which suggested that the recruitment/stock abundance curve for lobsters was asymptotic and that, except at very low stock densities, recruitment of stage IV juveniles was not proportional to stock abundance 5–7 yr later (Fogarty and Idoine, 1986). However, Bannister and Addison (1998) later concluded

Figure 3.27 (A) Larva of the American lobster *Homarus americanus* (photo: A. Mercier). (B) Morphological varieties of cultured European lobster *H. gammarus* in Norway (photo: E. Farestveit). (C) Adult European lobster from southwestern Norway (photo: E. Farestveit). (D) Blue colour morph of the American lobster (photo: A. Mercier).

that the level of natural variation in the Bridlington Bay stock indicated that the habitat there was capable of supporting much larger numbers of juveniles than are supplied naturally. Indeed, apart from the study of Fogarty and Idoine (1986), all other evidence suggests that lobster stocks are usually recruitment-limited. In particular, the burrowing behaviour of lobsters pre-disposes them to live at potentially high densities as juveniles (i.e., there should be little density-dependent mortality related to food supply because most feeding occurs within the burrow).

In principle, therefore, lobsters are suitable for stock enhancement, although the imperative for *Homarus gammarus* throughout much of its range is for restocking to rebuild spawning biomass to far more productive levels. However, important questions remain about how to produce the large numbers of juveniles needed for effective contributions to populations at reasonable cost.

3.7.2. Supply of juveniles

The development of hatchery methods is described well by Nicosia and Lavalli (1999) and Beal and Chapman (2001). Where permitted, females carrying fertilized eggs are collected from the wild and maintained in hatcheries until the eggs hatch. Alternatively, females can be mated and will spawn in captivity. The larvae (Figure 3.27) are relatively easy to rear, and techniques are now standardized (Chang and Conklin, 1983; Nicosia and Lavalli, 1999). The most recent developments are variations of the 44 l 'Hughes Pot' or 'kreisel' developed in Massachusetts, in which cannibalism is minimised by turbulent water flow in a round-bottomed system that keeps larvae in suspension and distributed evenly. Larvae kept at densities of 4,000 per kreisel have a survival rate of ~75% and a 16-kreisel unit can produce about 0.5 million stage IV larvae in a summer season of 120 d at 22°C. Similar systems have been developed in the United Kingdom, Canada, Ireland and Norway (Beard and Wickins, 1992; Knudsen and Tveite, 1999; Beal and Chapman, 2001). The larvae are usually fed on frozen mysids supplemented by live, algal-enriched *Artemia*. There is great variability in production rates for the various hatcheries.

Beyond stage IV, it is necessary to rear juvenile lobsters in separate compartments because they are cannibalistic (Chang and Conklin, 1983; Lee and Wickins, 1992). This behaviour, combined with relatively slow growth rates, makes production of juveniles expensive. In 1994, the cost of producing a 15-mm CL, stage X–XII larva of *Homarus gammarus* was estimated to be £1.50 in the United Kingdom (Addison and Bannister, 1994). Prices have now been reduced to £0.36 (Burton and Adamson, 2002). Overall, production methods struggle to supply large numbers of juveniles, and Nicosia and Lavalli (1999), in their major review of lobster hatcheries, concluded that current technology is limiting output for stock enhancement and restocking.

In an attempt to reduce production costs, Knudsen and Tveite (1999) reared stage IV postlarvae of *Homarus gammarus* in small, unattended, individual cages on the seabed in Norway at depths of 1.5–10 m for 3 mo. Survival was 66%, and growth was comparable to that in hatcheries. This approach has been modified in Ireland by Beal *et al.* (2002), who reared stage IV–V postlarvae individually in unattended cages of 360 cm^3 with 3.2-mm mesh, and in perforated 200-cm^3 plastic petri dishes placed on racks in the water column at a depth of 8–9 m. Survival rates in the petri dishes were 53–75% after 10 mo and seemed to be similar to those in the cages, which were unfortunately confounded by escapement. Growth was slower than for fed juveniles maintained in the laboratory. However, this system may well be a viable alternative to rearing lobsters for release. In particular, the limited space and mode of feeding in this rearing system conforms to the observations that early benthic-phase lobsters are confined to their burrows and subsist on infauna and plankton (see Section 3.7.1). Although 'offshore nursery' rearing systems can also be expected to have a limited capacity for production, they promise to reduce costs per juvenile further (Beal *et al.*, 2002).

3.7.3. Defects caused by rearing

If cultured stage VI lobsters are not offered suitable shell materials to manipulate, they fail to develop crusher claws and end up with two scissor claws instead (Govind and Kent, 1982). This places them at a disadvantage in agonistic interactions and could lead to injuries, because the scissor claws are broken easily (van der Meeren and Uksnoey, 2000). The exact implication of this conspicuous defect is not clear, however, as some *Homarus gammarus* with double scissor claws occur naturally in the United Kingdom (Addison and Bannister, 1994).

Comparisons of responses of naive and experienced lobsters to intra-specific contacts, shelter and presence of fish predators demonstrate that most behaviour is inherited. However, naïve lobsters are more prone to unrestrained combat that exposes them to predation (van der Meeren, 1993). Burrowing behaviour is instinctive, and hatchery-reared lobsters that have not previously been exposed to any suitable substratum construct burrows when released (Wickins and Barry, 1996). Overall, Svåsand *et al.* (1998) concluded that hatchery-reared lobsters had few problematical morphological or behavioural deficits and should be fit for survival in the wild.

3.7.4. Release strategies

The burrowing habits of juvenile lobsters, and the problems in relocating them, has made evaluation of optimal strategies for releasing cultured juveniles in the wild difficult. Losses caused by predation at time of release can be observed by divers, but subsequent survival can only be estimated when individuals marked with microwire tags, or with distinctive features resulting from hatchery rearing (see Section 3.7.3), eventually enter the fishery.

Predation of hatchery-reared juvenile *Homarus gammarus*, ranging in size from 12–15 mm CL in 1990–1994 to 22–32 mm CL in 1998, was investigated in Norway by van der Meeren (1991a,b). Most losses occurred in the first hour after release as the lobsters sought shelter. The lobsters were preyed upon principally by several species of wrasse in the summer, whereas cod, shorthorn sculpin *Myxocephalus scorpius* and the edible crab *Cancer pagurus* were the main predators in winter. It should be noted, however, that lobsters of these sizes are normally part-way through their 'in-burrow' phase and being exposed on the surface is not a normal situation. Similar observations were also made in Ireland by Ball *et al.* (2001), where sand gobies *Pomatoschistus minutus*, rocklings *Ciliata mustela* and green crabs *Carcinus maenas* attacked juveniles within minutes of release.

Predation seems to have been exacerbated during early attempts to release *Homarus gammarus* in Norway, which involved cooling 1-yr-old lobsters packed in wet wood shavings during transport to release sites. These lobsters were semi-moribund when liberated, or behaved erratically, and divers observed that >10% were eaten by wrasse within a few minutes. A suitable acclimation period in water at ambient temperature is necessary to ensure that released lobsters move rapidly into available cover. More recent release methods in Norway involve liberating juveniles at the surface, close to shore, in depths <10 m (Agnalt *et al.*, 1999). The preferred release season is spring when predatory fish are in low numbers (van der Meeren, 1991a,b). However, lobsters released in summer were also able to find shelter within a very short time (Tveite and Grimsen, 1995).

In the United Kingdom, *Homarus gammarus* of 11–15 mm CL have been released in spring and autumn using a variety of methods. Juveniles were either carried on trays to the seabed by divers and released onto suitable habitat, released down a pipe guided by divers or liberated at the surface. Recoveries of individuals marked with microwire tags and released using the latter two methods were lower than when juveniles were placed in trays. Release sites included cobblestones, bedrock, scree and boulders on sand (Bannister and Howard, 1991; Bannister *et al.*, 1994).

Overall, it seems that releases of *Homarus gammarus* at the surface are effective in areas with much suitable juvenile habitat and few predators, but releases by divers are more effective where the necessary habitats are patchy.

Little has been done to develop release strategies for *Homarus americanus*; however, fieldwork by Wahle and Steneck (1991, 1992) suggests that the greatest survival occurs in cobblestone or shell habitats with a high level of micro-scale heterogeneity. Predators of *H. americanus* are known to include the cunner *Tautogolabrus adspersus* and mud crab *Neopanope texana*. Shorthorn sculpin and white hake *Urophycis tenuis* have also been identified as predators, but their relative importance is unknown (Lavalli and Barshaw, 1986; Barshaw and Lavalli, 1988).

3.7.5. Use of artificial habitat

Interest in the use of artificial reefs to enhance productivity of lobster fisheries resulted from the perception that suitable habitat for adults and juveniles was in short supply. Wahle and Steneck (1991) considered availability of habitat to be a 'bottleneck' to recruitment of *Homarus americanus* and used artificial beds of cobblestones to create additional nursery habitat. These beds were colonized rapidly by juvenile lobsters (30–40 mm CL). However, there seems to be little evidence that habitat for newly settled lobsters is generally limiting.

The observation that lobsters require shelter from strong currents (Howard, 1980, 1988) suggests that there is scope for construction of artificial reefs to increase the carrying capacity for larger individuals in areas without many topographic irregularities. Such constructions would open unused feeding areas to larger lobsters derived from the wild or from release programmes. Accumulation of adults around artificial reefs in a variety of places supports this idea. For example, Scarratt (1968) recorded densities of 16 *Homarus americanus* 100 m^{-2} on an artificial reef in Canada. Another reef constructed on the south coast of the United Kingdom, more than 3 km from natural reefs, was colonized rapidly by *H. gammarus* at densities of 16–27 100 m^{-2} (Jensen *et al.*, 1994, 1999; Jensen, 1999). Many of the latter group showed strong site fidelity, but some tagged individuals were recaptured up to 16 km from the reef (Jensen *et al.*, 1994, 1999). Other evidence that shelter may be limiting for larger lobsters comes from Steneck (1991), who reported greater densities of *H. americanus* of 40–90 mm CL in response to increased availability of shelter. Paille and Gendron (2001) also concluded that fabricated reefs were effective habitat for adult lobsters if they met the biological and ecological requirements. The high level of injuries of older lobsters in some places, usually inflicted by crushing claws, also indicates that habitat availability might be limiting production (Sheehy *et al.*, 1999).

The indications are, therefore, that production of lobsters could be increased in places where the larger size classes are unable to find suitable shelter or where reefs are overcrowded. Under such conditions, provision of additional habitat should be the first step in a stock enhancement programme; release of juveniles should only be undertaken when it is clear that available habitat for adults is not fully used. Assessments of the need for adult habitat should also be tempered by the observation that other factors may also result in increased catches, as in the case of North America in recent years. Expensive interventions should only be made where thorough research has demonstrated that they will be cost-effective.

3.7.6. Minimising genetic impacts

Inbreeding has not been a problem in lobster hatcheries because larvae are normally obtained from wild females and each female is usually replaced after release of the young. However, the successful rearing of large numbers of siblings obtained from a single individual or a small number of females could lead to unnaturally large numbers of closely related lobsters being released and dominating the population at a site. The limited movement of animals after release makes this a possibility and could lead to inbreeding in the wild. For example, one study in the United Kingdom showed that most lobsters released as juveniles were recovered 4–8 yr later within a radius of 6 km (Bannister and Addison, 1998). No attempts seem to have been made to avoid this problem. The measures required are the use of larger numbers of broodstock, release of multiple cohorts derived from different females at each site and releases of the same cohort at multiple sites.

In larger-scale release programmes, the risk of producing relatively homogeneous cohorts is expected to be reduced by the low fecundity of lobsters; hatcheries would have to obtain and hold sufficient numbers of preovigerous females to produce the desired quantity of juveniles. However, the need for larger numbers of spawners can lead to multiple use of broodstock. Aiken and Waddy (1985) reported that a Canadian hatchery obtained eggs from 122 females over a 3-yr period, and there was retention of 403 females for a second spawning over the following 7 yr. In this case, the juveniles produced were destined for farming, but if such practices occurred for stock enhancement or restocking, the genetic diversity of the animals could be limited by repeated use of the same broodstock.

Regrettably, past stock enhancement programmes for lobsters have not always paid sufficient attention to the measures needed to maintain the genetic identity of stocks. The shortage of adult females during the early work on hatchery propagation in Norway resulted in use of berried females from Scotland, and approximately 80,000 released juveniles were of Scottish origin (Jørstad and Farestveit, 1999). Likewise, some planned releases in the Adriatic were to be based on berried females imported from the United Kingdom (Scovacricchi *et al.*, 1999).

Small numbers of *Homarus americanus*, including a berried female, have been captured on several parts of the Norwegian coast. These are thought to have been derived from illegal importations (Anon, 2000). Such introductions have potential consequences because *H. gammarus* and *H. americanus* hybridize (Hedgecock *et al.*, 1976, 1977; Anon, 2000). In North America, there have been repeated attempts to establish stocks of *H. americanus* on the west coast, but there are no reports of success so far. Both *H. americanus* and *H. gammarus* have been introduced to Japan, and small stocks of both species are reported to have become established (Nicosia and Lavalli, 1999).

The potential risks of transferring lobsters within their ranges are currently difficult to assess, because there have been very few population genetics studies (Tam and Kornfield, 1996, 1998; Harding *et al.*, 1997; Mykles *et al.*, 1998). Tracey *et al.* (1975) assessed genetic variability in *Homarus americanus* at several enzyme loci expressed in seven different tissues but did not observe large differences. Also, Jørstad and Farestveit (1999) analysed 2,580 individual *H. gammarus* from 22 locations along the Norwegian coast and found minor genetic variation in samples from the south of the country, although one isolated northern stock was significantly different. Consequently, they recommended that release programmes should only be implemented in areas with low levels of genetic differentiation. Ferguson *et al.* (2002), in a large-scale examination of multiple European lobster samples, found rather low levels of inter-population genetic differentiation. Mitochondrial DNA differentiation exceeded that for microsatellites or allozymes (respective F_{ST} values: 0.078, 0.018 and 0.016). Four distinct population groupings were evident: Mediterranean, northern Norway, The Netherlands, and the remaining Atlantic samples. Only the microsatellites revealed heterogeneity in the main Atlantic group. Until more definite studies are completed, the precautionary principle dictates that juveniles should be released in the area where broodstock were obtained to avoid changes to regional gene pools. Ferguson *et al.* (2002) concluded that transfers should be confined to local areas and that especially northern Norway and the Aegean should be protected from outside introductions.

3.7.7. Disease risks

In well-maintained hatcheries, there is a reduced risk of chronic disease problems arising among broodstock and then being transmitted to juveniles because pre-ovigerous or berried females are taken regularly from the wild. In fact, no significant diseases of lobsters have been reported in hatcheries, and no pathogens are known to have been spread through release of hatchery-reared animals. The problems that have occurred have been due to transfer of pathogens between species. In particular, there have been several reports of the occurrence of gaffkemia, a bacterial disease caused by *Aerococcus viridans* var. homari, in wild stocks of *Homarus gammarus* in Norway. These infestations are thought to have been derived from imports of *H. americanus* (Agnalt *et al.*, 2004).

This is not to say that problems cannot occur as the result of release programmes; bacterial infestation of egg masses in broodstock held in hatcheries has been reported, and is exacerbated by elevated temperatures (Fisher *et al.*, 1978). Also, a nemertean that infects egg masses of wild lobsters can spread rapidly if introduced to hatcheries (Aiken and Waddy, 1985). However, effects of both problems on the quality of larvae or juveniles reared in hatcheries can be avoided with diligent husbandry.

3.7.8. Systems for marking juveniles to assess effectiveness of releases

Various dart and T-bar tags, normally inserted dorsally between the carapace and the abdomen, have been used to mark relatively large wild lobsters. However, development of coded microwire tags (Jefferts *et al.*, 1963) permitted hatchery-reared 3- to 6-mo-old *Homarus gammarus* of 12 mm CL to be marked, released and recovered by divers after 1 yr (Walker, 1986). This technique has now been adopted by many research institutions working on lobsters. The microwire tags are inserted in the base of the fifth pereiopod. Initial retention rates were 85–100% for *H. gammarus* of 9–15 mm CL, and the tags were retained through 22–29 moults over a period of up to 47 wk (Wickins *et al.*, 1986). Minor improvements in tagging techniques were suggested by Uglem and Grimsen (1995).

Visible implant fluorescent elastomer tags are also an effective marking system for cultured lobsters. They are fully retained over at least 3 moults when injected below the epidermal layer in the abdomen of lobsters of 20–24 mm CL. An advantage of these tags is that they are clearly visible, even without use of ultraviolet light (Uglem *et al.*, 1996).

One 'advantage' of the morphological deficit of having two scissor claws (described in Section 3.7.3) is that such individuals have been used to track hatchery-reared lobsters in Norway (van der Meeren, 1993). Another morphological marker, blue juveniles, can be used to distinguish hatchery-reared *Homarus americanus* from wild animals (Figure 3.27). These colour morphs are the offspring of blue parental crosses that occur naturally at the rate of ~1 in every 6 million individuals (Wahle and Inze, 1997). Genetically blue, cultured juveniles do not behave differently from normal-coloured, hatchery-reared individuals and predators such as green crabs, winter flounder *Pleuronectes americanus* and longhorn sculpin *Myxocephalus octodecemspinosus* respond similarly to both normal and blue lobsters (Beal *et al.*, 1998).

3.7.9. Status of stock enhancement initiatives

The releases of lobsters fall into three categories: (1) early attempts to release larvae and postlarvae; (2) more recent releases of larger juveniles in Norway designed to overcome severe decreases in population density (ostensibly restocking) and (3) recent releases of larger juveniles in the United Kingdom aimed at determining whether existing catches can be increased.

3.7.9.1. Early attempts

Attempts to release larval and postlarval lobsters date back at least 100 years. More than 20 hatcheries were established in the United States, Canada, France, Norway and the United Kingdom between 1885 and 1954, including 14 in Canada (Nicosia and Lavalli, 1999). These hatcheries released very large numbers of stage I larvae and, as the technology improved, stage IV postlarvae. In some cases, eggs were stripped from berried females destined for canneries, which may also have helped avoid loss of larvae to some degree. However, interest declined because of lack of evidence of any beneficial effects from releases. Almost all of the hatcheries in North America and Europe were closed by 1955.

Failure of the early stock enhancement attempts is not surprising given what we now know about egg production and mortality rates. For example, Campbell and Pezzack (1986) calculated that lobsters from the Bay of Fundy and adjacent areas produced between 5.7 billion and 20.4 billion eggs in 1983. In contrast, a yearly average of 200 million stage I larvae were released in Canada over a 10-yr period, equal to no more than 1–3.5% of natural production.

Efforts have since been made to increase the likelihood of making a significant contribution to the population by rearing larvae through to stage IV (Aiken and Waddy, 1985). However, the numbers of stage IV larvae needed to have an impact in release programmes are still enormous, and there have been no perceptible results from these efforts. Nevertheless, one hatchery in Massachusetts has continued to produce and release postlarvae (Bannister *et al.*, 1994; Nicosia and Lavalli, 1999), and two hatcheries owned by fishermen's cooperatives in Ireland have been producing and releasing stage IV *Homarus gammarus* since 1994 (Browne and Mercer, 1999; Beal and Chapman, 2001).

3.7.9.2. Norway

The first releases of small numbers of hatchery-reared *Homarus gammarus* in Norway took place in 1979, followed by releases of >200,000 1-yr-old juveniles at four sites between 1983 and 1987 (Tveite and Grimsen, 1995). The animals were not tagged but were identifiable by their double scissor claws or by their lighter blueish colouration. Most lobsters did not move appreciable distances from release sites, and their contributions to catches were variable, ranging from zero at one site to 50% of the catch 4–6 yr after release. Overall, ∼5% of released lobsters survived to enter the catch, and there were almost as many cultured lobsters as wild ones in the vicinity of the release sites (Tveite and Grimsen, 1995).

Experimental releases were continued around the Kvitsøy Islands in southern Norway from 1990–1995. Each year, 17,000–30,000 lobsters marked with microwire tags were liberated. The initial batch of juveniles was 19.5 mo old (21.1 mm CL), but subsequent batches were considerably younger, 6.5–10.5 mo old (11.8–16.4 mm CL). The first recaptures were made after 2 yr. Numbers of cultured lobsters in the catch then increased steadily; by 1997, the tagged individuals accounted for 33% of the catch of ∼19,000 lobsters and 73% of a sample of sub-legal lobsters (Figure 3.28). Of the six year-classes released, 7.9% of the 1989 year-class had been recaptured at above the legal size by 1997. The islands are surrounded by water >80 m deep, and so outmigration was highly unlikely; landings in adjacent mainland areas contained no micro-tagged lobsters (A.-L. Agnalt, personal communication 2002). This programme demonstrated clearly that a depleted lobster population could be augmented substantially by the release of cultured juveniles (Agnalt *et al.*, 1999; 2004). Although the minimum legal size was changed several times, there were no other attempts to protect the released animals. The releases essentially amounted to a 'put-and-take' fishery, not a planned attempt to restore the natural spawning biomass. Considering the collapse of the lobster fishery in Norway (Moksness, 2004) (Figure 3.26), the releases could have delivered greater benefits if they had been used for restocking instead of stock enhancement. Borthen *et al.* (1999) developed a simulation model for the Norwegian lobster release programme that recognizes recruitment limitation as a factor and indicates that 23% of released lobsters need to be recaptured if the programme is to be profitable. However, they also argued that increasing the normal minimum size by 8% and allowing females to spawn twice would dramatically increase returns.

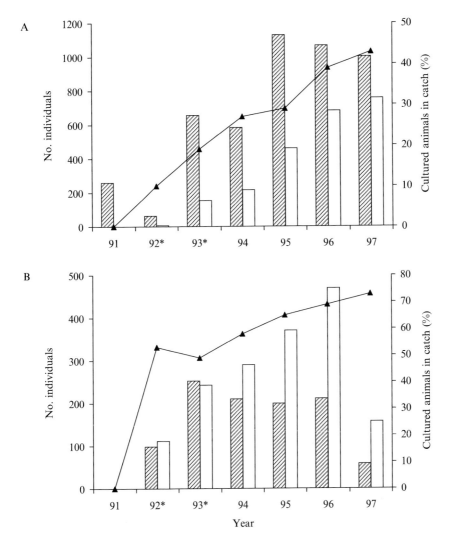

Figure 3.28 Autumn landings of wild (▨) and cultured (□) lobster *Homarus gammarus* of (A) legal size and (B) below legal size, at Kvitsøy Islands, Norway, from 1991–1997. Filled triangles show the percentage of cultured animals in the catch. Asterisks indicate years when the legal size was increased (redrawn from Agnalt *et al.*, 1999).

3.7.9.3. United Kingdom

The first attempt to release *Homarus gammarus* of larger sizes in the United Kingdom in stock enhancement programmes took place between 1983 and 1988, when ~50,000 microwire-tagged juveniles 3 mo old (stages X–XII with 11–15 mm CL) were released near Bridlington Bay, England (Bannister and Howard, 1991; Bannister *et al.*, 1994). Smaller numbers were also released in Wales (19,000) and Scotland (23,000). Tagged lobsters appeared in the commercial catches in significant numbers 4 yr after release and largely disappeared after 8 yr, although some returns were still being obtained when the 9-yr (1985–1993) monitoring programme terminated. Recapture rates varied with the cohort and location. However, returns from early cohorts fully recruited to the fishery varied from 1.6–5.5%. Most tagged lobsters were caught within 6 km of the release site. However, some individuals moved offshore for distances of up to 40 km (Addison and Bannister, 1994), lending support to the view of Sheehy *et al.* (1999) that older *H. gammarus* might move offshore to areas not usually targeted by fishermen and supply inshore areas with recruits.

In the Bridlington Bay area, a supplementary programme of tagging lobsters with T-bar tags was undertaken in 1990 and 1991 to estimate the exploitation rate in the fishery and the corresponding survival rates of microwire-tagged lobsters. From the time of release to attaining the minimum legal size of 85 mm CL, survival rate of microwire-tagged lobsters was 50–84%. Within release areas, from which most lobsters did not stray, cultured lobsters contributed 10–35% to the catch of the same size class (Bannister *et al.*, 1994). This is similar to results from a study site in Norway where ~26% of catches taken from 1992–1997 consisted of hatchery-reared lobsters.

In evaluating the extensive work in Bridlington Bay, Bannister *et al.* (1994) concluded that 'a worthwhile addition to the catch might be difficult to achieve' because of apparently low catchability. However, this is not fully borne out by the data on recaptures of tagged lobsters reported by Bannister and Addison (1998), which show that released lobsters were fully recruited to the inshore fishery at 5–6 yr, but lobsters older than 8 yr disappeared, as a result of exploitation, natural mortality or emigration. Based on the rate of attrition of microwire-tagged lobsters between ages 6 and 7 in the 1983–1986 cohorts, annual survival rate would be ~23%. This suggests that 5- to 8-yr-old lobsters were quite heavily exploited and that catchability could not have been low.

Bannister and Addison (1998) reported that the annual catch of *Homarus gammarus* in the Bridlington Bay fishery totaled 20,000–85,000 individuals between 1980 and 1990. If fishing effort was relatively stable, this large variation in catch must have been the result of variability in recruitment or catchability but, as explained previously, the latter seems improbable. It seems likely, therefore, that at most levels, populations of *H. gammarus* are recruitment limited. The final results of the tagging programme (Bannister and Addison, 1998) show that a large proportion of 4- to 8-yr-olds were removed by the fishery, and this suggests that annual releases of, for example, 10,000 stage X–XII juveniles, might increase the number of recruits of 85 mm CL by 6–42%, most of which would be caught in the following 4 years.

3.7.9.4. Overview

The current methods for producing juvenile lobsters, which are reliable but deliver relatively low numbers at moderately high cost, are better suited to restocking programmes designed to re-establish depleted populations than to stock enhancement. There is little evidence to suggest that stock enhancement of operational fisheries through release of cultured juveniles is economically feasible. The main problems are the current high costs of rearing juveniles to a size at which they will have good survival after release and the limits on producing the large numbers required to make the process worthwhile. Whether the increased efficiency promised by the new offshore nursery rearing methods developed by Beal *et al.* (2002) will make stock enhancement projects more feasible remains to be determined.

3.7.10. Management

The fisheries for both *Homarus americanus* and *H. gammarus* are well regulated. Prohibitions on the possession of 'berried' female *H. americanus* were introduced in Maine in 1872 and in Canada in 1874. These rules were soon repealed and replaced by closed seasons covering the summer months when eggs hatch. The possibility that stocks could be recruitment-limited was recognised in Maine in 1917, when legislation was introduced requiring that all ovigerous females be sold to the state. A hole was punched in the uropod, and the female was then released as state property. The hole was replaced by a V-notch in the telson in 1948. There is also a uniform minimum size of 81 mm CL for *H. americanus*.

In most parts of Europe, the minimum size for *Homarus gammarus* is 85 mm CL, except in Norway, where it was increased to 88 mm CL in 1993. However, as the size at maturity is 92–95 mm CL, the size limit does not directly protect immature individuals; it simply increases the probability that animals will reach sexual maturity before they are caught. The confirmed decline of Norwegian stocks indicates that greater restrictions on fishing effort for *H. gammarus* are required if stocks are to be restored.

Tighter restrictions would also help to increase spawning biomass of most other populations of *Homarus gammarus* in Europe, although other useful measures are in place in some countries. These include a mid-summer closed season, and V-notch telson marking in Ireland, where recruitment over-fishing has been recognized and females are protected for life (Browne *et al.*, 2001). For one fishery in Ireland managed this way, egg production increased by ~25%, and the number of undersized lobsters by 75%, within 4 yr of implementation (Tully, 1999).

Recent legislation in the United Kingdom permits groups such as co-operatives to apply for ownership and harvesting rights over certain species, including lobsters, in defined areas. Such legislation sets the stage to confer protection of hatchery-reared animals (Bannister and Addison, 1998; Burton, 1999) and provides incentive for fishermen to evaluate the potential bene-fits of restocking and bear the costs of such interventions where they are demonstrated to be profitable.

A new law has also been introduced in Norway that will encourage 'put-and-take' sea ranching of lobsters (Moksness, 2004). Under the new regula-tion, only license holders will be permitted to release cultured lobsters into certain areas, and only license holders will be able to harvest any lobster from the area. The government has also transferred management of lobsters in such areas to the license holders. Although this law provides a good incentive to bear the cost of producing and releasing lobsters, it remains to be seen whether it will also benefit recovery of wild populations. One concern is that unless the licensed areas are large, any larvae spawned there may be advected elsewhere. If so, there would be no incentive for license holders to allow a proportion of lobsters to reproduce. Application of a minimum legal size above the size of maturity in licensed areas would be in the national interest.

Overall, there is a reasonable framework for management in several countries, although much tighter controls on fishing will be needed to restore spawning biomass and to reap the maximum benefits from applying hatch-ery technology in restocking programmes where these are deemed necessary.

3.7.11. Problems to overcome

Despite recent progress, the high cost of hatchery and rearing operations is a major constraint to large-scale stock enhancement and restocking programmes. Consequently, production of juveniles is usually insufficient to make an impact at economically meaningful scales. In many countries, it is difficult to encourage solutions to these problems. There is a lack of legal protection of investments after the released stock enter the public domain, where they can be the targets of amateur fishermen, unlicensed divers and illegal fishermen (Scovacricchi *et al.*, 1999). Enforced ownership rights for fishermen will help pave the way for further innovations to increase production of juveniles at reduced cost (see Chapter 6, Section 6.5).

3.7.12. Future research

The key areas for future research centre around: (1) developing methods for greater production of juveniles at lower cost; (2) determining whether stocks are usually limited by the supply of juveniles; (3) assessing how and when artificial habitats can increase carrying capacity for adults; (4) developing appropriate protocols to maintain genetic diversity and (5) determining whether there are deepwater refugia for *Homarus gammarus*.

The marked difference in the size of the fisheries for *Homarus americanus* and *H. gammarus* and the ability to find juveniles of the two species indicate that recruitment limitation may be having a greater effect for *H. gammarus*. Further research is needed on the distribution and variation in abundance of juvenile *H. gammarus* before the full potential for restocking and stock enhancement of this species can be assessed (Linnane *et al.*, 2001).

Relatively rapid colonization of artificial reefs by larger lobsters indicates that stock enhancement programmes may be ineffective where adult habitat is limited unless additional refuge is provided, particularly at the local scales on which releases will need to operate. The recent increase in landings of *Homarus americanus* also suggests, however, that there is potential for greater harvests without such interventions.

It is important to identify the minimum number of female broodstock required to maintain gene frequencies of released cohorts representative of wild populations. Further large-scale surveys of gene frequencies of wild populations are needed to ensure that restocking and stock enhancement programmes maintain the natural genetic diversity.

The suggestions that *Homarus gammarus* in British waters occupy deepwater refugia as they get older and that large, fecund females are important in sustaining recruitment to local fisheries (Sheehy *et al.*, 1999) are intriguing. If these are true, then the behaviour of *H. gammarus* would be more similar to *H. americanus* than currently assumed. However, there are few records of large lobsters being taken as by-catch by trawlers, as might be expected if they move deeper in search of refugia. The degree to which stocks of *H. gammarus* might be replenished by old lobsters inhabiting deep water is an important question for managers. If verified, such refugia could be protected as marine reserves.

3.8. SEA URCHINS

3.8.1. Background and rationale for stock enhancement

Sea urchins are harvested commercially for their 'roe' (eggs) in most temperate regions of the world, although the major fisheries are focused in Japan, Chile and North America (see Andrew *et al.*, 2002 for review) (Figure 3.29). Japan dominates the market for sea urchin roe, currently consuming more than 80% of world production. The sea urchin fisheries in Chile and North America also exist largely to serve this enormous market. Over-fishing of several important fisheries, and an overall decline in world production (Andrew *et al.*, 2002), has prompted increasing interest in stock enhancement as a means of increasing production and, to a lesser extent, restocking to rebuild diminished resources. Stock enhancement of sea urchin fisheries is most developed in Japan (Agatsuma *et al.*, 2003), and the programmes there are large by any standard. Less developed efforts to restock sea urchins are also underway in South Korea and the Philippines.

The recorded history of Japan's sea urchin fishery dates back to 833 AD (Kawamura, 1969). Before 1985, the Japanese fishery was the largest in the world, but production from Chile and the United States has surpassed that from Japan at various times since then (Andrew *et al.*, 2002). The harvests of sea urchins from Japan were at their greatest in 1969, when a total of 27,528 t wet weight was landed. Production has declined since then, particularly in Hokkaido, and only 13,653 t was harvested in 1998 (Andrew *et al.*, 2002). Japan is increasingly reliant on imports to satisfy its market (Andrew *et al.*, 2002; Agatsuma *et al.*, 2003).

Of the six species harvested commercially, *Strongylocentrotus intermedius* and *S. nudus* are the most important (Agatsuma 2001a,b; Andrew *et al.*, 2002). These species are harvested mainly in Hokkaido, which produced 48% of total landings in 1997 (Andrew *et al.*, 2002), and in the northern prefectures of Honshu. *Hemicentrotus pulcherrimus, Pseudocentrotus depressus* and *Anthocidaris crassispina* contribute most of the remainder of the catch. These species are harvested from prefectures in southern Honshu, Shikoku and Kyushu. *Tripneustes gratilla* is caught in small quantities in Okinawa.

In Hokkaido, the sea urchin fishery began to decline in the mid 1980s, and landings fell sharply between 1988 and 1991. The reduced yield was attributed mainly to falling catches of *Strongylocentrotus intermedius,* but this was also true to some extent for *S. nudus.* For example, catches of *S. nudus* declined more gradually in Miyagi Prefecture but are now less than half of those in 1982 (Agatsuma *et al.*, 2003). Landings in the southern prefectures have also fallen approximately 50% since 1985.

Figure 3.29   (A) Sea urchins *Evechinus chlorotus* grazing on a reef in New Zealand (photo: M. Francis). (B) A sea urchin cut in half to show the roe (eggs) for which they are harvested (photo: C. Woods).

Stock enhancement plays a major part in the management of sea urchin fisheries in many Japanese prefectures and in Hokkaido in particular. Here, we summarise efforts to: (1) release cultured juveniles to augment the numbers of animals available for harvest; (2) provide additional habitat to increase the carrying capacity of inshore waters for sea urchins and (3) improve the size and quality of the roe. Our summary draws heavily on the information provided by Agatsuma *et al.* (2003).

3.8.2. Supply of juveniles

Hatchery techniques for propagating sea urchins are well developed in Japan, and competent juveniles can be produced relatively easily (Anonymous, 1992; Hagen, 1996; Kawahara, 1996, cited in Agatsuma *et al.*, 2003). For the major species, *Strongylocentrotus intermedius*, juveniles <5 mm are fed on diatoms or the green alga, *Ulvella lens*, after which they are reared in tanks on land or in sea cages until they are >10 mm dia. During this latter period, juveniles are fed on kelp or the terrestrial plant *Reynoutria sachalinensis* (Kawamura, 1993, cited in Agatsuma *et al.*, 2003).

At one stage, it seemed that juvenile *Strongylocentrotus intermedius* induced to settle in the wild on specially designed plates may also be available for use in stock enhancement programmes in Hokkaido (Anon., 1984; Tegner, 1989). However, although such techniques occasionally yielded catches of up to 1 million small sea urchins per year, natural variation in supply of juveniles was too great to make the methods commercially viable. All juvenile sea urchins used for stock enhancement programmes in Japan now come from hatcheries (Saito, 1992; Agatsuma *et al.*, 2003).

3.8.3. Release strategies

Two factors are critical to maximising survival and growth of juvenile sea urchins released in the wild in Japan: reducing predation and providing suitable substrata for grazing (Agatsuma *et al.*, 2003). Predation can be intense. For example, sea stars and crabs attack *Strongylocentrotus intermedius* in Hokkaido immediately after release at a size of 15–20 mm, and so it is common practice to reduce post-release mortality by removing crabs such as *Pugettia quadridens, Paralithodes brevipes* and *Telmessus cheiragonus* using baited traps (Agatsuma *et al.*, 1995; Kawai and Agatsuma, 1995). Some species of sea urchin (e.g., *S. nudus*) migrate between the subtidal and intertidal zones, once they reach an age of 1 yr and a size >20 mm, in search of preferred algae (Agatsuma and Kawai, 1997). Knowledge of such migrations and the habitat preferences of juveniles of different sizes are used to improve post-release survival and growth (Agatsuma, 2001b; Agatsuma *et al.*, 2003).

Other factors that affect survival of hatchery-reared sea urchins released in the wild are burial under sand moved inshore during heavy seas (Momma *et al.*, 1992) and 'stranding' on the shore with drift algae after storms (Agatsuma *et al.*, 1995). It is difficult to develop release strategies to avoid these chance events.

In areas of Japan where sea urchins remove kelp from reefs by grazing, the densities of the animals are manipulated to create kelp forests (Katada, 1963; Taniguchi, 1991; Taniguchi, 1996). In Aomori, Myagi and Nagasaki Prefectures, removal of urchins and transplantation of adult kelps are also reported to lead to rapid development of kelp forests (Kikuchi and Uki, 1981; Sawada *et al.*, 1981; Yotsui and Maesako, 1993). Management of the algal community to increase the food supply for sea urchins promotes their growth and development of the roe (see Section 3.8.5). Release of hatchery-reared juveniles into such areas can also result in localised increases in harvests (Agatsuma, 1999a,b).

3.8.4. Use of artificial habitat

Efforts to increase production of sea urchins by expanding areas of good habitat and promoting colonisation of algae on reefs to increase the supply of food are widespread in Japan (Agatsuma *et al.*, 2003). The use of reefs to provide shelter or create suitable habitat is concentrated in Hokkaido, Tohoku, Yamaguchi and Kyushu Prefectures and can be extensive. For example, reefs were constructed at 122 sites in Hokkaido from 1976–1992 (Kawamura, 1993) and, nationally, about 50 sites are enhanced this way each year. Construction of reefs varies considerably among prefectures and includes beds of stones surrounded by concrete blocks, small reefs with stones surrounded by iron rods and beds of stone protected from ice floes by concrete or iron breakwaters (Agatsuma *et al.*, 2003 and references therein).

Two examples of different approaches come from Fukui and Hokkaido Prefectures. In Fukui, where abundance of *Hemicentrotus pulcherrimus* is correlated with availability of shelter (Agatsuma, 2001c), stones have been introduced to enhance habitat since 1910 (Taki and Higashida, 1964, cited in Agatsuma *et al.*, 2003). Large catches have been recorded from areas 'enhanced' in this way, but it is unclear whether this was attributable to the introduction of stones (Taki and Higashida, 1964). In Hokkaido, additional structures are not always used to provide more shelter *per se*; in some cases they are used indirectly to provide suitable habitat. For example, large concrete blocks have been used as breakwaters to create areas protected from strong wave action, which are the preferred habitat of *Strongylocentrotus intermedius* (Agatsuma, 1991). Small stones are added behind these barriers to further augment the reefs to suit the habitat requirements of this species (Kawamura, 1973). However, annual levels of recruitment by sea urchins to these areas fluctuated in the same way as unenhanced reefs, and no significant differences have been found between natural reefs and those protected from wave action (Agatsuma, 1991).

3.8.5. Enhancement of roe

Because sea urchins are harvested for their roe (Figure 3.29), a major feature of the fishery in Japan is the movement of animals to locations where a better supply of food improves the size and quality of the eggs (Agatsuma *et al.*, 2003). This practice is successful because nutrients are stored in somatic cells within the gonad before being used for growth or reproduction (Giese, 1966; Holland *et al.*, 1967). In Japan, sea urchin roe is most valuable before, or in the early stages of, gametogenesis when the gonads are large but still firm. The value decreases when there are enough gametes in the lumen of the gonad to alter the texture of the roe (Agatsuma *et al.*, 2003).

Roe size and quality have been enhanced for wild populations of most commercial species of sea urchin by transplanting adults with poorly developed gonads from deep water and/or barrens to kelp forests (Agatsuma *et al.*, 2003 and references therein). Transplantations can, however, have an impact on the 'host' population, and so the collection of animals to be transplanted needs to be managed (see Section 3.8.10).

3.8.6. Minimising genetic impacts

There is relatively limited published information on the genetic implications of releasing hatchery-reared sea urchins in Japan or protocols for minimising loss of genetic diversity. Agatsuma *et al.* (2003) report responsible hatchery practices for *Strongylocentrotus intermedius* released in Hokkaido in 2000; the juveniles were derived from seven spawnings, involving almost 300 females and 300 males, and released on the same reefs where broodstock were collected. On the other hand, Natsukari *et al.* (1995) (cited in Agatsuma *et al.*, 2003) used allozyme techniques to show that released and wild *Pseudocentrotus depressus* were genetically heterogeneous and that release of cultured juveniles reduced the proportion of polymorphic loci and heterozygosity in the wild population.

Comprehensive research on population genetics of sea urchins on the west coast of the United States has revealed that, even for species with considerable potential for larval dispersal, genetic exchange may be limited over ecological and intergenerational time scales (Edmands *et al.*, 1996; Debenham *et al.*, 2000). Thus, local populations of sea urchins may be more isolated than larval dynamics suggest. The implications are that, until the genetic population structures of sea urchins have been studied thoroughly, cultured juveniles should only be released at locations where the broodstock were collected. There is an additional reason to have a sound understanding of the genetic structure of sea urchin populations in Japan; the incremental movements stemming from widespread transplantation of animals from barrens to more productive areas could result in loss of genetic diversity and localised adaptations.

3.8.7. Disease risks and other environmental impacts

Information on the possible risk of diseases and other ecological impacts caused by stock enhancement of sea urchin fisheries in Japan is scarce. Nevertheless, removal of predators to increase the survival of released sea urchins and large-scale introduction of concrete blocks and stones can be expected to cause changes to the ecology of reefs. It is also clear that release of large numbers of sea urchins will decrease algal biomass (Sakai et al., 2004) and will change the nature of interactions with other herbivores.

Sakai et al. (2004) report benefits to the commercial harvest of kelp Laminaria spp. as a result of stock enhancement of Strongylocentrotus intermedius in Hokkaido. Urchins as small as 6 mm diameter remove algae that compete with Laminaria. Once all sea urchins are harvested, the valuable kelp is able to recolonize the area.

3.8.8. Systems for marking juveniles to assess effectiveness of releases

Unlike many other species of marine invertebrates, there are no discernible differences between hatchery-reared and wild sea urchins, and no satisfactory external tag has been developed (Ebert, 2001). However, differences in size at age of released and wild animals, measured by the width of the first rings formed in the fifth genital plates (Kawamura, 1973; Agatsuma, 1987), have been used to assess percentages of released animals in populations at several localities in Hokkaido (Agatsuma et al., 1995, 2003; Sakai et al., 2004).

3.8.9. Status of stock enhancement initiatives

Release of sea urchins in Japan commenced in the late 1970s, increased sharply in the late 1980s, but has leveled out since 1994 at ~70 million per year (Figure 3.30). Stock enhancement is currently being attempted

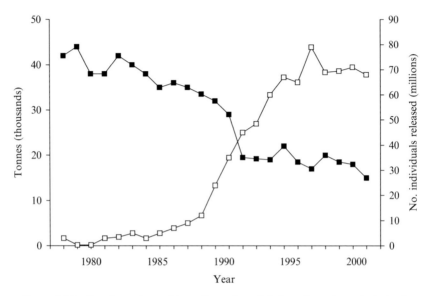

Figure 3.30   Total number of juveniles released (□) and total catch (■) of sea urchins (all species combined) in Japan from 1978–2000 (redrawn from Agatsuma *et al.*, 2003).

for all six species in the fishery, although most releases have been of *Strongylocentrotus intermedius* in Hokkaido (Table 3.10). These began in 1985, when >1 million sea urchins were released. Production of juveniles then increased steadily over the next decade until, in 1996, >60 million juveniles were released (Figure 3.31). In 2000, *S. intermedius* accounted for 84% of releases in Japan (Table 3.10).

The effectiveness of stock enhancement programmes for sea urchins in Japan has not yet been evaluated comprehensively at the national or prefectural scale (Saito, 1992; Kitada, 1999; Agastsuma *et al.*, 2003). At the national level, Kitada (1999) describes an index of 'stocking impact', which provides a measure of the relative numbers of juveniles released and the number in the catch. Contributions from cultured sea urchins compare poorly with other enhanced species, although the index is difficult to interpret as a measure of the effectiveness of stock enhancement in improving production.

Assessments of success at the prefectural level are restricted to *Strongylocentrotus intermedius* in Hokkaido. For the first 8 yr of this programme, there was no indication that the ever-increasing numbers of juveniles released had an impact on catches, which continued to decline (Figure 3.31).

Table 3.10 Numbers of sea urchins released (as thousands) in Japan in 2000 by prefecture and species.

Prefecture	Species				
	S. intermedius	*S. nudus*	*P. depressus*	*T. gratilla*	*H. pulcherrimus*
Hokkaido	55,202	1,750			
Aomori	250	655			
Iwate	701	3,930			
Miyagi		360			
Fukushima		280			
Mie			36		
Wakayama			55		
Ehime			151		
Oita			113		
Kagoshima			435	19	
Kumamoto			359		
Nagasaki			1,307		
Saga			242		321
Fukuoka			346		
Yamaguchi			98		
Shimane			45		100
Okinawa				5	
Total	56,153	6,975	3,187	24	421

Note that stock enhancement has been applied to *Anthocidaris crassispina* but none were released in 2000 (after Agatsuma *et al.*, 2003).

Since 1992, however, catches of *S. intermedius* in Hokkaido, in terms of roe weight, have stabilised and even risen slightly (Figure 3.31). Using the method described in Section 3.8.8 to differentiate hatchery-reared urchins from wild ones, Agatsuma (1998) estimated that released animals accounted for 62% of the catch in Hokkaido in 1994, 66% in 1995 and 80% in 1996. Of the animals harvested in 1996, 45% were released in 1994. These findings support the conclusion that the stabilization and subsequent modest increase in the landings of *S. intermedius* in Hokkaido after 1992 (Figure 3.31) was in large part an effect of the increase in number of juveniles released.

Similar results have been reported at smaller scales. Sakai *et al.* (2004) estimated that released sea urchins made up 62.3–80.2% of the catch of *Strongylocentrotus intermedius* at Tamori, southwestern Hokkaido, in the mid 1990s. They also showed that net profits were 100–200% greater than the costs of releasing juveniles in some years.

In contrast, there is no evidence that release of cultured juveniles has had any effect on the other major species, *Strongylocentrotus nudus* (Figure 3.31). Annual harvests of *S. nudus* have fluctuated between 350 and 550 t roe

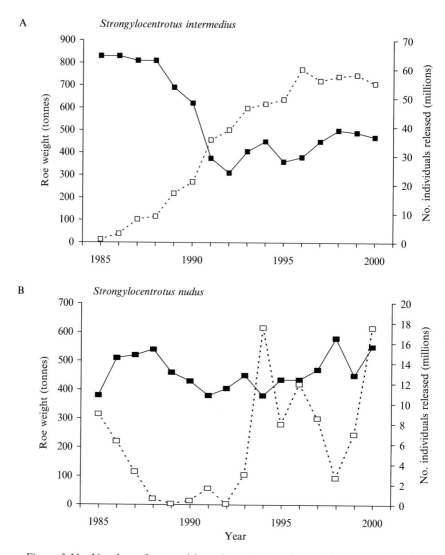

Figure 3.31 Number of sea urchins released (□) and caught as roe weight (■), from 1985–2000 for **(A)** *Strongylocentrotus intermedius* and **(B)** *S. nudus* from Hokkaido Prefecture, Japan (redrawn from Agatsuma *et al.*, 2003).

weight, even for those years immediately after the largest releases. The results from other prefectures where stock enhancement programmes for *S. nudus* are underway, particularly Aomori and Iwate Prefectures, are similar, i.e., there is no correlation between rate of catch and number of juveniles released (Agatsuma *et al.*, 2003). The same type of picture also emerges for the other four species released in the southern prefectures. In fact, annual catches of sea urchins in southern Japan have nearly halved since 1985, despite a steady increase in the numbers of juveniles released (Agatsuma *et al.*, 2003).

The only other countries currently releasing cultured sea urchins are the Philippines and South Korea. In the Philippines, releases of *Tripneustes gratilla* have been made on coral reefs to restock populations of this species where they have been depleted severely by over-fishing (Juinio-Menez *et al.*, 1998). This work has been done in conjunction with other management measures, such as fishery closures and development of cage-culture techniques. In South Korea, up to 700,000 cultured juvenile *Strongylocentrotus intermedius* and *Anthocidaris crassispina* have been released per year; however, the effects of this programme on augmentation of wild stocks is unknown (National Fisheries Research and Development Institute, 2000).

3.8.10. Management

Sea urchin fisheries in Japan are managed by Fisheries Co-operative Associations. The management of individual species is integrated with that of other resources, such as seaweeds and abalone (Andrew *et al.*, 2002). Such fisheries are managed with the aim of maximising production, and this is achieved by seeking to maintain the species mix that will achieve that goal. Indeed, Fisheries Co-operative Associations often intervene to 'manage' ecological relationships. For example, predatory crabs and starfish are removed from reefs (see Section 3.8.3) in 40% of sea urchin fisheries. Although the release of cultured juveniles, and manipulations of stocking density, habitat and predators, are central to the management of sea urchin fisheries in Japan, such programmes have met with mixed success in western countries (Andrew *et al.*, 2002). Also, the nature of property rights and a different philosophy of managing nearshore resources in western countries means that the strongly interventionist approaches used in Japan and South Korea are unlikely to be implemented there.

3.8.11. Future research

Much of the basic technical and ecological knowledge required to mass-produce juvenile sea urchins and to release them with a relatively high probability of survival is now in place (Andrew *et al.*, 2002; Agatsuma *et al.*,

2003). Future research on stock enhancement of sea urchins should therefore be targeted at: (1) the economics of large-scale releases; (2) additional strategies to improve survival of released juveniles; (3) the genetic population structure of stocks to determine whether parts of the resource should be managed separately; (4) the long-term effects of large-scale releases, manipulations of other species (e.g., predators), and introduction of additional habitat on the ecosystem and (5) management regimes that will allow maximum benefits to be obtained from investments in hatchery-production of juveniles.

Chapter 4

Overview and Progress Towards a Responsible Approach

4.1. RESTOCKING INITIATIVES

4.1.1. General trends

A key feature of the restocking initiatives for marine invertebrates described in this review is that the cultured juveniles have always been produced in hatcheries. This is not just a result of the parlous state of the stocks, which greatly reduces the scope for collecting juveniles from the wild. It is because methods for attracting and collecting large numbers of postlarvae of giant clams, sea cucumbers and topshell have not been developed and are unlikely to be cost-effective.

The only exception to the uniform use of hatchery-reared juveniles for restocking is the translocation of adults to form a critical mass of spawners. This is proposed for topshell because it seems to be more cost-effective than depending on survival of enough very small juveniles that are highly susceptible to predation, or rearing sufficient cultured animals to spawning size in cages, to form effective aggregations of adults. This begs the question 'should translocation of adults be examined more closely for restocking other sedentary species with limited larval dispersal?' The sea cucumber *Holothuria scabra* is a possible candidate for this approach in locations where its longer-lived larvae are retained by local currents (Chapter 2, Section 2.3.9). It has also been proposed for abalone in programmes where the primary need for release of juveniles is to rebuild severely depleted stocks (Chapter 3, Section 3.3.4).

ADVANCES IN MARINE BIOLOGY VOL 49
0065-2881/05 $35.00
DOI: 10.1016/S0065-2881(05)49004-5

Hatchery-reared juveniles may show morphological changes and some defects in behaviour, but this does not seem to be a major impediment for the emerging restocking initiatives described here. It is not a problem for giant clams because the 3- to 4-yr ocean nursery phase 'weeds out' unfit individuals before they are placed on coral reefs. There are some concerns with juvenile topshell, particularly with shell morphology and its effect on predation rates, but this problem may disappear if the conclusion holds that translocation of wild adults is the most effective way to establish critical spawning biomass. It is still too early to determine whether juvenile sea cucumbers produced in hatcheries are more vulnerable to predation than wild ones. The relevant release experiments have yet to be done, but the evidence to date suggests that any effects are likely to be small.

Two striking features of the restocking initiatives, compared with stock enhancement programmes (see Section 4.2), are that they are all at an early stage and that they are underway mainly in developing countries and Small Island Developing States (SIDS) with the assistance of international partners. Indeed, the numbers of juveniles released to date in restocking programmes are typically 3–4 orders of magnitude lower than for the major invertebrate stock enhancement programmes based on hatchery-reared juveniles (Table 4.1). Although this reflects the small-scale nature of many fisheries for invertebrates in developing countries and SIDS, it is curious to note that none of the developed countries has embarked seriously on restocking research. Part of the answer may lie in the need to first evaluate whether alternative management measures can be used to restore stocks, and the realisation that restocking programmes usually need to be accompanied by a complete moratorium on fishing until replenishment occurs (see Chapter 6). Such bans are often difficult to negotiate, legislate and implement.

A point of concern is that some of the developing countries and SIDS will not have the financial and human resources to implement restocking programmes over the long time frames necessary for recovery. This has already become evident in the case of most of the attempts to restock giant clams (Chapter 2, Section 2.1.7). Sustained assistance from development agencies will be needed to support these programmes until replenishment occurs (Bell, 1999c; Bartley *et al.*, 2004).

Another general feature of restocking initiatives is that, because of their early stage of development, they have yet to demonstrate that they will succeed in fast-tracking recovery compared with other forms of management. Much will depend on the use of genetic tags to determine the proportion of individuals derived from released animals and their progeny, perhaps three to four generations after release (Chapter 7, Section 7.1). A shortcoming of programmes based on hatchery-reared juveniles is that none of them has developed an inheritable tag. Without good evidence to demonstrate that restocking programmes succeed in meeting their goals, it will be difficult to encourage managers to consider and apply this intervention.

Table 4.1 Maximum number of hatchery-reared juveniles released per year for restocking and stock enhancement initiatives for marine invertebrates. Note that species released mainly as wild spat are not included.

Species/group	No. released	Location	Reference
Restocking			
Giant clams			
Tridacna derasa	17,000	Samoa	Bell (1999b)
Tridacna gigas	4,500	Australia	Bell (1999b)
Hippopus hippopus	4,800	Tonga	Bell (1999b)
Tridacna squamosa	>100,000	Japan	Bell (1999b)
Tridacna crocea	>100,000	Japan	Shokita *et al.* (1991)
Topshell			
Trochus niloticus	132,000	Japan	Isa *et al.* (1997)
Sea cucumbers			
Holothuria scabra	Small-scale experimental releases only	Pacific	Dance *et al.* (2003)
Stock enhancement			
Other bivalves			
Mercenaria mercenaria	>6 million	Long Island, New York	Malouf (1989)
Abalone			
Haliotis discus hannai	150 million	Japan	Seki and Taniguchi (2000)
Queen conch			
Strombus gigas	750,000	Bonaire	Iversen and Jory (1997)
Shrimp			
Penaeus japonicus	500 million	Japan	Kitada (1999)
Penaeus chinensis	>2 billion	China	Wang (personal communication 2002)
Lobsters			
Homarus gammarus	30,000	Norway	Agnalt *et al.* (2004)
Sea urchins			
Strongylocentrotus intermedius	60 million	Japan (Hokkaido)	Agatsuma *et al.* (2003)
Sea cucumbers			
Apostichopus japonicus	2 million	Japan[a]	Yanagisawa (1998)
	>700 million	China[b]	Chen (2004)

[a]See Chapter 2, Section 2.3.11 for details.
[b]Releases were for 'put-and-take' sea ranching (see Chapter 2, Section 2.3.11 for details).

4.1.2. Progress towards a responsible approach for restocking

Increased awareness of the value of marine resources, and the importance of maintaining biodiversity at all levels, has led to a code of conduct for responsible fishing (FAO, 1995a) and an ecosystem approach to managing fisheries (FAO, 2003b). Blankenship and Leber (1995) have set out clear principles for restocking and stock enhancement programmes to help ensure that releases of cultured juveniles conform with the codes and guidelines developed by FAO. In this section, we compare progress against the 10 components elucidated by Blankenship and Leber (1995) that are now widely accepted to represent a responsible approach for releasing cultured juveniles to replenish or enhance marine fisheries (Table 4.2). Because most of the restocking initiatives are at the early stage of determining whether aquaculture can be applied to produce large numbers of animals for release in the wild, the emphasis here is on examining whether a responsible approach is being applied to development of the technology rather than to restocking programmes *per se*. Progress for each of the three restocking initiatives (giant clams, topshell and sea cucumbers) against the 10 components of the responsible approach is summarised in Table 4.2, and across all three species/groups in the following.

4.1.2.1. Component 1: Select priorities and target species

A process for rational selection of species was included by Blankenship and Leber (1995) in the responsible approach to ensure that funds available for research were directed toward the most appropriate candidate species. However, this particular criterion is of limited relevance for restocking programmes, because they tend to suggest themselves; they involve fisheries in a dire condition where release of cultured juveniles promises to aid recovery more quickly than would have been possible otherwise. The question therefore becomes 'can the species be restored more effectively through release of cultured juveniles?' rather than 'is this the best species to consider for restocking?' All species reviewed in Chapter 2 have been over-fished severely in several (but not all) locations. The state of the resource alone was the main driver to consider restocking, although lack of impacts on other parts of the ecosystem was also a factor in favour of restocking as a potential management tool in the case of these tropical invertebrates (Munro, 1989; Nash, 1993; Battaglene and Bell, 1999).

4.1.2.2. Component 2: Develop a management plan

Various measures are used to manage the species involved in restocking initiatives (Chapter 2). However, in no case has a plan been developed that seeks to identify whether restocking is needed in addition to other forms of management.

An aspect in which most initiatives also need to improve is determining the genetic structure of the stocks and identifying whether all, or only some, of the populations that make up the stock require restoration (this is linked to Component 4 in the following, see also Chapter 6, Section 6.1 and Chapter 7, Section 7.2). With the exception of placement of giant clams in marine protected areas in the Philippines, no action has been taken to maximise the benefits of releasing cultured animals in restocking programmes (see Chapter 6, Section 6.2 for details of the approaches recommended). This partly reflects the fact that in many cases the potential for restocking invertebrates has not yet been assessed fully, and that mature technologies have not yet been delivered to managers.

4.1.2.3. Component 3: Define measures of success

Because none of the restocking initiatives has proceeded to the stage where the methods have been adopted on large scales, no quantitative measures of success have been established. The closest thing to setting a measure of success is the study in Vietnam by Strehlow (2004) that estimated the anticipated rate of recovery in a limited fishery for the sea cucumber *Holothuria scabra* over a 10-year period after release of cultured juveniles and appropriate companion management (Chapter 2, Section 2.3.8).

Defining measures of success for restocking programmes will require careful analysis. It will depend on: (1) the size, distribution and reproductive mode of the remnant population(s); (2) the number and size of cohorts that can be released; (3) the estimated rate of population increase each generation, expected to be a logistic (S-shaped) function given that habitat and food will not be limiting at low abundance unless a competitor species is present in large numbers and (4) the number of generations needed to reach the desired target spawning biomass before re-opening the fishery (say 40% of original spawning biomass). For example, setting the defined measure of success at a 50% contribution of animals derived from hatchery releases to the restored population, an achievement that would meet the goal of 'fast-tracking' recovery, would require the total number of animals released in multiple cohorts to exceed the initial remnant stock, especially if the time to sexual maturity is prolonged. However, the numbers released could be reduced if the cultured animals had greater spawning success than wild individuals, which is likely for sessile species if the remnant population is dispersed. Clearly, the potential contribution of restocking will need to be modelled carefully in each case to set a realistic measure of success. As indicated in Section 4.1.1, evaluation of success will depend on development of a genetic tag.

4.1.2.4. Component 4: Use of genetic resource management

For giant clams and sea cucumbers, it is now apparent that gene frequencies differ markedly across the distributions of species (at relatively small scales

Table 4.2 Summary of how the various restocking initiatives have fulfilled the 10 components of a responsible approach advocated by Blankenship and Leber (1995).

Recommendation	Giant clams	Topshell	Sea cucumbers
1. Prioritize and select target species	Primary driver was severe depletion, sometimes local extinction, of some species.	Restocking investigated because several stocks are over-fished and hatchery production of juveniles is relatively easy.	Sandfish selected because of high value, wide distribution and relative ease of larval rearing.
2. Develop a management plan	Incomplete for most countries. Some countries ban export for CITES purposes.	Local management plans focusing on sustainable harvests developed for several countries. Most plans do not yet involve restocking.	Management measures identified in many countries, but management plans involving restocking have not been included, because the technology is not yet mature.
3. Define measures of success	Not specified for any country.	Restocking targets not yet specified for any country.	Restocking targets not specified for any country.
4. Use genetic resources management	Hatchery protocols to maintain genetic diversity developed. Gene frequencies of some natural populations identified. Releases in the Philippines have maintained regional genetic structure.	Genetic structure of topshell populations not studied. Natural stock structure probably altered by numerous cross-border translocations.	Natural variation in gene frequencies identified for sandfish in several countries and under investigation in others. Protocols to produce genetically diverse juveniles in place.
5. Use disease and health management	Protocol to transfer larvae among countries to reduce risk of introducing pathogens and parasites applied by some countries.	Risk of transferring diseases deemed to be low. Fouling on shells of adults poses biosecurity risk unless epibiota are destroyed before transfer	Diseases are rare among hatchery-reared juveniles, but quarantine protocols have been identified.

6. Develop restocking objectives and tactics	Importance of predation quantified. Grow-out methods developed to reduce mortality. Effects of restocking on other species considered minimal because of sessile, autotrophic habits of giant clams.	Intensity of predation on juveniles recognised and is rationale for transferring adults instead of releasing juveniles. No adverse impact of releases on other species.	Measures to overcome severe predation of juveniles under investigation. Releases have minimum impact on ecosystem, because sea cucumbers are low in the food chain and benefit sediments through bioturbation.
7. Identify released individuals and assess stocking effects	Contribution of progeny from cultured clams to wild populations unknown—requires a genetic tag.	Methods for mark and recapture developed for use in experiments. Genetic tag needed to determine contribution of hatchery-reared juveniles to replenishment.	Attempts to develop a long-term physical tag are still in progress. A genetic tag is needed; colour morphs are a possibility but patterns of inheritance are unknown.
8. Identify optimum release strategies	Effective methods of husbandry developed to rear clams in ocean nurseries until they are large enough to escape predation.	Cost-benefit of releasing juveniles of different sizes still unclear despite considerable experimentation. Translocation of adults seems to be most effective.	Some information available on suitable release habitats, but release strategies are still under development.
9. Identify economic and policy objectives	Not developed, except in Philippines where many clams have been placed in protected areas.	Importance of topshell widely appreciated in island communities. Policies to support cost-effective ways of increasing harvests should be accepted readily.	Feasibility of hatchery dedicated to restocking assessed for Vietnam. Institutional arrangements to protect released animals until replenishment occurs need to be developed.
10. Use adaptive management	Model developed to reduce costs by linking grow-out to 'escape' size with small-scale farming.	Poor survival of released juveniles has led to use of adults instead of juveniles for restocking. Steps to increase success of restocking identified.	Experiments underway to determine whether the cost of mass-producing juveniles can be defrayed by co-culture with shrimp.

in the case of *Holothuria scabra*). Information is readily available to guide managers about how to release animals in a way that maintains the integrity of the genetic population structure of the species. This involves application of protocols for maintenance of genetic diversity in hatchery-reared juveniles and releasing cultured animals at sites where broodstock were collected (Chapter 2, Sections 2.1.5 and 2.3.5, see also Chapter 7, Section 7.2).

4.1.2.5. Component 5: Use disease and health management

Reasonable attention has been paid to the need to quarantine hatchery-reared juveniles before release during development of restocking technology for giant clams, topshell and sea cucumbers. Fortuitously, none of the species currently under investigation for restocking is particularly suscepti-ble to diseases, and there have been no serious problems stemming from release of animals in the wild at the small scales of operations undertaken so far. Nevertheless, because most restocking programmes rely on use of hatch-ery-reared juveniles, and poorly managed hatcheries can become 'hotbeds' for development of pathogens and parasites, restocking has potential to cause harm to the remnant wild stock (and other species) unless quarantine protocols are implemented correctly (see Chapter 7, Section 7.3).

4.1.2.6. Component 6: Develop restocking tactics

This component of the responsible approach requires a sound understanding of the broader ecological setting into which juveniles will be released and in which the fishery operates. For example, managers need to know the abun-dance and type of predators, food availability, accessibility of critical habi-tat, competition for food and space, and environmental carrying capacity for the target species. Armed with this information, managers can then commis-sion experiments to determine how to release juveniles in ways that result in high rates of survival (see Section 4.1.2.8).

All restocking initiatives reviewed here have devoted considerable time and resources to an understanding of the broader ecosystem into which juveniles are placed. The released animals have encountered severe predation, which is the single most important factor affecting survival of hatchery-reared juve-niles in the wild. For giant clams, and probably topshell, acceptable levels of survival can only be achieved with relatively long-term use of small sea-cages to rear juveniles to 'escape' size. This is expensive. It remains to be seen whether high rates of survival of cultured sea cucumbers can be achieved by releasing them into habitats and at times, sizes and densities that minimize predation. In general, restocking programmes have not attempted to use artificial substrata to increase the carrying capacity of the habitat for the target species. Rather, the emphasis has been on restoring abundances to higher levels within natural habitats.

4.1.2.7. Component 7: Identify released individuals and assess stocking effects

Considerable effort has gone into developing marks for sea cucumbers and topshell to track individuals during experiments. Sea cucumbers have proved difficult to tag, and current methods are essentially restricted to physical tags suitable for estimating survival of small numbers of individuals for limited periods (months) after release (Chapter 2, Section 2.3.7). Topshell are easier to mark, because physical tags can be glued to the shell. However, even at the experimental scale, detection and tag loss have posed problems. No attempt has been made to tag large numbers of hatchery-reared giant clams, because the large size and growth form of the shell prevents use of an easily detectable chemical or physical mark.

Overall, the preoccupation with developing a physical tag for experiments has not equipped restocking programmes to assess the long-term contribution of releases to rebuilding spawning biomass. At best, physical tags can be used to estimate survival of released individuals during their lifetime but cannot tell us anything about the contribution of the progeny of released animals to the target population size. As mentioned earlier, this can only be done with the use of a genetic tag that will enable the proportion of the restored stock derived from released animals and their progeny to be determined (see Chapter 7, Section 7.1 for details).

4.1.2.8. Component 8: Use an empirical process to define optimum release strategies

In general, this has been part of most restocking programmes reviewed here. Large-scale replicated experiments have been used to test different methods of rearing giant clams in ocean nurseries to sizes where they escape predation. These trials have provided reliable estimates of expected rates of growth and survival (see Chapter 2, Section 2.1.4). A rigorous empirical approach has also been used for evaluating the consequences of releasing cultured topshell at a range of sizes and in different ways (see Chapter 2, Section 2.2.4). Such research underway for sea cucumbers is not yet complete.

4.1.2.9. Component 9: Identify economic and policy objectives

Economic objectives and analyses are lacking for most restocking programmes, with the exception of recent work on sea cucumbers in Vietnam. Perhaps this is because the potential value of restocking is largely self-evident; a valuable resource has been lost and management wishes to see it restored. Nevertheless, such analyses should be conducted so that governments, and supporting development agencies, have an accurate picture of the beneficiaries, time-scales and costs involved in restoration. These can be expected to vary from species to species and from place to place, and specific evaluations should be done for each programme. In some cases, the costs will be prohibitive. However, it will be important to avoid rejecting a restocking proposal just because the cost of producing an individual is greater than its market value. On the contrary, relatively high costs for hatchery-reared juveniles could be considered, because a well-managed and carefully planned restocking programme should deliver long-term financial benefits from a restored and sustainably managed fishery far in excess of the initial cost of producing the juveniles (see Chapter 6, Section 6.1).

Few policies are currently in place to protect animals released in restocking programmes to ensure that maximum benefit is gained from the investment (see Section 4.1.2.2). Such policies should be implemented before restocking programmes begin in earnest. This will require changes to institutional arrangements (see Chapter 6). Policies should also be developed to ensure that a responsible approach is always taken to the production and release of animals and that production costs are not reduced at the expense of changes to the genetic structure of the stock or undue risk of introducing diseases (see Chapter 7).

4.1.2.10. Component 10: Use adaptive management

Although restocking programmes are still at an early stage, three examples of adaptive management have occurred, albeit during the development and refining of methods for release rather than by managers of the fisheries. The first is the realization that developing countries and SIDS will have difficulty meeting the costs of rearing giant clams to the size at which they escape predation. In this case, there has been a proposal to link restocking programmes to small-scale farming activities to defray the costs (see Chapter 2, Section 2.1.8). The second is the proposal to move from the use of hatchery-reared juveniles to the translocation of adults to restock topshell (see Chapter 2, Section 2.2.9). This occurred in response to intense predation of hatchery-reared juveniles and limits the number of animals that can be reared in cages to a size where they escape predation. The third is the experimentation underway to determine whether the costs of mass-producing sandfish for restocking programmes can be reduced by rearing the juveniles in shrimp ponds (Chapter 2, Section 2.3.2).

4.2. STOCK ENHANCEMENT INITIATIVES

4.2.1. General trends

In contrast to restocking initiatives, several different methods have been used to supply juveniles for marine invertebrate stock enhancement programmes. These include: (1) collection and nursing of wild 'spat'; (2) translocation of excess juveniles that have settled naturally; (3) hatchery-reared juveniles and (4) provision of suitable habitat to enable spat to settle in greater numbers or in areas where settlement would not otherwise be possible.

The most successful method, in terms of the number of juveniles available for release and the magnitude of increased production, has been the collection, nursing and release of wild spat. This has been the basis for sustainable landings in the scallop fishery in northern Japan well in excess of previous maximum catches and dramatic improvements in the productivity of the scallop fishery in the South Island of New Zealand (see Chapter 3, Section 3.1.8). Collection and nursing of wild spat has so far generally failed to work well for scallops outside Japan and New Zealand and has not proved feasible for several other species. For example, spat collection looked promising for sea urchins in Japan but was eventually too unreliable to enable regular large-scale releases of juveniles. On the other hand, collection and nursing of spat seem to be among the few viable options for stock enhancement of some species. Spiny lobsters are a case in point. For these valuable animals, the likelihood of cost-effective mass-production of large numbers of juveniles in hatcheries is a long way off, if possible at all, whereas capture and culture of puerulus larvae can improve rates of survival dramatically during the first year of benthic life. Such intervention provides more juveniles both for farming and for supplementing the wild stock (see Chapter 3, Section 3.6 for details and also information on the use of artificial habitats to improve survival and recruitment of juvenile spiny lobsters).

Overall, it is evident that capture and culture of wild spat is one of the simplest and most attractive options for stock enhancement; it reduces the cost of supplying large numbers of environmentally fit juveniles and can avoid effects of releases on the genetic structure of stocks. Collection and nursing of spat should, therefore, be investigated thoroughly before considering other options for stock enhancement. But a word of caution is needed here: capture and culture of spat can only result in stock enhancement if the juveniles are collected at or very shortly after settlement (i.e., before they have been exposed to the intense predation that occurs at the beginning of benthic life). If juveniles are collected once these high rates of mortality have run their course, the nursing and release process will do little to increase the numbers of individuals recruiting to the fishery.

A variation on the theme of capturing and culturing spat is translocation of settled juveniles from places where their abundance exceeds the
carrying capacity of the habitat, i.e., where density has an adverse effect on
growth, reproduction and survival. The scope for such translocations seems
to be limited to a variety of bivalves (see Chapter 3, Section 3.2). It will also
depend on being able to collect the juveniles *en masse* cost-effectively and
re-establish them in suitable habitats that received few juveniles.

The vagaries of larval supply, the inability to locate concentrations of
settling larvae or difficulties in developing methods to collect and nurse them
in large numbers, means that many stock enhancement programmes have
had to depend on hatchery-reared juveniles. This option has been made
readily available for some species through development of cost-effective
culture methods for aquaculture. Even so, some stock enhancement initiatives have struggled to produce the numbers of juveniles needed to make a
meaningful difference to natural population size (e.g., geoduck and hard
clams in the United States and queen conch in the Caribbean). On the other
hand, stock enhancement programmes for shrimp in Japan and China have
released hundreds of millions, and billions, of hatchery-reared juveniles per
year, respectively (Table 4.1), and often had significant effects on catches
shortly afterwards (Chapter 3, Section 3.5.8). Overall, however, there is
considerable doubt that the scale of releases of hatchery-reared juveniles
has been great enough to make a difference to the status of the stocks. The
release of up to 30 million abalone per year (Chapter 3, Section 3.3.9) and
up to >60 million sea urchins per year (Chapter 3, Section 3.8.9) in Japan
is testimony to this. At best, these efforts sometimes arrested further
declines in the fishery but did not enhance yields nationally. Estimates are
needed of the numbers of juveniles to be released each year to achieve the
desired increase in harvest. Only then can managers determine whether
such production is possible and evaluate whether a more limited supply of
juveniles from the hatchery will have a beneficial effect (see Chapter 6,
Section 6.4 for details).

The difficulties producing sufficient juveniles in hatcheries are not improved by the fact that cultured animals often have morphological and
behavioural deficits such as thinner shells and slower rates of burrowing
(see examples throughout Chapter 3). This leads to greater rates of predation
than occur for wild individuals of the same size and so higher numbers are
needed to compensate for the larger losses. Although experiments have
shown that cultured juveniles can learn natural behaviour by 'acclimating'
them to predators before release, it is difficult to see how this could be done
for the millions of individuals needed for effective stock enhancement.

The problem of defects in morphology and behaviour needs to be kept
in perspective, however. The increases in mortality rate that result can
be completely overshadowed by the naturally high levels of predation on

juveniles, especially if they are placed in the wrong habitat (see examples throughout Chapter 3). Provided any defects of cultured animals are caused by rearing conditions, not adverse genetic effects, and the juveniles are reasonably fit for survival in the wild, a more important challenge for stock enhancement programmes is to minimise predation rather than fine-tune the quality of juveniles. The focus of such strategies should be the period immediately after release, because this is when most predation occurs.

Stock enhancement projects also stand out from restocking initiatives because there has been widespread use of artificial habitats in the former. Such habitats have been used in two ways: to increase survival of settling larvae and to increase the carrying capacity of the habitat for older life history stages. Deployment of a variety of substrata has 'harnessed' wild spat of several species of clams in the United States and resulted in harvests that may not otherwise have occurred in those places. It is not out of the question, however, that construction of the habitats simply moved the location of spatfall from one site to another. The postlarvae may have remained competent to settle in the plankton until they encountered suitable surfaces elsewhere. For some species, there is no doubt that the creation of additional suitable habitat for juveniles has resulted in a net increase in productivity. The best example is the use of simple additional shelters to improve settlement success of the spiny lobster *Panulirus argus* in the Caribbean. In this case, the shelters reduced the high rates of predation typical of the first benthic year of life and, in one case, resulted in a six-fold increase in survival (see Chapter 3, Section 3.6.2).

For herbivorous species, like abalone and sea urchins, there is a direct link between production and surface area colonized by suitable algae, and so it is not surprising that stock enhancement programmes for these animals have put much effort into providing additional habitat for shelter and foraging. Interestingly, the large-scale construction of such habitats in Japan has not yet resulted in increased production of these two species. Assessments are needed to determine the cost–benefit of such measures. There is also a case for using artificial habitat to improve the carrying capacity of fishing areas for lobsters (see Chapter 3, Section 3.7.5) where the shelter needed by adults is limiting. However, before embarking on such investments, it would be critical to verify that a good supply of the smaller size classes is available to colonise the additional habitats regularly.

Many marine invertebrates feed low in the food chain and do not interact directly with other fisheries species, so few negative impacts have been reported from releases of cultured juveniles. Nevertheless, spread of diseases and the removal of predators and competitors can cause potential adverse effects. The majority of species reviewed here are relatively resistant to disease, and outbreaks among wild populations have been reported only for abalone, scallops and shrimp. Not all diseases associated with species managed by stock enhancement can be attributed to transfers from hatcheries; some originated from introduction of exotic species for aquaculture and are now spread by hatchery-reared endemic animals released in the wild. Also, 'clean' cultured animals released in the wild often succumb to diseases that occur there naturally. Even so, great care is needed to ensure that release of hatchery-reared juveniles does not spread diseases to other species or reduce the size or value of the populations. The disease risks associated with stock enhancement programmes are discussed in more detail in Chapter 7, Section 7.3.

Removal of predators and competitors by resource managers or fisherman will not necessarily be seen as an adverse effect by other stakeholders in the enhanced resource. For example, they may make value judgments that manipulation of one species in favour of another is warranted to increase production of the more valuable one. Nevertheless, increased interest in an 'ecosystem approach to fisheries' (FAO, 2003b) means that managers will need to consult widely before incorporating the culling of predators or competitors in management plans for stock enhancement programmes. Indeed, the broader stakeholder community in some countries may wish to see a more balanced ecosystem rather than increased production of particular species at the expense of reduced biodiversity. Such decisions will not always be straightforward, because several competing species may be of commercial value. For example, removal of sea urchins can increase production of abalone (see Chapter 3, Section 3.3.5). The effects of stock enhancement on other components of the ecosystem are discussed in more detail in Chapter 7, Section 7.4.

4.2.2. Progress towards a responsible approach to stock enhancement

The stock enhancement initiatives described for the eight species/groups provide more opportunities to assess progress toward a responsible approach than the work done so far on restocking. In particular, they encompass established programmes that have already delivered benefits, as well as projects at the earlier stages of research, development and assessment. Progress for each of the eight stock enhancement initiatives against the 10 components of the responsible approach is summarised in Table 4.3 and across all eight species/groups in the following.

4.2.2.1. Component 1: Select priorities and target species

This component suggested by Blankenship and Leber (1995) has not generally been considered. The main reason is that they were encouraging careful allocation of research funds for stock enhancement among several candidate species, whereas managers have usually been presented with clear choices (e.g., requests to increase production of a particular fishery to former levels). With the exception of Japan, where stock enhancement has been attempted for >90 species (Honma, 1993; Imamura, 1999), the need to give priority to stock enhancement efforts among species has seldom arisen, because requests from the commercial or recreational sectors to intervene have been limited to just a few, and sometimes only one species. Nevertheless, some initiatives could have benefited from a more objective analysis of the potential of stock enhancement to redress the problem. Attempts to increase catches of hard clams on the east coast of the United States come into this category (see Chapter 3, Section 3.2.9).

4.2.2.2. Component 2: Develop a management plan

The ideal of integrating stock enhancement with other forms of management has only been achieved comprehensively for scallops, and then only in Japan and New Zealand (see Chapter 3, Section 3.1.9). In part, this is because technology for stock enhancement of several species is still under development, so managers are not yet in a position to apply the methods with confidence. Where hatchery methods do make mass-release of juveniles a possibility (e.g., for shrimp and, to lesser extent, sea urchins and abalone), attention has not always been paid to determining the structure of the stock. It is necessary to identify whether all or just some of the populations comprising the fishery would benefit from the release of juveniles and to determine the level of supplementation required to optimize yields (see Chapter 6 for details).

There are two other noteworthy examples of good management. The first is assessment of the potential to increase production of spiny lobsters through capture and culture of puerulus larvae. In New Zealand and Australia, managers successfully balanced for the use of the resource for fishing, aquaculture and possible stock enhancement (see Chapter 3, Section 3.6.2). The second is the comprehensive plan to assess the potential benefits of large-scale releases of the shrimp *Penaeus esculentus* in Exmouth Gulf, Australia. There, a logical step-wise approach was used involving the industry partner and a broad array of scientists to determine whether releases would meet the prescribed objectives (see Chapter 3, Sections 3.5.9 and 3.5.11).

4.2.2.3. Component 3: Define measures of success

Most of the stock enhancement initiatives have had reasonably clear goals for releases of cultured juveniles. These goals have included: reaching the carrying capacity of the habitat for scallops in Hokkaido, Japan; decreasing

Table 4.3 Summary of how the various stock enhancement initiatives have fulfilled the 10 components of a responsible approach advocated by Blankenship and Leber (1995).

Recommendation	Scallops	Other bivalves	Abalone	Queen conch
1. Prioritize and select target species	Not applicable in most countries, because one species dominates catches.	No formal process used. Programmes launched in response to public requests.	Not applicable in most countries, because one or two species dominate catch.	Not applicable; only one species involved.
2. Develop a management plan	Comprehensive plans in place in Japan and New Zealand, based on release of cultured spat and incorporating companion management (e.g., rotational fishing).	Many management measures in place in the United States, but limited analysis of the role of stock enhancement.	Well-developed plans for most countries, but large-scale stock enhancement incorporated only in Japan.	Management plans developed for many countries but do not include stock enhancement, because the benefits have not yet been demonstrated.
3. Define measures of success	Specified in Japan and New Zealand (e.g., harvests at carrying capacity).	Specified for geoduck clams on the west coast of United States for subtidal commercial and intertidal recreational fishery.	Often not specified but usually involve restoring production to previous higher levels.	Not specified, although a density of 50 adults ha^{-1} is needed to overcome 'Allee effect'.
4. Use genetic resource management	Simplified by the use of wild spat. Loss of local adaptions by translocation of spat recognised as a potential problem and caution advised.	Population genetic studies undertaken for some species. Hatchery protocols to obtain desired progeny developed.	Population genetics described for several species. Awareness of need to overcome selective breeding in hatcheries.	Genetic variation of wild conch quantified. Use of egg masses from wild adults reduces effects of hatchery rearing on gene frequencies.
5. Use disease and health management	Red tides and ectoparasites cause problems for wild and enhanced fisheries. Movements and introductions pose threats to receiving stocks.	High awareness of risks. Protocols developed for distribution of juveniles to reduce introduction or spread of pathogens.	Risk of damage to stocks by pathogens and parasites is high and well understood. Regulations in place in the United States to reduce risk of spreading disease.	No disease problems encountered.

6. Develop enhancement objectives and tactics	Ecosystem well understood; spat released at sizes where predation is reduced. Predators removed in Japan.	Biological requirements of released animals well understood. Measures implemented to reduce predation.	Ecology well understood, and requirements of released juveniles identified. Predators removed in some cases.	Excellent understanding of ecological requirements of juveniles.
7. Identify released individuals and assess stocking effects	Marks on shells of cultured animals used as natural tags. Survival of released juveniles well known.	Many hatchery-reared species identified easily by distinctive marks on the shell. Objective analyses of stocking effects are rare.	Batch marking easy by controlling diet in hatchery phase. Genetic tagging is now possible but still relatively expensive.	Methods for mass-marking still to be developed. Stock enhancement programmes are not cost-effective.
8. Identify optimum release strategies	Release strategies have been 'fined-tuned' in Japan and New Zealand.	Release strategies developed and depend on protective measures (e.g., artificial shelter) for several species.	Release strategies identified and improved by use of protective release 'chambers' or habitats.	Releases must be made in known nursery areas. Survival increased 10-fold by placing juveniles in protective pens.
9. Identify economic and policy objectives	Access or property rights established in Japan and New Zealand.	Objective is to improve recreational catch for many species. Cost–benefit of releases rarely assessed beforehand.	Limited entry fisheries for abalone in most countries mean that any benefits will accrue to existing participants.	Not clearly defined, apart from increasing yields from this valuable fishery in the Caribbean.
10. Use adaptive management	Stocking density adjusted in Japan to reduce disease risk. Releases in New Zealand based on annual surveys of natural abundance.	Ecological knowledge used to increase likelihood of success of releasing hard clams.	Release of larvae is a response to high cost of larger hatchery-reared juveniles. A shift is underway from stock enhancement to restocking.	No clear examples, but the need to shift from investigating stock enhancement to restocking is recognised.

(Continued)

Table 4.3 (Continued)

Recommendation	Shrimp	Spiny lobsters	Lobsters	Sea urchins
1. Prioritize and select target species	Depleted local stocks are primary driver, but availability of hatchery technology for aquaculture has also been a factor.	Research on stock enhancement driven by desire to stabilize and improve production.	Not applicable because only one species fished in Europe and one in North America.	Enhancement attempted for all species in Japan, but greatest effort is in Hokkaido.
2. Develop a management plan	Plans exist in many countries but include stock enhancement only for particular areas in China, Japan and Australia.	Major fisheries all have well-developed management plans. Provisions for stock enhancement under development in only a few countries at present.	Plans long established in North America and U.K. but do not yet include enhancement. Plans in place in other countries, but tighter regulations and enforcement needed.	Management plans in Japan determined by Fisheries Co-operative Associations and include stock enhancement.
3. Define measures of success	Clearly defined in Japan and Australia (e.g., decreased inter-annual variation in catch and increased long-term average catch).	Identified in some places (e.g., Tasmania), where removal of pueruli for aquaculture must be matched by release of 20% more juveniles than expected to survive naturally.	Not clearly defined for stock enhancement. Need for restocking in Europe not widely recognised and should be promoted.	Not specified but general aim is to return production of roe to historical average catches.
4. Use genetic resource management	Protocols developed to maintain gene frequencies of cultured shrimp, but large-scale releases in China seem to have reduced genetic diversity of wild population(s).	Population genetics studied in some countries. Use of wild pueruli avoids risks of producing juveniles with limited gene frequencies.	Population genetics studies undertaken in Europe. Regular replacement of wild fertilized broodstock helps maintain gene frequencies. Transfers of eggs between countries pose problems.	Responsible release practices reported for parts of Japan, but there are differences in gene frequencies of hatchery-reared and wild animals. Population genetics well known in the United States.

5. Use disease and health management	High risk because of widespread use of excess juveniles from aquaculture for stock enhancement. Protocols to reduce risk developed.	Low incidence of diseases reported, but severe mortalities have occurred in cage culture in Vietnam because of poor husbandry.	Disease risk in hatcheries is low. Imports of *Homarus americanus* to Europe have spread diseases to *H. gammarus*.	Disease risks not assessed.
6. Develop enhancement objectives and tactics	Ecology well known, and nursery habitats identified for major species. Predation is intense.	Ecology well known. On-growing in aquaculture and artificial habitats used to improve survival.	Ecology well known for larger sizes but knowledge is poor for juveniles.	Habitat requirements and effects of predation well understood. Crab and starfish predators actively removed.
7. Identify released individuals and assess stocking effects	Effective tags for tracking large batches of juveniles still needed. Genetic tags show promise. Assessing effects of releases in Japan depends on differences in size frequencies.	Marking systems exist at experimental scale. No large-scale releases made yet.	Released animals identified by morphological differences. Recovery of juveniles released as stages IV–V well documented in U.K. and Norway.	No satisfactory external tag developed. Released animals distinguished by size and contribution of releases to total harvested roe weight.
8. Identify optimum release strategies	Effects of release size and release habitat well known. Effects of release season and stocking density are less well understood and under investigation.	Still under development for reared juveniles, but provision of additional suitable shelter greatly increases survival of wild juveniles.	Difficult to determine because of cryptic behaviour of juveniles for 2–3 yr. Immediate post-release survival best when juveniles placed in trays on seabed.	Good knowledge of ecology has resulted in effective release strategies. Artificial structures used to improve quality of release areas.
9. Identify economic and policy objectives	Policies to devolve the costs of stock enhancement to fishers have been developed in China, Japan and Australia.	Objectives well identified in Australia and New Zealand.	Property rights introduced in U.K. This is expected to promote investment in restocking.	Clear goals to increase production of roe and integrate fishery with other important resources.
10. Use adaptive management	Cost–benefit of releases at different sizes identified.	Transfer of technology for puerulus collectors from recruitment studies to aquaculture to supply juveniles.	No clear examples. Change of emphasis from stock enhancement to restocking needed.	This is an integral part of measures to increase production of sea urchin roe in Japan.

inter/annual variation in catches and increasing long-term average catch of shrimp in Exmouth Gulf, Australia; doubling the catch of geoduck clams on the west coast of the United States, and using some of the puerulus larvae of spiny lobsters collected for aquaculture to increase the number of benthic juveniles entering the fishery in Tasmania, Australia. Overall, many of the current stock enhancement initiatives have had reasonably clear objectives that can be translated readily into testable hypotheses and measures of success. However, with the exception of scallops in Japan and New Zealand, some bivalves in the United States, abalone and sea urchins in Japan and shrimp in China, Japan and Australia, the methods for stock enhancement have not yet developed to the point where the objectives can be tested on a large scale.

4.2.2.4. *Component 4: Use of genetic resource management*

Use of wild spat and modifications or additions to settlement habitats to increase the supply of juveniles has reduced the potential range of genetic consequences for some stock enhancement initiatives. Nevertheless, more attention needs to be given to the effects of relocating wild spat on dilution of gene frequencies and the possible implications for important local adaptations.

For many of the stock enhancement initiatives that rely on hatchery-reared juveniles, due regard has been given to describing the genetic structure of populations and to using protocols to maintain normal genetic diversities among cultured animals. However, questions remain about the sensitivity of some of the genetic methods in use, and a more sophisticated approach may be needed in some cases (e.g., use of microsatellites) to delineate the spatial structure of self-replenishing populations adequately to provide the basis for responsible and effective patterns of releases (see Chapter 6, Section 6.1 and Chapter 7, Section 7.2).

4.2.2.5. *Component 5: Use of disease and health management*

There is good awareness of the risks posed to populations of scallops and other bivalves from toxic dinoflagellate algal blooms (red tides), diseases and parasites. Protocols have been developed to reduce the risks of spreading pathogens and parasites from hatcheries to the wild, and they seem to be applied well in some countries (e.g., the United States). Apart from legislation for oysters and clams in the United States, it is not all that clear what regulations are in place elsewhere to reduce the risk of spreading red tides and diseases in the wild by restricting translocation of shellfish from one site to another. This needs to be examined because it is evident that severe losses of production can ensue from careless movements of animals.

Diseases have caused havoc in shrimp aquaculture and have sometimes been transferred to wild populations. Pathogens and parasites have also plagued fisheries for abalone as the result of introductions and translocations (Chapter 3, Section 3.3.7). Recent occurrence of debilitating diseases among farmed, wild-caught juvenile spiny lobsters in Vietnam (see Chapter 3, Section 3.6.5) provides yet another warning that pathogens can flourish in situations of poor husbandry. The implications for stock enhancement programmes, and other species in the ecosystem, are described in detail in Chapter 7, Section 7.3. Overall, more attention needs to be given to minimising the potential risks from diseases and parasites in management of invertebrate fisheries. In some cases this will involve stronger legislation and enforcement of the protocols already developed for releases of hatchery-reared juveniles.

4.2.2.6. Component 6: Develop stock enhancement tactics

Most stock enhancement programmes have developed a sound understanding of the broader ecosystem into which cultured juveniles are placed. And, like restocking, predation is the greatest hurdle for stock enhancement programmes to overcome once environmentally fit juveniles can be produced *en masse*. Because of their larger scale, and the priority currently given to them compared with restocking initiatives, stock enhancement programmes have used a much greater range of interventions to control predation. These include purpose-built shelter for released sea urchins and abalone and the naturally settling puerulus larvae of spiny lobsters, as well as active removal of starfish and crab predators of scallops and sea urchins.

An important principle is that releases are usually more effective when made at low densities. Such releases reduce density-dependent effects on growth and the chances that predators will converge on high concentrations of juveniles. Distributing juveniles at low density can be more expensive, however, and managers need to resist the temptation to release animals in the easiest way. Otherwise, much of the investment in hatchery production will be lost.

4.2.2.7. Component 7: Identify released individuals and assess stocking effects

There has been mixed success in developing economical ways to mark the very large numbers of individuals needed to test the effectiveness of stock enhancement programmes. It has been relatively easy to identify released scallops, several species of bivalves and abalone because of distinctive marks that result from culture or that can be bred into hatchery-reared animals. In some cases (e.g., for shrimp and sea urchins), it has not yet been feasible to mark juveniles physically *en masse,* and the contribution of cultured animals has been gauged instead by estimating differences in size distributions of released and wild individuals. This has often involved releasing animals out of synchrony with natural recruitment so that cultured and wild cohorts can be distinguished. Development of genetic tags promises to help solve this potential problem (see Chapter 7, Section 7.1).

Ironically, where marking has been relatively easy (e.g., for hard clams), good estimates of survival have not always been made. With the exception of stock enhancement programmes for scallops in Japan and New Zealand, more attention needs to be given to accurate assessments of survival of released animals and their contribution to total catches. Where it is not possible to tag large numbers of cultured animals at relatively modest cost, the impact of releasing juveniles on the total abundance of the population can be estimated using 'impact assessment' sampling strategies (see Chapter 7, Section 7.1 for details).

4.2.2.8. Component 8: Use an empirical process to define optimum release strategies

For many of the species/groups reviewed here, experiments and trials have been done to identify how, where and when to release cultured juveniles, and at what size to release them. This is helping to obtain acceptable rates of survival in the face of often intense predation. Even so, relatively low survival, and/or severe limits on mass-production of juveniles, currently restricts the scope for using stock enhancement to increase productivity of several marine invertebrates. Release strategies for wild scallop spat have been fine-tuned to obtain levels of survival that make enhancement profitable and remain the outstanding example. However, scallop stock enhancement based on hatchery-reared spat continues to struggle to release a critical mass of juveniles and this typifies the problems encountered for many other species.

4.2.2.9. Component 9: Identify economic and policy objectives

With the exception of the detailed planning and analysis of the proposed stock enhancement of the fishery for the shrimp *Penaeus esculentus* in Exmouth Gulf, Australia, and the scallop fishery operated by the Challenger Scallop Company in New Zealand, detailed descriptions of the economic and policy objectives of stock enhancement programmes are not common. However, as described under Component 3, it is evident that several of the stock enhancement initiatives have set out to meet clear objectives centred around increasing productivity of target fisheries. These initiatives have often considered the policies that will be necessary to make them work (e.g., property rights for investors in lobster stock enhancement programmes in Europe and access for recreational clam fishermen in the United States).

Policies have not been so well developed, however, for assessing the success of stock enhancement trials. Such policies might be expected to include, for example, decision points about when to terminate initiatives that have had equivocal results or to continue to invest until it is clear whether releases of cultured juveniles are cost-effective. Also, little thought has gone into policies to determine how enhancement programmes may affect harvests of other species in mixed fisheries. This may be in part related to the fact that many marine invertebrates are low in the food chain, and increased abundances are not expected to have pronounced effects on many other species. Nevertheless, this aspect of stock enhancement will need increased attention given the growing interest in an ecosystem approach to fisheries (FAO, 2003b).

4.2.2.10. Component 10: Use adaptive management

Good examples of adaptive management have resulted mainly from the well-established enhanced scallop fisheries. In Japan, stocking densities were reduced to five to six individuals m^{-2}, including scallops in hanging culture above the seabed, after unexplained mortalities and reduced rates of growth occurred at higher stocking densities. There was also an attempt to reduce the numbers of scallops released each year in Sarufutsu, Hokkaido, once the fishery had been augmented to historically high levels for several years. However, when catches fell dramatically, releases were adjusted, albeit to considerably lower levels than used previously, to re-attain record harvest levels (see Chapter 3, Section 3.1.9).

In New Zealand, catches of scallop spat are adjusted each year to meet the shortfall between availability of natural juveniles and the desired stocking density. This has lowered the cost of collecting spat in several years. The number of cultured juveniles required has been reduced further by using a full immersion system for transporting spat from collection to release sites (see Chapter 3, Section 3.1.9).

For several stock enhancement initiatives, particularly for abalone, queen conch and sea urchins, it is now recognised that although the current technology is unlikely to achieve the original aims of the releases, it could well be suitable for restocking programmes in particular situations. Such adaptive management would also benefit fisheries for the lobster *Homarus gammarus* throughout much of Europe.

Chapter 5

Lessons Learned

The broad range of case studies in this review provides a clear message for fisheries agencies considering the use of restocking and stock enhancement as management tools, that is, there are no generic methods for such interventions. Even relatively small differences in life-history traits among species (e.g., larval duration, nursery habitats, vulnerability to predation, age at first maturity, feeding areas) can have major effects on the approach required, and costs involved, for releasing marine invertebrates in restocking and stock enhancement programmes. These differences have necessitated approaches as diverse as: (1) hatchery production of juveniles; (2) collection of wild spat; (3) provision of more settlement habitat; (4) redistribution of settled juveniles and (5) translocation of adults. Associated measures include construction of additional habitat to increase carrying capacity, removal of predators and installation of temporary artificial shelters to reduce the high levels of predation that can occur during release of juveniles.

The variations in life history among species, the range of possible ways of increasing the numbers of juveniles, and the numerous failures, lead to the inescapable conclusion that there are no shortcuts to targeted research to identify whether a particular form of restocking or stock enhancement is likely to be cost-effective. A number of key lessons emerge from the accounts presented for the 11 species/groups described in Chapters 2 and 3. We summarise these lessons below to provide guidance for the design, development and implementation of future restocking or stock enhancement programmes.

0065-2881/05 $35.00
DOI: 10.1016/S0065-2881(05)49005-7

5.1. LESSONS FOR RESTOCKING

Lesson 1: The costs and time frames involved in restocking programmes for some species can be prohibitive. These costs will be unpalatable enough for developed countries, but beyond the capacity of developing countries unless they receive long-term assistance. Restocking cannot, therefore, always be relied on to correct abuse of the resource, so management agencies should take great care to avoid depletion of populations to chronically low levels. Long-lived sessile species are particularly vulnerable to depletion to low densities, and such mismanagement is likely to result in local extinction because the remnant population may be dispersed too widely to reproduce effectively.

Lesson 2: The costs of restocking can be reduced greatly for some species simply by relocating a proportion of adults to form a viable spawning biomass. This method is likely to be effective only for species with multiple, largely self-replenishing populations at relatively small spatial scales and/or very limited larval dispersal (e.g., topshell, abalone and sea cucumbers, but also scallops and other bivalves in some situations). Even then, this measure will probably only be successful at sites with good larval retention. It will also depend on transferring sufficient adults to establish a viable breeding population and a moratorium on fishing until the spawning biomass reaches a level that can support regular, substantial harvests (see Chapter 6, Section 6.2).

Lesson 3: The high costs involved in releasing some species at a size where they escape most predation can be defrayed by combining the culture of animals intended for restocking with other forms of aquaculture. For example, giant clams for release in the wild could be reared on farms supplying the aquarium trade (Chapter 2, Section 2.1.8), and the sea cucumber *Holothuria scabra* could possibly be produced in combined culture with shrimp in ponds (Chapter 2, Section 2.3.2).

5.2. LESSONS FOR STOCK ENHANCEMENT

Lesson 4: Very large numbers of juveniles are needed for effective stock enhancement. Even where hatcheries can produce large numbers (tens of millions) of fit individuals at reasonable cost, releases will not have an economic impact on catches unless the supply of recruits is generally limiting, the cultured juveniles represent a large proportion of recruitment, and fishing is regulated appropriately. The release programmes for hard clams on

Long Island, New York (Chapter 3, Section 3.2.9), and for abalone and sea urchins in Japan (Chapter 3, Sections 3.3.9 and 3.8.9) help illustrate this point. It is important not to underestimate the scale of releases needed for stock enhancement programmes. The two groups of marine invertebrates for which stock enhancement has been most successful, scallops and shrimp, have been released in hundreds of millions to billions of juveniles each year. The releases of abalone and sea urchins in Japan, which have been an order of magnitude lower, have not met the goals of increasing harvests, although they may have stabilised catches at lower levels by arresting further declines.

Lesson 5: Excessive releases of juveniles cause density-dependent mortality. This is evident at both large and small scales. In Japan and China, the population size of shrimp seems to have been reduced after releases of very large numbers of juveniles (Chapter 3, Section 3.5.8). For abalone, releases of low numbers of juveniles have been as successful as releases of higher numbers at limited spatial scales (Chapter 3, Section 3.3.11). This highlights the fact that a sound understanding of the ecology of the target species, and the carrying capacity of the ecosystem, is needed to optimise the number of animals released (see Chapter 6, Section 6.4).

Lesson 6: Wild spat can provide an abundant, low-cost source of juveniles for some species. The most successful form of marine invertebrate stock enhancement (scallops) is based on the collection of wild spat (Chapter 3, Section 3.1.2). However, caution is needed here, because variation in abundance of spat may be too great for establishment of stock enhancement programmes. This was the case for sea urchins in Japan (Chapter 3, Section 3.8.2) and is a firm reminder that development of methods for collection of spat (or postlarvae) cannot always be translated into a cost-effective way of supplying juveniles for stock enhancement.

Lesson 7: Artificial habitats can be used to increase the carrying capacity for target species by providing space and/or food for additional released animals. Such man-made structures have provided additional shelter and feeding surfaces for abalone (Chapter 3, Section 3.3.5) and increased areas for development of the algal communities needed to improve roe weight in sea urchin fisheries (Chapter 3, Section 3.8.5). There is also a strong indication that habitat is limiting for adult lobsters in some places. Where this occurs, any increase in the supply of juvenile lobsters will have little effect without provision of additional habitat for the adults (Chapter 3, Section 3.7.5).

Lesson 8: Yields of some species can be increased simply by providing suitable settlement habitat. This has been demonstrated for clam species on the west coast of the United States where gravel beds have 'harnessed' juveniles that otherwise would not have settled at the same locality (Chapter 3, Section 3.2.2). Provision of artificial shelter with crevices of the appropriate size has also increased settlement success of spiny lobsters up to six-fold in the Caribbean (Chapter 3, Section 3.6.2).

Lesson 9: Harvests of some species can be improved by redistributing some juveniles from areas of heavy settlement to areas where spat failed to settle. The twin benefits of this intervention are: (1) that growth and survival are likely to improve in areas with abundant juveniles because of reduced intraspecific interactions and (2) harvests can be taken from areas that would otherwise have yielded little because larvae generally failed to arrive or settle there. This has been demonstrated for razor clams on the west coast of the United States (Chapter 3, Section 3.2.9). However, great care must be taken not to alter the conditions needed for successful future settlement when juveniles are collected and transplanted. The destruction of oyster reefs in Chesapeake Bay is a salutary reminder of the economic losses that can occur if the underpinning ecosystem is degraded (Chapter 3, Section 3.2.1).

Lesson 10: Large areas are needed for stock enhancement of some species of marine invertebrates. Access to such areas will need to be negotiated before investment in large-scale releases and then protected if the stock enhancement programme is to be successful and sustainable in the long term. Scallops are a case in point (Chapter 3, Section 3.1.11).

Lesson 11: The most successful stock enhancement programmes for marine invertebrates are run by co-operatives and the private sector. This has occurred because fishermen have been given access or property rights to the resource, and therefore the incentive to invest in application of the technology. The best examples of this are the scallop stock enhancement programmes in Japan and New Zealand (Chapter 3, Section 3.1.9, see also Chapter 6, Section 6.5). The corollary to this is that, ultimately, large-scale releases for stock enhancement commenced by governments will not be sustainable unless the beneficiaries bear the costs involved. This is evident from the massive reductions in the releases of shrimp in China since the implementation of new economic policies (Chapter 3, Section 3.5.9).

Lesson 12: Successful stock enhancement may change the marketing conditions for some species. This has occurred for scallops in Japan, where increased production from stock enhancement and aquaculture meant that fisheries co-operatives had to add value to the product, engage in wholesaling and make concerted efforts to raise awareness of the quality and availability of their scallops to maintain prices (Chapter 3, Section 3.1.10). By contrast, the great increase in production of the sea cucumber *Apostichopus japonicus* in China through 'put-and-take' sea ranching has not yet lowered prices. Rather, the price has increased because of greater prosperity in the country, which has resulted in many more consumers and higher demand (Chapter 2, Section 2.3.11).

5.3. LESSONS FOR BOTH RESTOCKING AND STOCK ENHANCEMENT

Lesson 13: Establishment of a successful restocking or stock enhancement programme at one site is no guarantee that the methods can be transferred to other countries or to other sites in the same country. For example, few nations have been able to implement the methods for collecting scallop spat that are so beneficial in northern Japan, and the technology has only been successful in one of the areas where it has been applied in New Zealand (Chapter 3, Section 3.1.8).

Lesson 14: Small-scale experiments to test methods for restocking and stock enhancement can give misleading results. Increases in the scale of releases often reveal problems not apparent in initial experiments (e.g., for scallops in Australia and Canada [Chapter 3, Section 3.1.8] and geoduck clams in subtidal habitats in the United States [Chapter 3, Section 3.2.4]). Therefore, decisions to invest in large-scale releases of juveniles should not be made on the basis of small-scale, albeit well-designed, experiments. Large, 'commercial' scale experiments should be done to determine whether the assumptions stemming from initial experiments hold over sizable management areas.

Lesson 15: Predation is the single greatest hurdle to the successful establishment of released juveniles in restocking and stock enhancement programmes. The need for a thorough understanding of the ecology of the species and the ecosystem at the release site, and field experiments to identify the best way to release animals to maximise survival, are essential steps to reduce predation. In general, the chances of success can be improved greatly by releasing animals at low density in areas where predators are in relatively low abundance and at a size where the juveniles escape most predation. For some species, culling predators, and/or providing structures that impede predation, can also improve survival of released animals substantially.

Lesson 16: Aquaculture will create opportunities for restocking and stock enhancement. Although restocking and stock enhancement programmes were among the prime reasons for development of hatchery technology for many species (e.g., giant clams, sea cucumbers, topshell, abalone, sea urchins), large-scale releases of other species (e.g., shrimp) have only been made possible through investments in aquaculture. 'Spin-off' benefits from aquaculture, in terms of methods for culturing juveniles for release, can be expected to continue. One of the most promising developments in this regard is the capture and culture of puerulus larvae of spiny lobsters (Chapter 3, Section 3.6.2).

Lesson 17: Availability of methods to rear larvae in hatcheries does not necessarily mean that they can be applied cost-effectively to produce the large numbers of juveniles needed for effective restocking and stock enhancement programmes. This problem has been particularly severe for giant clams, topshell, queen conch and lobsters, where the costs involved in rearing juveniles to a size large enough to escape most predation have been punitive. However, it has also occurred to some extent for most other species except shrimp. When there are severe limits on production of juveniles because of high costs and restrictions on capacity, use of hatchery-reared juveniles should be considered only for restocking, when the initial high cost of rebuilding spawning biomass may be justified. Even then, managers need to confirm that restocking will add value to other forms of management and that the supply of cultured juveniles is adequate to meet the aim of the programme (see Chapter 6, Section 6.1.2).

Lesson 18: Large-scale releases of hatchery-reared juveniles can affect genetic diversity of wild populations. The evidence for this is still equivocal for some species. However, it has been reported for sea urchins in Japan (Chapter 3, Section 3.8.6), and the reduced diversity of *Penaeus chinensis* in the coastal waters of China, compared with Korea, is thought to be due to very large releases of cultured shrimp for stock enhancement (Chapter 3, Sections 3.5.5 and 3.5.8). Whether the shrimp released in China were of more limited genetic diversity than the wild population because of intentional or unintentional selective breeding, or whether there were differences in diversity between *P. chinensis* in China and Korea originally, remains to be determined. Only then will it be possible to gauge the full impact of stock enhancement on the gene frequencies of these shrimp.

Lesson 19: The potential benefits of releases of hatchery-reared juveniles will be reduced unless the releases address the most urgent needs of the fishery and appropriate measures are used to manage the intervention. The release of lobsters in Norway is a case in point (Chapter 3, Section 3.7.9). There, legislation to prevent fishing for the released animals, and the remnant wild population, until replenishment occurred would have delivered greater benefits than allowing the cultured animals to be caught when they reached the minimum legal size. Given the current status of the fishery, releases of lobsters in Norway funded from the public purse would be managed better as restocking rather than stock enhancement.

Lesson 20: Adaptive management can reduce the cost of restocking and stock enhancement. This is epitomised by the Challenger Scallop Enhancement Company in New Zealand, where the quantity of spat collected, grown and released each year is based on accurate annual estimates of the size of the population and the natural supply of juveniles (Chapter 3, Section 3.1.9). The quantity of cultured scallops needed has also been reduced further by improved ways of preparing collected spat for release (Chapter 3, Section 3.1.4).

Chapter 6

Management of Restocking and Stock Enhancement Programmes

Throughout this review, we have stressed that restocking and stock enhancement are not the same process. It should come as no surprise then that, although these interventions often share technology for the propagation and release of juveniles, restocking and stock enhancement need to be managed differently. However, management decisions about restocking and stock enhancement should not just be about how to make the best use of the animals once they have been produced and released. On the contrary, the most important management decisions need to be made well beforehand (Hilborn, 1998; Leber, 1999; Molony *et al.*, 2003; Bell, 2004; Lorenzen, 2005). In particular, managers need to answer the question: 'Are investments in hatchery-reared juveniles likely to add value to existing and alternative forms of management?'

Unfortunately, decisions to propagate and release juveniles have not always been preceded by the necessary thorough analysis (Leber, 1999; Lorenzen, 2005). Instead, they are often precipitated by the broad appeal of stock enhancement programmes with politicians (Malouf, 1989; Hilborn, 1998, 1999; Leber, 2002; Molony *et al.*, 2003). Rather than being tempted to

0065-2881/05 $35.00
DOI: 10.1016/S0065-2881(05)49006-9

use the technology for releasing cultured juveniles just because it is available, we reiterate that managers need to be confident that: (1) restocking programmes are likely to deliver additional value to other ways of restoring spawning biomass, and (2) stock enhancement programmes have a high probability of stabilizing yields and that the gains will exceed the costs.

In this Chapter, we outline the basic information required to determine whether restocking is likely to help rebuild severely depleted stocks or whether stock enhancement could help overcome recruitment limitation in operational fisheries and return a profit on investment. Where it can be demonstrated that either restocking or stock enhancement is likely to be beneficial, we then identify various ways in which the released animals should be managed to obtain optimum benefits from these two types of interventions.

6.1. INFORMATION TO EVALUATE THE NEED FOR RESTOCKING

To determine whether release of cultured juveniles will add value to other measures normally used to rebuild spawning biomass to more productive levels, managers need sound information on stock delineation, i.e., the biological and genetic structure of the population(s) comprising the fishery, the projected time frames for recovery with and without various levels of restocking, and the capacity of hatcheries to produce juveniles.

6.1.1. Stock delineation

Stock delineation is a key first step in evaluating the potential benefit of restocking programmes. Managers need to know whether they are dealing with a single population or several largely self-recruiting populations. If a fishery is based on more than one population, assessments will need to be made about whether all populations need to be restored and, for those that do need to be rebuilt, how best to go about it. Some populations may benefit substantially from the addition of cultured juveniles, whereas others will not. Simply adding more juveniles with the aim of eventually increasing spawning biomass in the absence of knowledge about the population(s) comprising the fishery runs the severe risk of mis-using the investment by failing to place the juveniles where they are needed most. There are also genetic consequences (see Chapter 7, Section 7.2), particularly the loss of local adaptations (e.g., spawning at the times and places that result in good survival of larvae).

A good indication of spatial stock structure can be provided by a thorough analysis of the population genetics of the species (see Chapter 7,

Section 7.2), although some stocks can still be divided into relatively isolated population units even when gene frequencies are generally homogeneous (see Chapter 3, Section 3.1.5 for examples). Thus, other tools to help determine stock delineation, such as multivariate comparison of morphometrics (Chapter 3, Section 3.2.6) or species composition of parasites, may also be important in some situations.

6.1.2. Time frames for recovery with and without restocking

Once the spatial structure of the stock has been ascertained, it is then necessary to identify which population(s) is likely to require restocking. This can be done by modelling how long it will take the spawning biomass of the population(s) to recover with and without the addition of cultured juveniles (Lorenzen, 2005; Nash and Catchpole, in press), assuming that management measures like a total moratorium on fishing or use of networks of marine protected areas are implemented to protect the remnant wild stock and released animals (Figure 6.1). Managers should first identify the desired level of spawning biomass (e.g., 40% of the virgin level) and ensure that data are collected that will enable the potential contribution of restocking to this target to be assessed (Heppell and Crowder, 1998). To estimate the time needed for recovery under management regimes such as a moratorium on fishing, without restocking, these data include: remnant population size; age structure; generation time; fecundity; larval dispersal patterns; natural

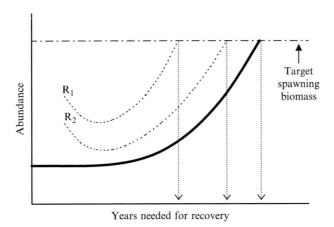

Figure 6.1 Hypothetical time taken to reach target spawning biomass under a fishing moratorium alone (solid curve) and at two levels of restocking, R_1 and R_2 (broken curves). Curves for restocking reflect expected mortality after release.

mortality and growth rates of different life history stages and behaviour of the species that may affect spawning success or survival at low population density. Caddy and Defeo (2003) provide a comprehensive guide to the tools and methods available for modelling the dynamics of discrete populations of marine invertebrates.

To estimate the value that could be added under different restocking scenarios (variations in the magnitude and frequency of releases), additional data are required on: the cost of juveniles; survival rate of the released animals and the subsequent survival of their F_1, F_2, F_3,..... progeny. Restocking can be considered to be worthwhile if the value gained from reopening the fishery earlier as a result of the release of cultured juveniles substantially exceeds the cost of producing the juveniles.

In cases where it is not possible to make a relatively accurate estimate of remnant population size, life-history characteristics can be used as a guide to assess the scope for restoration with and without release of cultured juveniles. Heppell and Crowder (1998) illustrate how different life history traits among species may increase or decrease the relative effectiveness of restocking programmes. For example, species with high fecundity and sexual maturity at a young age have greater potential for replenishment than those with limited fecundity and delayed maturity.

Nevertheless, the research required to identify the most appropriate form of management to restore spawning biomass and provide a regular supply of recruits should not be underestimated. Research examining the potential for using marine protected areas (MPAs) to manage a fishery for the spiny lobster *Panulirus argus* in the Bahamas helps to illustrate this (Lipcius *et al.*, 1997; Stockhausen *et al.*, 2000; Lipcius *et al.*, 2001; Stockhausen and Lipcius, 2001). To assess the benefits of various management options, these authors had to take into account the fact that postlarval abundance is decoupled from adult abundance by physical transport and that adult abundance is decoupled from postlarval supply by variation of habitat quality. They concluded that a single large reserve (20% of coastal habitat) would deliver greater benefits than several small reserves (totalling 20% of coastal habitat) and that both configurations were better than reducing effort by 20%. Stockhausen and Lipcius (2001) also concluded that the location of the MPA was important and that good information on larval transport was essential to maximising the benefits of a fishing reserve to recruitment to the meta-population.

6.1.3. Capacity of hatcheries to produce sufficient juveniles

If the modelling indicates that restocking will be beneficial, provision then needs to be made to ensure that hatcheries have the capacity to produce the required numbers of juveniles. Problems can be expected to arise when managers have limited access to cultured juveniles because of the size of the national economy (e.g., for Small Island Developing States) or because the spatial structure of the stock means that only relatively small numbers of animals can be produced for release within each population (when several populations need restocking). In such cases, at least two solutions are available. First, multiple smaller batches of juveniles could be produced and released. Second, the progeny from relatively low numbers of released animals could be moved progressively to uncolonised areas of suitable habitat to form new spawning aggregations. This ensures that replenishment occurs throughout the range of the population (see Chapter 2, Section 2.3.9). Such limitations can be expected to extend the time required for replenishment and should be incorporated into the modelling described previously.

Proceeding with restocking programmes in the absence of the necessary information can be futile. An attempt to restock hard clams in New York (Chapter 3, Section 3.2.9) provides a sobering example of the need for information on spawning biomass and rate of replenishment in evaluating the potential contribution of hatchery releases. In that case, the 125,000 relocated adults contributed only 0.005% to larval production, demonstrating that funds available for management could have been used more effectively by strengthening other measures such as limitation of catches.

6.2. MANAGEMENT OF RESTOCKING

One overriding rule should be used to manage the restoration of spawning biomass to productive levels as soon as possible. That is, the remaining wild individuals, the released animals and their collective progeny should be protected until the number of adults has reached the desired level (Bell, 2004). Such protection may be required for several generations.

In most cases, a moratorium on fishing will be needed to provide the necessary levels of protection. However, this may be difficult to implement if the species forms part of a mixed fishery and individuals are taken as by-catch. In such situations, fishing gear could be modified to reduce the impact on the species, fishing for the other species could be done at times when the restocked species is least vulnerable, and fishing closures or MPAs could be established to safeguard a substantial proportion of the spawning biomass. Lorenzen (2005) provides valuable insights into the potential costs and benefits of different combinations of restocking and fishing on the recovery of a fishery.

An effective moratorium on the capture of the species is likely to cause some short-term hardship for fishermen, although their revenues from fishing for the species are presumably already low because of the impoverished status of the stock. Nevertheless, it will be important to explain the longer-term benefits of the restocking programme to fishermen, and the need for restraint until the spawning biomass has recovered to the point where the population can once again yield sustainable harvests. Fishermen also need to be informed that future harvests will have to be set at lower levels; otherwise over-fishing and stock reduction will recur. Implementation of property or access rights for fishermen before the moratorium will provide the incentive to comply with such management, because they will be the ultimate beneficiaries. In the short to medium term, however, other incentives may need to be provided to transfer to alternative livelihoods.

6.3. INFORMATION TO EVALUATE THE NEED FOR STOCK ENHANCEMENT

Stock enhancement should only be contemplated where there is good evidence that production is often limited by recruitment limitation (Munro and Bell, 1997; Grimes, 1998; Doherty, 1999; Bell, 2004; Lorenzen, 2005). Great annual variation in abundance of juveniles settling from the plankton, continued rapid growth of new recruits, and persistence of intermittent strong year classes are good indicators that the carrying capacity of the habitat for juveniles is seldom reached (Munro and Bell, 1997; Doherty, 1999). Fisheries with these characteristics are candidates for stock enhancement, provided relatively large numbers of juveniles can be released cost-effectively, and their survival to minimum legal size is high. However, as discussed later, the need for stock enhancement will vary from year to year, depending on annual variations in the natural supply of juveniles and the carrying capacity of the habitat.

6.4. MANAGEMENT OF STOCK ENHANCEMENT

Whereas managers of restocking programmes are aiming for longer-term benefits, managers of stock enhancement programmes need to demonstrate, often over a shorter term, that the value of the additional harvest exceeds the cost of producing and releasing the juveniles (Hilborn, 1998). In general, they need to pay more attention to development of regulations to optimise the biological, social and financial sustainability of the interventions made to increase the productivity of the fishery. They need to pay particular attention to the number of juveniles released and to identifying and implementing measures that will maximise the benefits from the process (Bell, 2004).

6.4.1. Estimating how many juveniles to release

The number of juveniles released in a stock enhancement programme each year should be based on the natural supply of juveniles and the carrying capacity of the habitat (Munro and Bell, 1997; Grimes, 1998; Doherty, 1999; Caddy and Defeo, 2003; Bell, 2004). If at all possible, the natural supply of juveniles should be measured as early as possible in the settlement period to provide the opportunity to produce and release the required number of cultured animals that year. Where it is difficult to assess the abundance of wild juveniles, and where the natural supply of juveniles is highly erratic, managers can take the alternative approach of initiating large-scale production of juveniles at the time of spawning each year, then proceed with rearing the number required when the annual estimates of carrying capacity and abundance of wild juveniles are available.

Carrying capacity is difficult to measure and may vary with season and time. Indeed, until recently, much of the literature on the rationale and application of stock enhancement paid little attention to carrying capacity, except to say that it is seldom met for a given species because of recruitment limitation. Too often it is inferred that release of cultured animals provides the opportunity to 'fill' the habitat with juveniles, thereby increasing productivity. It is now evident that considerable thought needs to go into how any vacant carrying capacity is used if maximum benefits are to be obtained from stock enhancement (Caddy and Defeo, 2003; Bell, 2004). It is also important that the concept of 'carrying capacity' is clearly understood. Carrying capacity should not be thought of as the number of individuals the environment can support. Rather, it should be the number of individuals that can be supported at the optimal stocking density (i.e., the maximum density at which growth and survival are not affected). Estimates of carrying capacity will, therefore, need to take into account the productivity of the ecosystem and the abundance of predators and potential competitors.

Decisions about how many juveniles to release are relatively simple for 'annual' species, such as shrimp, because basically only one age class is involved; sufficient juveniles should be added so that the carrying capacity of the habitat for adults will be reached once the animals are at harvest size and natural mortality has run its course.

The calculations are more complex for species with multiple year classes. For such species, managers need to identify the most desirable harvest regime for the fishery and then plan for stock enhancement to deliver the required quantity and frequency of yields. Adding juveniles to an over-exploited stock in numbers that would use the carrying capacity fully by the time the released animals reach harvest size will not necessarily be in the best interest of the fishery. Such action would lead to an age structure dominated by that age class and a 'bumper' harvest followed by negligible returns for several years (Figure 6.2, see also Caddy and Defeo, 2003; Lorenzen, 2005). More regular harvests are usually desired by the industry.

As described by Bell (2004), a preferable stock enhancement regime would be to use average natural mortality and growth rates to allocate carrying capacity among age classes in the way that delivers the greatest regular harvest each year (Figure 6.3A). The desired optimum age structure can then be compared with that of the population (Figure 6.3B) and cultured juveniles added to provide the number of 1-yr-olds required to produce the optimum harvest by the time the animals grow to minimum legal size (Figure 6.3C). Provided the principle is applied each year, the fishery will remain on a path of more or less regular harvests at the desired level.

This approach is easiest to follow for species that occupy the same general habitat throughout their life cycle. However, it also applies to those species where the juveniles occupy separate (nursery) habitats from the adults (Bell, 2004). Although the survival of juveniles is not dependent on the density of adults in such populations, it is still preferable to have a consistent number of individuals contributing to the 'first' age class in the adult stock rather than an intermittent strong age class followed by relatively low numbers in subsequent years. For such species, it is most important that stock enhancement programmes succeed in producing juveniles that are competent to migrate from the nursery to adult habitat.

No matter whether the stock enhancement programme involves 'annual' species, or long-lived species that use one or more habitats during their life cycle, a clear understanding is needed of density-dependent growth and mortality processes during the first year of life so that managers can release the desired number of animals each year in a cost-effective way (Caddy and Defeo, 2003; Hilborn, 2004; Leber, 2004; Lorenzen, 2005 and references therein). As mentioned in Chapters 4 and 5, it is now apparent that many species should be released at low densities; otherwise, they will compete among themselves and with naturally recruited juveniles. This will result in reduced growth rates and, in the worst-case scenario, undue levels of mortality, resulting in failure to establish the desired year class strength and harvest regime.

Tailoring the releases of hatchery-reared juveniles as described in Figure 6.3 will also address one of the key questions frequently asked about stock enhancement programmes: do the released animals simply replace wild ones (Svåsand *et al.*, 2000; Hilborn, 2004; Leber, 2004; Lorenzen, 2005 and references therein)? Clearly, this would happen if there were no recruitment limitation and the habitat was already supporting a population at optimal density. As we have said earlier, there is no justification for considering stock enhancement in such situations. However, where there is recruitment limitation, a clear understanding of: (1) the natural supply of juveniles; (2) carrying capacity; (3) the number of animals to be added to provide the

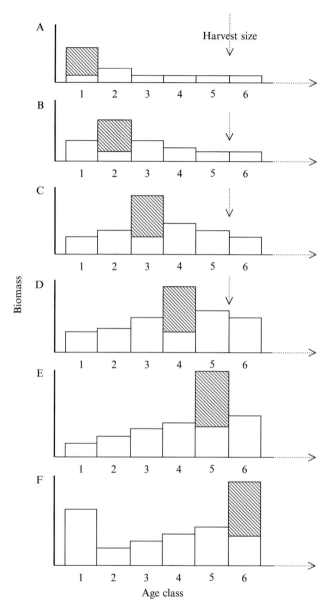

Figure 6.2 The structure of a generalised stock with multiple age classes that has been 'fished down', showing how addition of excessive cultured juveniles (hatched column) during a year of poor natural recruitment (A) could create an age class with a biomass that dominates the carrying capacity of the habitat and limits recruitment in subsequent years (B–E) until that age class reaches harvest size. Removal of much of the dominant age class by fishing would then provide vacant carrying capacity to once again support substantial numbers of wild and/or cultured juveniles (F) (redrawn from Bell, 2004).

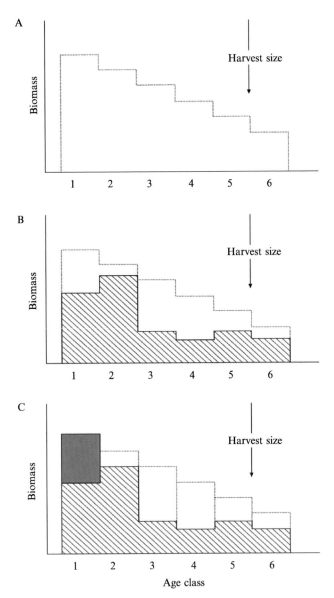

Figure 6.3 (A) Hypothetical distribution of biomass among age classes that will produce the highest possible yield each year (where the total biomass depicted by the histogram represents the carrying capacity of the habitat). (B) Differences in the optimum (blank column) and 'actual' (hatched column) distribution of biomass among age classes. (C) Quantity of juveniles (solid column) to be released in year 1 to yield optimum harvests; this process should be repeated each year, although the number to be released will vary depending on the natural supply of juveniles (redrawn from Bell, 2004).

desired age structure and harvest regime and (4) any density-dependent effects resulting from different release strategies will help ensure that released animals do not replace wild ones.

A problem that could arise in using stock enhancement to stabilize production between years as described in Figure 6.3 is that one of the existing age classes in the population may be very strong and well exceed its optimum biomass. In such cases, addition of cultured juveniles could be deferred until the strong age class has been harvested, although it may be better to prematurely 'fish down' the age class to the optimum level so that the stock enhancement programme could begin earlier (Bell, 2004).

As previously noted, the carrying capacity of a habitat can be expected to change spatially and from year to year (Grimes, 1998; Caddy and Defeo, 2003; Hilborn et al., 2003; Molony et al., 2003; Beamish and Noakes, 2004; Collie et al., 2004). Adjusting releases in response to changes in carrying capacity will increase variation in annual harvests and work against the concept of regular optimum yields. Managers will need to weigh the relative merits of setting optimum harvests at a constant, conservative level to accommodate years of low carrying capacity against catches that are greater on average but more variable. Sometimes, decreases in carrying capacity will be so extreme that it will be nonsense to release animals. For example, in Exmouth Gulf, Australia, where the harvests of *Penaeus esculentus* are correlated with the amount of seagrass available to provide essential habitat for juveniles, the area of seagrass varied from 3 km^2 to 50 km^2 between 1999 and 2001 as a result of damage by Cyclone Vance (Kenyon et al., 2003a,b). Clearly, there would be little sense in implementing a stock enhancement programme for *P. esculentus* in years when the required juvenile seagrass habitat was severely damaged. Caddy and Defeo (2003) describe several other examples of bottlenecks in the supply of essential habitat that limit the scope for stock enhancement.

Overall, the tool of stock enhancement holds promise for improving and stabilizing production of some species of sedentary invertebrates. The prime concern is that managers use it appropriately. Mis-application may squander resources, cause undesirable variation in harvests, replace wild, fitter individuals and/or result in increased impacts on other species (see Chapter 7).

6.4.2. Increasing productivity further through manipulations of the ecosystem

Interventions to increase yields from fisheries through stock enhancement are not limited to the release of cultured animals to overcome shortfalls in the natural supply of juveniles. As the accounts for sea urchins, spiny lobster and scallops in Chapter 3 demonstrate, production can be increased further by providing additional habitat and removing predators. The range of artificial substrata developed in Japan (Grove *et al.*, 1994; Okaichi and Yanagi, 1997; Morikawa, 1999) indicates that this approach may be feasible for an even wider range of species. A key question about the use of artificial habitats, however, is the cost of construction: will the cumulative value of the additional harvests exceed the expense of the installations?

In highlighting the possible role of artificial structures, we are referring mainly to those provided to create more habitat for released juveniles of sedentary invertebrates as opposed to those constructed to increase productivity of wild populations of more mobile species like fish. The 'production–attraction' debate still needs to be resolved for mobile species before managers can be confident that artificial habitats do not simply attract animals and make them more vulnerable to fishing (see Chapter 3, Section 3.6.2). Nevertheless, artificial habitats are credited with increasing natural settlement success of abalone (Chapter 3, Section 3.3.5), sea urchins (Chapter 3, Section 3.8.4) and spiny lobsters (Chapter 3, Section 3.6.2) under certain conditions. Caddy and Defeo (2003) and Leber (2004) correctly point out that some of the most important uncertainties about stock enhancement will be addressed by research that combines manipulations of both habitat and stocking density.

A concern exists among the advocates of an ecosystem approach to managing fisheries that removal (culling) of predators may favour production of one species at the expense of other, possibly useful or important, species (Caddy and Defeo, 2003; FAO, 2003b; Leber, 2004). Indeed, in some countries, existing fisheries and environmental legislation could be used to prohibit culling. Clearly, managers will need to consult widely before endorsing proposals to remove predators (see also Chapter 7, Section 7.4).

6.4.3. Other considerations for management of enhanced fisheries

In general, enhanced populations do not require specialized management because the cultured animals simply add to the population available for capture and should be managed together with the wild individuals to maximise yields using the conventional measures described by Caddy and Defeo (2003). However, some aspects of management of stock enhancement programmes should receive particular attention to ensure that the most is made of the investment and that the costs of future releases are minimised. These aspects are described briefly in the following.

6.4.3.1. Replenishment

Any perception on the part of fishermen that releases of juveniles will correct excess harvesting should be combated by sound capture regulations and rigorous enforcement. Otherwise, excessive harvests of adults will reduce the scope for natural replenishment and increase the need for, and cost of, producing cultured animals to increase the supply of juveniles. Well-managed stock enhancement programmes will leave sufficient spawning biomass to fulfill the potential for natural replenishment each year and use adaptive management to release only the number of cultured juveniles needed to obtain the desired annual yield. The enhancement of scallops in New Zealand is one of the few programmes currently operating this way (Chapter 3, Section 3.1.9).

Rotational fishing, as described in Chapter 3, Section 3.1.9 (see also Caddy and Defeo, 2003), is one measure that can be used to assist natural replenishment in some fisheries. It provides improved opportunity for reproduction when the number of areas fished sequentially exceeds the number of years it takes the species to reach maturity.

6.4.3.2. Integration with other uses of the resources and habitat

As more pressure is placed on inshore marine resources by increasing human populations, managers will be required to find equitable solutions to multiple demands. This will involve promoting beneficial synergies in some cases and resolving conflicts in others. Judicious use of wild juveniles can promote synergies between aquaculture and stock enhancement for some valuable species (Hair *et al.*, 2002). For example, proposals to use the settling (puerulus) larvae of spiny lobsters for aquaculture (Chapter 3, Section 3.6.2) has led to arrangements where a greater proportion of pueruli than would normally survive are to be returned to the wild at a larger size by the farmers, thus enhancing the stock.

The problems encountered by stock enhancement initiatives for scallops in France, where insufficient seabed was available because of other uses of the area (Chapter 3, Section 3.1.11), is an example of the conflicts that may be encountered in future stock enhancement programmes. Even when access rights are negotiated, there is no guarantee that other demands for use of the habitat will not emerge, as in the recent claims by mussel farmers for a substantial part of the area allocated to the Challenger Scallop Enhancement Company in New Zealand (Chapter 3, Section 3.1.11).

6.5. MEASURES TO OPTIMISE SOCIAL AND FINANCIAL BENEFITS OF RESTOCKING AND STOCK ENHANCEMENT

Important questions facing managers assessing the potential value of restocking and stock enhancement are: 'who will bear the cost?' and 'who will be permitted to benefit from the increased catch?' To date, much of the cost involved in developing methods for propagation and release of juveniles has been met by governments and non-governmental, non-profit organizations. Although governments are expected to take responsibility for common property resources, it is increasingly difficult to justify spending public money for new forms of management, like stock enhancement, that benefit users who make little or no contribution to the costs involved. Added to this is the realisation that, like other forms of management in many open access fisheries, there may be little incentive for fishermen to comply with the regulations required for successful stock enhancement, leading to the well-described 'tragedy of the commons' (Arbuckle and Metzger, 2000). Another point to consider is that successful publicly-funded stock enhancement can be expected to attract additional effort into an open access fishery (Hilborn, 1998; Lorenzen *et al.*, 2001; Leber, 2002, 2004). If this occurs, fishermen will be no better off than before and unable, or unwilling, to contribute to the cost of stock enhancement. In short, stock enhancement programmes are unlikely to be long-lived under open access conditions.

Stock enhancement is far more likely to be implemented successfully where access to the fishery is limited through catch, access or ownership rights (Lorenzen *et al.*, 2001; Caddy and Defeo, 2003). Then, there is a direct incentive to develop and comply with the new form of management because the participants are the direct beneficiaries. Strong government institutions, capable of selecting the most appropriate groups of users and overseeing compliance with the regulations, are required to set the stage for effective stock enhancement. In the absence of such institutional arrangements, there is no framework for transferring the existing technology to the private sector and paving the way for them to adopt, devise and implement sustainable stock enhancement programmes.

The development of successful enhanced scallop fisheries in northern Japan by fishing co-operatives, and in New Zealand by the Challenger Scallop Enhancement Company (Chapter 3, Section 3.1.9), are prime examples of how the major costs of stock enhancement can be borne by private enterprise when fishers have the rights to the increased harvests.

Caddy and Defeo (2003) present a variety of co-management and community-based scenarios for management of stock enhancement of marine invertebrates at small spatial scales in developing countries. These local, area-based management systems hold much promise, particularly where they have uniform regulations and collectively cover the distribution of the self-replenishing population supporting the fishery, because they allow the fishing effort of all stakeholders to be identified and managed appropriately.

Restocking is in a different category than stock enhancement in this regard. Programmes to release cultured juveniles to rebuild fisheries that are at such a low ebb that catches have not enticed fishermen to seek property rights will probably remain the province of governments. This situation may change if fishermen see sufficient successful examples of how restocking has been used to rebuild populations to convince them to bear the costs of restoration in exchange for property rights.

6.6. INDEPENDENT ASSESSMENTS

Because stock enhancement programmes have so many production and environmental standards to meet, industry, managers and scientists need a framework for co-operation on all the relevant issues. Managers also need to arrange regular independent assessments to verify that stock enhancement programmes are operating efficiently and in a responsible way. These assessments should cover such issues as whether: (1) the desired population age structure has been achieved to deliver consistent harvests; (2) the carrying capacity has been calculated adequately; (3) the value of the incremental harvests exceeds the costs of producing and releasing the juveniles; (4) the interests of the investors are being met and (5) the gains have not been made at the expense of decreases in abundance of other valuable species, damage to wild conspecifics through reduced genetic diversity or introduction of diseases to wild populations (see Chapter 7).

The criteria for assessing the effectiveness of restocking programmes differ from those for stock enhancement in the following ways. First, there can be less emphasis on the need to identify age structure or monitor the ability of the habitat to support the population because of the assumption that releases will fall well short of the carrying capacity of the habitat. The only caveat here is that releases should be made in a way that does not result in immediate post-release density-dependent mortality. Second, the cost of producing and releasing juveniles is of less concern, and the market value of the released animals (once they reach adult size) should not be expected to exceed the cost of producing them. Rather, the funds required for restocking should be viewed as an investment that will be returned once the replenished stock yields regular, substantial harvests. As mentioned in Chapter 6, Section 6.1, two provisos are that: (1) the recommended modelling shows that restocking will reduce the time required for restoration significantly compared with other forms of management; and (2) the value gained from early reopening of the fishery exceeds the cost of restocking.

Chapter 7

Other Important Considerations for All Initiatives

Although there are important differences between restocking and stock enhancement, several considerations apply equally to both types of interventions where they are deemed to be beneficial to the management of invertebrate fisheries. These considerations are the need to: measure whether the intervention has been as successful as expected; minimise the potential negative impacts of releasing cultured juveniles on the genetic integrity of the wild populations; reduce the risk of introducing diseases to the target species and broader animal community and avoid undesirable changes to the relative abundances of other valuable species in the ecosystem.

7.1. MEASURING SUCCESS

Where thorough analysis has indicated that restocking or stock enhancement should be beneficial, assessing whether releases of cultured juveniles have achieved their objective is essential to responsible and adaptive management

0065-2881/05 $35.00
DOI: 10.1016/S0065-2881(05)49007-0

(Blankenship and Leber, 1995; Munro and Bell, 1997; Leber, 1999). Indeed, how else will managers know whether their investments have been fully effective or whether methods need to be improved? The need to measure success is appreciated widely and, in the case of stock enhancement initiatives, a variety of indicators have been proposed to estimate the contribution of cultured juveniles. These indicators include: (1) the ratio of cultured juveniles to the estimated recruitment from the wild population; (2) the proportion of released animals in the commercial catch; (3) the survival rate of released animals at the size of harvest and (4) increases in total catch after enhancement (Munro and Bell, 1997 and references therein). In the past, however, such indicators have often been used to assess the success of releases on the assumption that the releases were necessary and that the greater the rate of survival or representation of tagged individuals, the greater the success of the intervention. From the arguments that we have put forward in Chapter 6, we suggest that the purposes of measuring success need to be re-evaluated. It is now clear that success should be measured at two separate stages. The first stage involves determining what proportion of released juveniles survives to the desired stage (to maturity for restocking and to minimum legal size for stock enhancement) so that the information can be used in the models needed to assess whether release of cultured juveniles will add value to other forms of management. The second stage applies only for those populations where modelling indicates that releases are likely to deliver additional benefits over and above other forms of management. In such cases, measures need to be taken to identify the contribution of: (1) released animals and their progeny to the target spawning biomass identified for restocking programmes and (2) released juveniles to strengthening year classes to provide optimum yields in stock enhancement initiatives.

Marking the cultured juveniles is the most powerful way of estimating these contributions (Blankenship and Leber, 1995). This is not always straightforward for marine invertebrates because both the tagging procedures and the extensive sampling programmes needed to retrieve tagged animals can be a significant cost during the research for, and application of, restocking and stock enhancement. Moreover, different approaches are needed for restocking and stock enhancement. Genetic tags are needed to measure the success of restoring severely depleted populations through hatchery releases by tracing the contribution of progeny derived from the released animals to the target spawning biomass. Additional tagging options are available for stock enhancement, because success is usually measured within the life span of the released animals.

In the following, we summarise the current range of tags available for marine invertebrates and the issues that need to be considered when selecting a tag for a restocking or stock enhancement programme. We also describe briefly an alternative approach for measuring the success of hatchery releases when it is not practical or cost-effective to tag large numbers of individuals.

7.1.1. Availability of tags for marine invertebrates

Tags that can be used to distinguish released marine invertebrates from wild conspecifics fall into four broad categories: physical, chemical, biological and electronic. Physical tags can be applied externally or internally; external physical tags include discs, darts, anchors and streamers that are glued to hard parts (e.g., the shell) or impaled through the integument and visible outside the animal's body. Internal physical tags are inserted (injected or sprayed) beneath the integument and include monofilaments, coded microwires, and fluorescent elastomers. Chemical tags are naturally occurring substances or introduced chemical components that include stable isotopes, stains, metals, rare earth elements and activatable tracers. Biological tags involve a wide variety of options and fall into three groups: genetic tags, conspicuous marks and distinctive parasites. Genetic tags vary from colour morphs and hybrids with phenotypic characteristics to DNA-based markers that assess parentage or subpopulation affinity. Conspicuous marks are alterations to the animal's appearance that arise from maintaining them in culture or altering their diet, or from clipping or branding body parts. Distinctive parasites can be used as tags where unique geographical host–parasite relationships and loads occur. Electronic tags are devices attached or implanted into animals that emit signals designed to be detected by a remote receiver without recapturing the animal.

Although most tags currently available have been developed for fish (Nielson, 1992), most of the techniques can also be applied to marine invertebrates. However, there are some particular problems involved with tagging invertebrates (e.g., moulting in crustaceans and tag shedding in echinoderms). The lack of structures such as scales and large otoliths also precludes the use of some popular batch staining techniques (e.g., tetracycline) with invertebrates. Conversely, some options are available for invertebrates but not fish (e.g., adhesive tags in shelled molluscs).

We have not attempted to review all the types of tags used for restocking and stock enhancement of marine invertebrates, although several examples have been highlighted throughout Chapters 2 and 3. Instead, we have provided a guide to the extensive literature on tagging marine invertebrates (Table 7.1) to indicate which techniques have been applied to date. This information allows a rapid assessment of the range of species that have been marked with a particular tag and the different types of tags that have been used for major taxonomic groups or species.

Table 7.1 A summary of information available on different methods for tagging a wide range of marine invertebrates.

Tag type	Taxa	Species	Reference
Reviews			
General			Emery and Wydoski (1987)
			Nielson (1992)
Genetic tags			Keenan and MacDonald (1988)
Parasites as tags			Kabata (1963)
Taxonomic			
	Crustacea		Farmer (1981)
	Shrimp		Neal (1969); Ishioka (1981); Glaister (1988); Rothlisberg (1998); Rothlisberg and Preston (1992)
	Abalone		McShane (1988)
	Lobsters	*Homarus americanus*	Haakonsen and Anoruo (1994)
		H. gammarus	Linnane and Mercer (1998)
	Spiny lobster	*Panulirus ornatus*	Trendall (1988)
	Marine larvae		Levin (1990)
Targeted studies (by tag type, taxa, species)			
Physical (external)			
Plastic (glued)	Topshell	*Trochus niloticus*	Castell *et al.* (1996); Crowe *et al.* (1997)
Aluminium (glued)	Abalone	*Haliotis laevigata*	C. Dixon, personal communication, 2004
Petersen disc	Topshell	*Trochus niloticus*	Crowe *et al.* (2001)
Dart, anchor or Tbar	Queen conch	*Strombus gigas*	Glazer *et al.* (1997b)
	Shrimp	*Penaeus setiferus*	Marullo *et al.* (1976)
	Spiny lobsters	*Panulirus homarus*	Mohamed and George (1968)
		P. cygnus	Melville-Smith and Chubb (1997)
	Lobster	*Homarus americanus*	Krouse and Nutting (1990)
	Shrimp	*Penaeus aztecus*	Neal (1969)
		P. setiferus	
		P. duorarum	
		P. setiferus	Bearden and McKenzie (1972)

Table 7.1   (Continued)

Tag type	Taxa	Species	Reference
(Sphyrion)	Lobsters	*Homarus americanus*	Scarratt and Elson (1965); Scarratt (1970); Ennis (1972; 1986); Moriyasu *et al.* (1995)
(Spaghetti)	Queen conch	*Strombus gigas*	Appeldoorn and Ballentine (1983); Stoner and Davis (1997)
Vinyl streamer (Floy tag)	Shrimp	*Penaeus duorarum* *P. aztecus* *P. setiferus* *P. monodon*	Marullo *et al.* (1976) Su *et al.* (1990); Primavera and Caballero (1992)
	Spiny lobsters	*Panulirus cygnus*	Melville-Smith and Chubb (1997)
Vinyl streamer (glued)	Queen conch	*Strombus gigas*	Appeldoorn and Ballentine (1983); Stoner and Davis (1997)
Loop (Atkins tag)	Shrimp	*Penaeus indicus* *P. semisulcatus*	Maheswarudu *et al.* (1998)
Carapace	Crab	*Callinectes sapidus*	Fisschler and Walburg (1962)
Eyestalk rings	Shrimp	*Penaeus monodon*	Su & Liao (1999)
Physical (internal)			
Monofilament	Shrimp	*Penaeus monodon*	Teboul (1993)
PVC+stain	Shrimp	*Penaeus* spp.	Neal (1969)
Coded microwire	Lobsters	*Homarus americanus*	Krouse and Nutting (1990); Cowan (1999); James-Pirri and Cobb (1999)
		H. gammarus	Wickins *et al.* (1986) Uglem and Grimsen (1995)
	Spiny lobsters	*Panulirus argus*	Sharp *et al.* (2000)
	Shrimp	*Penaeus monodon*	Su and Liao (1999)
	Crab	*Callinectes sapidus*	van Montfrans *et al.* (1986)
(Bergman-Jefferts)	Shrimp	*Pandalus platyceros*	West and Chew (1968); Prentice and Rensel (1977)

(Continued)

Table 7.1   (Continued)

Tag type	Taxa	Species	Reference
Fluorescent elastomer (injected)	Shrimp	*Penaeus aztecus* *P. setiferus* *P. duorarum*	Klima (1965)
		P. vannamei	Godin *et al.* (1996)
	Lobsters	*Homarus gammarus*	Uglem *et al.* (1996)
Fluorescent elastomer (sprayed)	Shrimp	*Penaeus aztecus* *P. setiferus*	Benton and Lightner (1972)
Chemical			
Stable carbon isotope	Shrimp	*Penaeus aztecus*	Fry (1981); Fry *et al.* (1999)
		P. monodon	Preston *et al.* (1996)
Gold bit tags		*Penaeus japonicus, Metapenaeus ensis*	Ariyama *et al.* (1994)
Staining	Clam	*Ruditapes philippinarum*	Koshikawa *et al.* (1997)
	Shrimp	*Penaeus aztecus* *P. setiferus* *P. duorarum*	Klima (1965)
Rare earth elements	Marine larvae		Levin (1993)
Activatable tracers (Europium)	Shrimp	*Penaeus japonicus*	Ishioka (1981)
Biological – genetic			
Phenotypic (colour morphs)	Lobsters	*Homarus americanus*	Irvine *et al.* (1991); Wahle (1991); Wahle and Inze (1997)
(claw shape)		*H. gammarus*	van der Meeren (1993)
Genotypic (allozymes)	Scallops	*Argopecten irradians*	Krause (1992); Tettelbach *et al.* (1997)
	Giant clam	*Tridacna gigas* *T. derasa* *T. maxima*	Benzie (1993a,b)
Mitochondrial DNA	Shrimp	*Penaeus monodon*	Benzie *et al.* (1993)
	Scallops	*Argopecten irradians*	Seyoum *et al.* (2003)
Microsatellites	Abalone	*Haliotis rufescens* *H. rubra* *H. asinina*	Kirby *et al.* (1998) Evans *et al.* (2000) Selvamani *et al.* (2001)
	Shrimp	*Penaeus esculentus*	Meadows *et al.* (2003); Lehnert *et al.* (2003) Bravington and Ward (2004)
		P. monodon	Brooker *et al.* (2000)

Table 7.1   (Continued)

Tag type	Taxa	Species	Reference
Amplified length polymorphisms (AFLPs)	Shrimp	*Penaeus japonicus*	Moore *et al.* (1999)
Cross-breeding Hybrids	Hard clam	*Mercenaria mercenaria*	Manzi (1990)
Biological–marks			
Tail clipping	Shrimp	*Penaeus japonicus*	Miyajima and Toyota (2002)
Eyestalk ablation	Shrimp	*Penaeus japonicus*	Tanaka and Nakamura (1971)
Epidermal implants	Spiny lobsters	*Panulirus cygnus*	Melville-Smith *et al.* (1997)
	Lobsters	*Nephrops norvegicus*	Shelton and Chapman (1995)
Heat/cold branding	Crabs	*Birgus latro*	Fletcher *et al.* (1989)
Feeding	Abalone	*Haliotis rufescens*	Leighton (1961)
		H. discus hannai	Sakai (1962)
		H. rufescens	Olsen (1968a)
		H. corrugata	Olsen (1968b)
		H. sorenseni	
		H. assimilis	
Stress bands	Scallops	*Pecten maximus*	Dao *et al.* (1999)
Growth checks	Softshell clam	*Mya arenaria*	Beal (1993)
Biological–parasites			
	Shrimp	*Penaeus merguiensis*	Owens (1983, 1985)
		P. semisulcatus	Mathews *et al.* (1988)
Electronic			
Electromagnetic	Spiny lobsters	*Panulirus cygnus*	Phillips *et al.* (1984); Jernakoff (1987, 1988)
		Jasus novaehollandiae	Ramm (1980)
	Crab	*Cancer pagurus*	Smith *et al.* (2000)
	Lobsters	*Homarus gammarus*	van der Meeren (1997) Smith *et al.* (2000)
Acoustic	Lobsters	*Nephrops norvegicus*	Newland and Chapman (1993)
	Spiny lobsters	*Jasus edwardsii*	Mills *et al.* (2003)
	Crab	*Cancer magister*	Prentice (1990)

7.1.2. Factors to consider when selecting a tag

Although the information provided in Table 7.1 provides a good guide to the types of tags that may be suitable for a particular species, a number of factors need to be considered carefully when selecting a tag for research on restocking or stock enhancement or for any subsequent large-scale release programmes. These include: the objective of the study; the age (or size) of the individuals to be marked; the numbers of animals to be tagged; the ease of marking and/or likelihood of recovering the tag (including any propensity for the tag to increase mortality); and the costs and public/consumer acceptance. No tag will satisfy all demands but, with care, costs and benefits can usually be optimized to suit the objectives of the study. Table 7.2 illustrates the comparative advantages and disadvantages of different types of tags for shrimp.

7.1.2.1. Restocking programmes

In general, there are fewer factors to be considered for restocking than for stock enhancement programmes because, as described previously, a genetic

Table 7.2 Tag type × attribute matrix example for penaeid shrimp.

		Tag type			
			Genetic		
Attributes	Physical	Genotypic	Phenotypic	Altered	Chemical
Mark early stages	−	+	−	+	+
Detect all life history stages	−	+	−	+	+
Local population	+	−	−	+	+
Multiple marking	+	−	−	+	+
Cheap to mark	−	+	+	+	+
Cheap to detect	+	−	+	+	+
Rapid marking	−	+	+	+	+
Rapid detection	+	−	+	+	+
Transmitted	−	±	±	±	−
Benign	−	0	0	+	+
Public acceptance	±	+	+	±	±
Score (+)	4–5	5–6	5–6	8–11	9–10

+, Positive attribute; −, negative attribute; and 0, neutral attribute; ±, either positive or negative attribute depending on circumstances (after Rothlisberg and Preston, 1992).

tag is needed to assess the contribution to the target spawning biomass of F_1, F_2, F_3 etc. generations derived from released animals. For restocking programmes, the most important issues are the need to select genetic tags that are cost effective and non-deleterious. The latter issue is essential; by definition, successful restocking programmes will propagate the genetic tag so that it is represented in most individuals in the population. Thus, there is potential to harm the recreated stock, and perhaps consumers, if the genetic tag has undesirable attributes. The onus is on managers to ensure that the changes to the genetic makeup of the target population that occur from the use of a genetic tag are benign.

One way of developing genetic tags is to use rare mtDNA or allozyme alleles produced by selecting broodstock with rare genotypes. Then, an increased frequency of such markers in the population after release of cultured juveniles indicates that restocking has made a contribution, whereas no change in frequency denotes failure. This approach has been used for a limited number of fish and invertebrates to date, for example in cod *Gadus morhua* (Jørstad *et al.*, 1994), salmonids (Crozier and Moffett, 1995; Hansen *et al.*, 1995), the bay scallop *Argopecten irradians* (Seyoum *et al.*, 2003) and abalone (Gaffney *et al.*, 1996). None of the studies has reported any deleterious effects to the recipient populations.

Discovery of hypervariable markers has provided an alternative approach that considerably reduces the risk of changes to the genetic structure of the population, provided the hatchery protocols described in Section 7.2.5 are applied. With this method, broodstock are taken directly and randomly from the wild, and the hypervariability of genetic markers is used to determine the stocking success of their progeny. Microsatellites (see Section 7.2.1.4) provide an abundant supply of hypervariable genetic markers, but development of a reliable suite of microsatellite markers for a new species can take time. Companies specialising in such work can be contracted to do this. Bravington and Ward (2004) describe a statistical procedure for assessing the proportion of released progeny in subsequent recaptures, given that the genotypes of both (or only one) of the hatchery parents are known. Multiple single pair matings can be used in the hatchery if enough hypervariable markers are available.

The ultimate need for a genetic tag to measure the success of restocking does not preclude the use of other types of tags during the preliminary research needed to generate data for modelling the cost–benefit of releasing cultured juveniles to help restore spawning biomass. At this stage of the process, a wide variety of tags may be suitable for obtaining robust estimates of survival to maturity. Such tagging studies will be very similar to those needed for the first stage of measuring the success of releasing animals for stock enhancement.

7.1.2.2. Stock enhancement programmes

Research for stock enhancement, and for stock enhancement programmes themselves, is likely to provide opportunities to use a broad range of the tags listed in Table 7.1. We would expect, however, that the diversity of tags used will be greater during the first of the two stages at which success should be measured (i.e., determining what proportion of released individuals survive to the predetermined target harvest size) (see Section 7.1). During such research, relatively expensive tagging systems can be used, because the short-term aims should be to acquire reliable information on survival, compare different release strategies etc., rather than to mark individuals *en masse* at low cost. In some cases, it may be necessary to identify animals individually (e.g., when growth rates need to be estimated). On the other hand, measuring success of mass releases during stock enhancement programmes themselves, which should only be launched once the modelling described in Chapter 6 has indicated that release of cultured juveniles will be cost-effective, should be possible with a subset of simpler batch-marking methods. During such releases, it is only necessary to distinguish between cultured individuals and wild ones, and between different year classes of cultured animals.

7.1.3. Alternative approaches

For some restocking or stock enhancement programmes, it may not be possible to find a suitable tag for the target species, or the scale of releases may preclude marking most individuals. This is currently the case for sea cucumbers (Purcell, 2004b). Where this occurs, the sampling designs developed by Underwood (1992, 1995) for detecting changes to abundance of a species after manipulations of the environment can be applied to assess the impact of hatchery-reared juveniles on abundance of individuals in the fishery (see also Hilborn, 2004).

Application of Underwood's sampling designs in the context of restocking and stock enhancement programmes depends on estimating the abundance and size structure of the self-recruiting population(s) to be restocked or enhanced with that of several 'control' populations (i.e., those of a similar fished status) on several occasions before and after release of juveniles. Significantly greater rates of increase in abundance for the target population(s) than for control populations can then be attributed to releases of hatchery-reared juveniles, rather than to other processes influencing replenishment (Figure 7.1). In the case of sedentary species with very limited larval dispersal that have been over-fished to the point where animals are scarce or absent throughout the distribution of the self-recruiting population, this

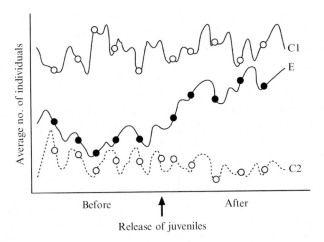

Figure 7.1 Sampling needed to detect impact of hatchery releases for a restocked or enhanced population (E) compared with two unenhanced (control) populations (C1 and C2). In this example, samples were taken on five occasions before and five occasions after release of juveniles (redrawn after Underwood, 1995).

approach could be applied at the scale of sites within this distribution. In such cases, however, control sites would eventually be expected to receive recruits from sites where released individuals were established, and so there is a limit to the duration over which this approach could be applied. The 'Before vs. After, Control vs. Impact' (BACI) sampling approach would also be suitable for measuring success of other forms of enhancement (e.g., provision of additional settlement habitat at the scale of sites within the distribution of a population).

In the absence of information from tagging studies, or the BACI sampling designs described here, correlations between hatchery releases, or other forms of stock enhancement, and increased population size cannot be attributed unequivocally to these interventions. Rather, they could be due to other coincidental events (e.g., recruitment of an unusually strong year class derived from the remnant wild individuals).

7.2. GENETIC CONSIDERATIONS

7.2.1. Measuring genetic population structure

It is evident from this review, and increasingly widely appreciated, that a good understanding of spatial population structure is needed as the basis for management of marine invertebrate fisheries. In particular, fisheries managers need to know the number and size of the largely self-recruiting populations in a fishery before they can identify which ones may be in need of restocking or stock enhancement (Chapter 6, Section 6.1, see also Carvalho and Hauser, 1995; Caddy and Defeo, 2003). Fortunately, a variety of population genetics methods is now available to help identify the basic population structures for marine invertebrates, although much variation in the reliability of the information can be expected among species, and groups of species, depending on such factors as larval duration, larval behaviour, habitats where gametes are typically released, local topography and current patterns. In this section, we describe and compare the methods currently used to estimate genetic affinities among populations of the same species. These methods involve either working directly with DNA or with proteins that reflect variation in DNA.

The oldest of these methods involves enzyme-specific histochemical staining after gel electrophoresis of proteins to separate allozymes or isozymes. It dates back 40 years and has been used extensively to examine genetic variation in a wide range of marine fish and invertebrate species. The molecular DNA techniques used in population genetics started with mitochondrial DNA (mtDNA) analysis using restriction fragment length polymorphisms (RFLP). More recently, considerable attention has focused on the use of the polymerase chain reaction (PCR) for amplification of specific DNA fragments. The amplified fragment can then be examined by DNA sequencing, by restriction fragment analysis or by size assessment. There is a bewildering array of molecular analytic tools now available, with acronyms such as EPIC (exon-primed intron-crossing PCR), RAPD (randomly amplified polymorphic DNA), AFLP (amplified fragment length polymorphism of DNA), VNTR (variable number of tandem repeats) and SNP (single nucleotide polymorphisms). The widely used microsatellite analysis is a VNTR application. We have chosen to concentrate on allozyme, mtDNA, PCR and microsatellite approaches because, together, they have been used in >90% of studies dealing with genetic issues associated with restocking and stock enhancement. Because an understanding of population genetics is critical to restocking and stock enhancement and is a highly technical field, we have provided more details than elsewhere in this review. Even so, the coverage we provide is limited, and further information on methods, principles and results of DNA analyses can be found in Avise (1994), Hillis *et al.* (1996), Caetano-Anollés and Gresshoff (1997), Hoelzel (1998), Ferguson and Danzmann (1998) and Goldstein and Schlötterer (1999).

7.2.1.1. *Allozyme analysis*

The mechanics of allozyme analysis are quite simple (Richardson *et al.*, 1986; Manchenko, 1994; Hillis *et al.*, 1996). A relatively crude tissue extract is prepared and placed on a starch or acrylamide gel or on cellulose acetate strips. Electricity is passed through the gel for a fixed period at a controlled level. Histochemical stains specific for particular enzymes are then applied. Some enzymes (isozymes) are encoded by multiple loci, so several discrete zones may be apparent after staining. Allozymes are the products of genetic variation at an enzyme-encoding locus, whereas isozymes are alternative forms of a protein encoded at different loci.

Although the products of most enzyme loci in most species are invariant (monomorphic) and levels of electrophoretically detectable variation vary widely among species (Nevo *et al.*, 1984; Ward *et al.*, 1992), on average ~10% of loci in a typical species (Table 7.3) will be heterozygous and ~30% of loci will be polymorphic. This results in detectable variation in banding patterns for the loci in question, allowing genotypes to be assigned, and provides a solid basis for estimating genetic affinities among individuals. For a number of reasons, however, such as the fact that not all amino acid substations are detectable electrophoretically, the true levels of DNA variation among individuals will be higher than detected by this method.

The relative simplicity and low cost of allozyme methods are balanced by some disadvantages, however. Chief among these is limited resolving power, with even the polymorphic enzymes rarely having more than three or four relatively common alleles. In addition, fresh or frozen tissue needs to be available, and because many enzymes are tissue-specific, different tissues might need to be accessed. Also, the debate over whether the variation revealed by allozyme electrophoresis is biologically significant (the selection

Table 7.3 Comparison of allozyme diversity levels among different classes of vertebrates and invertebrates.

Group	Species	H_T	H_S	F_{ST}
Insects	46	0.138 ± 0.009	0.122 ± 0.009	0.097 ± 0.020
Crustaceans	19	0.088 ± 0.016	0.063 ± 0.011	0.169 ± 0.061
Molluscs	44	0.157 ± 0.020	0.121 ± 0.019	0.263 ± 0.036
Amphibia	49	0.146 ± 0.012	0.097 ± 0.008	0.308 ± 0.033
Birds	28	0.059 ± 0.006	0.054 ± 0.006	0.078 ± 0.021
Mammals	83	0.077 ± 0.005	0.057 ± 0.004	0.207 ± 0.023
Reptiles	33	0.115 ± 0.014	0.086 ± 0.010	0.222 ± 0.036
Fish	113	0.062 ± 0.004	0.053 ± 0.003	0.132 ± 0.016
Marine	57	0.064 ± 0.004	0.059 ± 0.004	0.062 ± 0.011
Anadromous	7	0.057 ± 0.007	0.052 ± 0.008	0.108 ± 0.044
Freshwater	49	0.062 ± 0.007	0.046 ± 0.005	0.222 ± 0.031

H_T (total heterozygosity), H_S (average subpopulation or sample heterozygosity) and F_{ST} (proportion of total genetic diversity attributable to sample differentiation, see Wright, 1943 and Nei, 1973 for equations to derive F_{ST}). Invertebrate data from Ward *et al.* (1992), fish data from Ward *et al.* (1994), data from other vertebrates are updates of information in Ward *et al.* (1992). Requirements for inclusion were that 15 or more loci had been examined in 15 or more individuals per sample or subpopulation.

hypothesis) or not (the neutral hypothesis) has not been resolved (Lewontin, 1974; Avise, 1994).

7.2.1.2. Mitochondrial DNA analysis

The smaller size of mtDNA relative to nuclear DNA, and the fact that there can sometimes be several hundred copies of the mtDNA in one cell, have made it relatively easy to extract and purify. Another advantage of mtDNA is that it is haploid and almost always only inherited from the maternal parent, whereas nuclear DNA is diploid and inherited from both parents. This gives mtDNA a genetically effective population size of 25% of that of nuclear DNA, other factors being equal.

The mtDNA genome is composed of ~37 genes comprising 22 tRNAs, 2 rRNAs and 13 mRNAs encoding for proteins. In addition, mtDNA has a non-coding section usually known as the control region (or d-loop) that is the initiation site for mtDNA replication and transcription. Different regions of mtDNA evolve at different rates, depending on the degree of functional constraint; the control region has the highest rate of evolution and is typically the most variable region. Overall, the rate of evolution of mtDNA is about 5–10 times that of single copy nuclear DNA (Brown *et al.*, 1979; Wilson *et al.*, 1985).

The analysis of mtDNA variation often relies on restriction enzyme deployment. Restriction enzymes cleave DNA sequences at particular oligo-nucleotide sequences, usually of four, five or six bases. The sequence recognition and cutting is highly specific, and several hundred such enzymes are now available with different sequence specificities. RFLP analysis of mtDNA involves extracting the mtDNA, cutting it with one or more restriction enzymes and separating the fragments by gel electrophoresis. The position of the fragments can be visualised on the gel by radioactive labelling or by staining with ethidium bromide followed by ultra-violet visualisation. The different mtDNA patterns are termed haplotypes (e.g., five individuals might have haplotype A and 10 might have haplotype B). MtDNA is a non-recombining molecule, and generally there is only one type throughout the cells of an individual, although some instances of variation within individuals (heteroplasmy) have been recorded.

7.2.1.3. Polymerase chain reaction

PCR analysis permits the rapid and repeated amplification of one or more specific regions of DNA. To do this, DNA primers are required that are complementary to the flanking sequences of the targeted region. Usually, the primers are about 20 base pairs (bp) long, and so-called universal primers are available that target conserved regions of DNA from most animal species and allow the amplification of potentially variable regions between the primer sequences (Kessing *et al.*, 1989). These are especially useful in mtDNA studies. Typically, the amplified region is some hundreds of base pairs long, although regions of several thousand base pairs can now be amplified successfully. The amplified fragment can be examined for size variation, cut with restriction enzymes and the resulting fragments assessed for size variation, or the nucleotides can be sequenced.

The PCR approach is very powerful and flexible and, once primers have been developed, facilitates examination of any mtDNA or nuclear DNA region. Amplification also enables examination of very small amounts of tissue, permitting non-invasive and non-lethal sampling. Samples can be stored at room temperature in alcohol or other chemicals such as dimethyl sulfoxide, simplifying logistic and storage processes. Even old and dried museum specimens, including historic collections of fish scales (Nielsen *et al.*, 1999), can yield enough DNA for analysis.

7.2.1.4. Microsatellite analysis

Microsatellites are found in nuclear DNA. They are repeated sequences of DNA in which the repeated sequence is short (2–6 bp). For example, the sequence ACACACACAC would be a microsatellite region, with the dinucleotide repeat AC repeated five times. The total length of the microsatellite is usually less than 300 bp (Tautz, 1989) and is flanked by unique (non-repeat) sequences. These unique sequences are generally conserved within a species, making it possible to design primers to the flanking regions that permit the PCR amplification of the microsatellite. The microsatellite can then be sized by gel electrophoresis. Variation in the number of repeats is reflected in fragment size variation. At a particular microsatellite locus, a heterozygous individual might have, for example, eight repeats on one chromosome and 12 on the homologous chromosome. Mutation rates are high for microsatellites, estimated at $\sim$0.05–0.2% (Huang *et al.*, 1992; Kwiatkowski *et al.*, 1992), and heterozygosity is therefore high. Dozens of alleles per microsatellite locus are not uncommon. Allele data are most readily collected using primers with fluorescent labels and sized using automated DNA sequencers.

Generally, microsatellite primers tend to be specific to a particular species, or species group, because mutations that accumulate in the flanking region over time reduce the binding ability of the primers, so new primers might have to be developed. Non-amplifying or 'null' alleles are not uncommon. These can arise from mutations in the flanking region so that one of the primers does not bind and the fragment is not amplified, or from PCR processes not amplifying a long allele as readily as a short allele. Visible microsatellite (and allozyme) alleles are expressed in a co-dominant fashion, whereas null alleles are recessive alleles. In a somewhat analogous manner, the A and B alleles of the human ABO blood group system are co-dominant alleles, whereas the O allele is a recessive allele. Microsatellites are generally considered to be selectively neutral markers, evolving primarily by mutation and genetic drift.

7.2.1.5. *Merits of the different methods*

Given the range of methods for estimating the frequency of genotypes summarised earlier, and the considerable expense of several of these methods, it is important that careful thought is put into which technique to use when designing research to identify the population structure of marine invertebrate species supporting fisheries. The advantages and disadvantages of allozymes, mtDNA and microsatellites are summarised in Table 7.4.

Differentiation of populations is promoted by genetic drift and opposed by gene flow. Because genetic drift is more marked in small populations, the smaller effective population size of mtDNA than nuclear DNA makes data from mtDNA analysis a potentially more sensitive indicator of genetic affinities than data derived from nuclear DNA. However, mtDNA is a non-recombining molecule, meaning that even if several variable sites are examined in the mtDNA genome, these sites are not independent characters. Thus, mtDNA is best treated as a single character, whereas nuclear DNA assessment can be based on many independent characters or loci.

This ability to examine many independent loci, be they allozymes or microsatellites, is an important advantage of nuclear DNA and compensates for the slower rate of evolution of many nuclear DNA genes (especially allozyme genes) and for the larger effective population size of nuclear DNA markers. Another factor in favour of nuclear DNA is that random genetic drift can vary at different loci, and testing multiple independent loci allows this effect to average out. Among the DNA markers, microsatellites evolve faster than allozymes and, therefore, might be expected to provide more sensitive measures of weak differentiation among populations. Furthermore, microsatellites are thought to be neutral markers and, therefore, are not affected by selective forces that might confound interpretations of differentiation in allozymes. Certainly, allozyme analysis seems to give less resolution of population structure than mtDNA or microsatellite analysis (although the reverse is sometimes observed, see Ward and Grewe, 1994). Several instances of this are given in Section 7.2.2; another comes from the New Zealand greenshell mussel *Perna canaliculus*, where allozyme studies

Table 7.4 The major population genetic tools currently available, with some general comments and rankings of their properties. For example, allozymes are quick and inexpensive. The comments are necessarily partly subjective and sometimes over-simplified. For example, the costs of microsatellite analysis are significantly less when primer sequences are already available than when they have to be developed *de novo*. See Section 7.2.1 for definition of acronyms (after Ward, 2002).

Tool	Basis	PCR-based	Speed	Expense	Genetics	Tissue required	Number of markers	Variability of markers	Coding/ noncoding	Comments
Allozymes	Electrophoretically detectable protein variation	No	*****	*	Co-dominant	Fresh, frozen	***	**	Coding	Still an excellent tool if suitable tissue available
mt DNA-RFLP	Fragment size variation after restriction enzyme digestion	No	**	***	Haploid, maternally inherited	Fresh, frozen, preserved in ethanol	*	**	Mostly coding	Effectively a single marker
mt DNA-PCR-RFLP	As DNA-RFLP, but using PCR products	Yes	***	***	Haploid, maternally inherited	Fresh, frozen, preserved in ethanol	*	**	Mostly coding	Effectively a single marker
mt DNA-sequencing	Nucleotide sequence variation	Usually	***	****	Haploid, maternally inherited	Fresh, frozen, preserved in ethanol	**	**	Mostly coding	Effectively a single marker
Microsatellite	Variation in number of tandem repeats of di-tetranucleotide motifs	Yes	***	***	Co-dominant (some recessive null alleles)	Fresh, frozen, preserved in ethanol	*****	*****	Non-coding	Primer development required; scoring problems can occur

Rankings are: *, lowest; *****, highest.

indicated an absence of genetic structuring, whereas mtDNA and RAPD analyses showed a pronounced north–south split (Apte and Gardner, 2002; Star *et al.*, 2003).

Clearly, caution is needed before inferring population structure and levels of gene flow from analyses of a single marker, or even a single class of marker. Different markers have different mutation rates and vary in their likely sensitivity to natural selection and in the speed at which they react to changes in population size and migration rate. It will always be better to combine approaches than rely on a single method alone. However, because nuclear DNA markers with their larger effective population size take longer to reach demographic equilibrium than mtDNA markers (Neigel, 1994), the former may be better for revealing historical patterns of genetic population structure and the latter for assessing present-day patterns of gene flow. This was suggested by studies on the shrimp *Alpheus lottini* and the starfish *Linckia laevigata* in the Indo-Pacific region (Williams *et al.*, 2002). To date, most of our knowledge of the population structure of marine invertebrates is based on either allozyme or mtDNA analysis, although microsatellites are now beginning to be used widely.

7.2.1.6. Methods for analyzing the data

The data generated from allozymes, mtDNA and microsatellite analyses can be used to estimate genotype frequencies for the populations sampled and to identify the genetic structure of species across their distributions in a number of ways. Perhaps the best known of these is the Hardy-Weinberg expectation that genotype frequencies accord with those that should occur within a randomly mating population. Once allele or haplotype frequencies have been estimated in each sample, the extent of genetic differentiation among subpopulations (F_{ST}, where S denotes subpopulation and T the total population) can be quantified (Wright, 1943) as:

$$F_{ST} = V_P / (_P(1-_P)) \tag{1}$$

where V_P is the variance of the allele frequencies among subpopulations and $_P$ is the mean allele frequency of one of the alleles across subpopulations. F_{ST} was initially developed to deal with a simple two-allele polymorphism but later was extended to deal with multiple alleles and termed G_{ST} by Nei (1973):

$$G_{ST} = (H_T - H_S)/H_T = 1 - (H_S/H_T) \tag{2}$$

where H_T is the total heterozygosity (genetic diversity for mtDNA) across the pooled subpopulations and H_S is the average heterozygosity (genetic diversity) across subpopulations.

In practice, F_{ST} and G_{ST} are equivalent and are usually estimated, together with standard errors, by software packages such as Arlequin (Schneider et al., 2000) or GenePop (Raymond and Rousset, 2000). An F_{ST} value of, say, 0.02, means that 2% of the detected variation arises from inter-subpopulation differentiation, whereas 98% of the variation is found within subpopulations. Two populations fixed or homozygous for different alleles at a locus will have the maximum achievable F_{ST} value of 1.0 for that locus.

Hierarchical analyses of genetic diversity may also be carried out at different population levels, for example, by assigning all subpopulations in a particular region to that region and estimating levels of divergence not just across all subpopulations but also among regions.

Other approaches are now in use as well (e.g., computer-based coalescent-based models that take into account the genealogy of alleles or haplotypes [Beerli and Felsenstein, 1999, 2001; Bahlo and Griffiths, 2000; Nielsen and Wakeley, 2001] and techniques such as nested clade analysis based on evolutionary relationships among alleles or haplotypes and their spatial frequencies [Templeton, 1998]).

To identify self-replenishing populations or functional fishery management units, a current recommended genetic approach, taking into account time and cost factors, would be to examine ~100 individuals from each putative subpopulation for sequence variability of a known variable mtDNA region and for genetic variability at ~10 microsatellite loci. Clearly, the more individuals and loci examined, the more power any subsequent statistical analysis will have. Statistical analysis should include standard exact methods for assessing levels of population differentiation using an F_{ST} or allied approach, and incorporation of a nested clade analysis might well strengthen any subsequent conclusions.

7.2.2. Patterns of genetic population structure in marine invertebrates

A general trend in the animal kingdom is that population differentiation is related to dispersal. For example, there are much lower levels of population differentiation in marine fish than in freshwater fish (see F_{ST} values in Table 7.3). The low average F_{ST} value of marine fish is assumed to reflect a high degree of genetic connectivity of populations, presumably resulting from widespread dispersal of larvae, recruits, juveniles and/or adults. Similarly, insects and birds have average F_{ST} values that are much lower than for other land-based taxa (Table 7.3).

There is, however, increasing evidence that limited genetic differentiation is by no means typical of all marine species. On the contrary, recognition of multiple self-replenishing populations is increasing for marine species, especially those associated with islands or reefs (Jones *et al.*, 1999, Swearer *et al.*, 2002). Indeed, it is now evident that marine larvae should not be considered as passive particles that are displaced easily. Instead, larvae can exhibit vertical migrations that, when coupled with vertically stratified currents, help to retain them near shore. The likely prevalence of self-recruitment among many marine populations is also suggested by dispersion modelling. Simulations show that dilution and mortality effects are such that there simply are not enough larvae produced to sustain recruitment of 'downstream' distant populations (Cowen *et al.*, 2000).

There is great variation in life-history strategies among inshore marine species, including demersal eggs and larvae that might be expected to promote retention of juveniles in the natal area. Indeed, the expected negative correlation between dispersal ability and F_{ST} was demonstrated for 10 marine shorefishes and 7 coral reef fishes by Waples (1987) and Doherty *et al.* (1995), respectively. In short, the traditional simple picture of panmictic marine populations with widespread gene flow is now being challenged increasingly.

The variation in population differentiation described for marine fish is also reported for marine invertebrates. There are examples of genetic uniformity over long distances, populations isolated by distance, chaotic patchiness and abrupt genetic differentiation among populations.

A good example of extensive gene flow across a wide geographical range is described by Lessios *et al.* (1998). They studied allozyme and mtDNA variation in populations of the echinoid *Echinothrix diadema* separated by more than 5400 km of uninterrupted deep water and found clear evidence of massive recent gene flow across this region.

'Chaotic patchiness' has been recorded for some invertebrates with pelagic larvae and occurs where there is significant but ephemeral differentiation among local populations, with distant populations often no more differentiated than local populations. Examples include allozyme studies of limpets *Siphonaria* sp. (Johnson and Black, 1982, 1984), urchins *Echinometra mathaei* (Watts *et al.*, 1990) and *Strongylocentrotus franciscanus* (Moberg and Burton, 2000), crabs *Callinectes sapidus* (Kordos and Burton, 1993) and barnacles *Balanus glandula* (Hedgecock, 1994a). Genetic heterogeneity on such micro-geographical scales presumably reflects temporal variation in the genetic composition of recruits. This temporal variation itself may reflect recruitment from different source populations at different times, depending on current patterns (Kordos and Burton, 1993), selection on early life stages (Johnson and Black, 1984) or the possibility that only a few adults are responsible for much of the recruitment in any one season and location (the 'sweepstakes' hypothesis, Hedgecock, 1994b). In such situations, differences in genotype frequencies cannot be attributed to the existence of isolated self-replenishing populations.

Abrupt differentiation of populations in a widespread species is illustrated by the American oyster *Crassostrea virginica*. There is a pronounced genetic break between the Atlantic coast and the Gulf of Mexico, represented by allele frequency shifts of up to 50–75% for mtDNA (Reeb and Avise, 1990) and nuclear DNA loci (Hare *et al.*, 1996, Hare and Avise, 1998). However, allozyme loci show approximate allelic uniformity (Buroker, 1983) as do some other nuclear loci (McDonald *et al.*, 1996). The null hypothesis of population homogeneity might have been accepted (or at least not rejected) from one molecular approach but rejected from another. A history of vicariant separation between Atlantic and Gulf populations seems likely, but the data also suggest pronounced contemporary barriers to gene flow along the east coast of Florida (Hare and Avise, 1998). The allozyme uniformity might arise from stabilising selection across all populations, the other markers being selectively neutral and showing the effect of gene flow restrictions. Whatever the explanation, pronounced spatial differentiation exists despite the potential for long-distance dispersal by planktonic oyster larvae. Some other coastal species, including the horseshoe crab *Limulus polyphemus*, show a similar genetic break in this region (Saunders *et al.*, 1986).

Genetic breaks in marine invertebrates have also been recorded from the northern part of Australia. Examples include mtDNA studies of the crab *Scylla serrata* (Gopurenko and Hughes, 2002), and microsatellite studies of the shrimps *Penaeus monodon* (Brooker *et al.*, 2000) and *P. esculentus* (unpublished observations). The population differentiation of *P. esculentus* was not evident from an earlier allozyme study (Mulley and Latter, 1981). These genetic breaks may have arisen as a consequence of sea level changes in the Pleistocene, because eastern and western populations were isolated by a landbridge between Australia and New Guinea around 70,000–10,000 BP (Chivas *et al.*, 2001).

Not surprisingly, the extent of genetic differentiation among populations of marine invertebrates also often reflects differences in life history patterns. Sea cucumbers are a case in point; *Cucumaria miniata* showed no significant population structuring (mtDNA analysis), whereas *C. pseudocurata* showed extensive differentiation over a very similar geographic range in the northeast Pacific. The former has pelagic larvae, whereas the latter is a brooding species (Arndt and Smith, 1998).

MtDNA sequencing of the COI gene revealed that populations of the mantis shrimp *Haptosquilla pulchella* in Indonesian waters, a benthic crustacean with a suspected 4- to 6-wk planktonic larval period, showed F_{ST} values as high as 0.821 (Barber *et al.*, 2000). These striking differences may reflect past isolation during low sea level periods up to about 10,000 BP. However, the three deeply divergent mtDNA clades have levels of divergence similar to cryptic species of alpheid shrimp (Knowlton and Weigt, 1998), suggesting cryptic speciation may be involved (Barber *et al.*, 2002). Nonetheless, fine-scale genetic structuring within clades still indicates limited dispersal. Similarly, mtDNA sequencing of the COI gene of populations of the marine bivalve *Macoma balthica* showed the very high overall F_{ST} of 0.669 (Luttikhuizen *et al.*, 2003). In this case, however, pairwise F_{ST} values varied from 1.0 to zero. Some areas were connected by the high levels of gene flow anticipated from a 2–5 wk pelagic larva, whereas others were strongly isolated.

7.2.3. Genetic population structure in marine invertebrates targeted for restocking and stock enhancement

The message that populations of marine invertebrates can be more highly differentiated than expected is confirmed by data for some of the 11 species/ groups of species considered in this review.

Most population genetic surveys of abalone species have shown low but significant population structuring (Brown, 1991; Jiang *et al.*, 1995; Hamm and Burton, 2000; Del Rio-Portilla and Gonzalez-Aviles, 2001; Conod *et al.*, 2002; Evans *et al.*, 2004), presumably because of the limited dispersal power of a relatively short-lived (2–5 d) planktonic dispersing stage.

An allozyme survey of Norwegian populations of *Homarus gammarus* found low levels of variation within populations. Minor but statistically significant genetic population differentiation was most distinct in the northerly population (Jørstad and Farestveit, 1999). Another assessment of population structure of this species over a wider area, Norway to the Mediterranean Sea, was based on allozymes, mtDNA and microsatellites (Ferguson *et al.*, 2002). The three molecular approaches gave essentially concordant results, with low but significant differentiation, although the microsatellite analysis also revealed heterogeneity in the main Atlantic group of populations. Ferguson *et al.* (2002) concluded that the 'island population structure' of the European lobster revealed from their analysis implies that any over-exploited population is unlikely to be replenished from an adjacent one and that management of European lobster fisheries needs to be done at the local level.

Four of the eight species of giant clams under consideration for restocking have been subjected to allozyme population analysis, although a potentially useful DNA marker has also been described recently (Yu *et al.*, 2000). Many of these studies demonstrated relative homogeneity within island or reef groups but differences between well-separated groups (Macaranas *et al.*, 1992; Benzie and Williams, 1995; Benzie and Williams, 1997; Laurent *et al.*, 2002). Interestingly, the inferred patterns of gene flow are not consistent with present-day current patterns and may reflect past dispersal episodes at a time of lower sea level (Benzie and Williams, 1995, 1997).

Most studies of the population genetics of shrimp have been based on allozymes, but a few more recent studies have used mtDNA or microsatellites (see review by Benzie, 2000). Results have been generally concordant, showing little genetic differentiation over thousands of kilometres for several species, but with relatively abrupt changes over hundreds of kilometres for other species that are often associated with Indo-Pacific disjunction.

Holothurians show a wide range of reproductive strategies that can be associated with differences in population structure. Earlier it was noted that *Cucumaria miniata* has pelagic larvae and no significant spatial genetic heterogeneity, whereas *C. pseudocurata* has brooded larvae and extensive spatial differentiation (Arndt and Smith, 1998). Populations of *Holothuria nobilis* on the Great Barrier Reef separated by up to 1300 km could not be differentiated (Uthicke and Benzie, 2000), whereas those of *H. scabra* from Indonesia, Torres Strait, northeast Australia, the Solomon Islands and New Caledonia were more differentiated and showed less gene flow (Uthicke and Benzie, 2001; Uthicke and Purcell, 2004). Some species of holothurians can reproduce both sexually and asexually, leading to pronounced deviations from Hardy–Weinberg equilibria within populations. Allozyme analysis of *Stichopus chloronotus* and *H. atra* gave estimated sexual contributions of ~60% and 85%, respectively, and indicated that whereas asexual reproduction was important for maintaining local population size, the sexually produced larvae had high dispersal powers that led to low population differentiation (Uthicke *et al.*, 2001).

Perhaps one of the most telling pieces of research on measuring population structure comes from the European scallop *Pecten maximus*. Interestingly, despite insignificant allozyme and mtDNA differences between French and Scottish populations (Beaumont *et al.*, 1993; Wilding *et al.*, 1997), differences in reproductive cycles persist after transfer of juveniles (Cochard and Devauchelle, 1993; Mackie and Ansell, 1993). This suggests that there are genetic differences between these populations not detected by the molecular studies and raises the possibility of greater levels of population differentiation among marine invertebrates than we realize. Experimental reciprocal transfers and monitoring of reproductive behaviour for a greater number of species would help to identify the extent to which this may be true across marine invertebrate taxa.

7.2.4. Population structure and its implications for restocking and stock enhancement programmes

As described earlier, studies of marine invertebrates have revealed an enormous variety of genetic population structures: homogeneity, chaotic patchiness and multiple, largely self-replenishing populations often reflecting isolation by distance. It is also apparent that the high dispersal powers of many species are often not realized, and this may promote self-recruitment at different locations. As we explain in Chapter 6, Section 6.1.1, knowledge of population structure is pivotal to efficient management. We stress here that identifying population structure will depend heavily on a genetic study. It cannot be predicted reliably from existing knowledge of a species' biology.

The great variety of patterns in genetic structure of populations described previously, and the realization that population sizes can be smaller than expected, dictates that genetic analysis of spatial population structure should precede any assessment of the potential for improving management through restocking or stock enhancement. It is also evident that genetic studies should use a variety of methods, and that the results should perhaps be interpreted as revealing the minimum number of largely self-replenishing populations in the fishery. If the genetic study, and the complementary approaches described in Chapter 6, Section 6.1.7 produce convincing evidence for population homogeneity, then it may be acceptable to augment the target population with the offspring of broodstock from anywhere within the distribution of the species. However, if there is spatial heterogeneity in gene frequencies, implying restricted gene flow, then it is possible that different co-adapted genotypes might have evolved within the genetically different populations. In such circumstances, release of cultured juveniles with non-local genotypes carries the risk of introducing genotypes into the target population that are poorly adapted to that environment, possibly leading to a reduction in resilience. Therefore, wherever possible, broodstock for production of hatchery-reared juveniles should be taken from the target population. Where over-fishing has been so severe that there are insufficient broodstock available from the target population, and it is necessary to use spawners from elsewhere for a restocking programme, broodstock should be taken from the population that is likely to be genetically most similar to the recipient population (IUCN, 1998). The protection of genetic diversity in the target population should always be a primary aim of any restocking or stock enhancement programme. These issues are considered further by Cross (2000), Johnson (2000) and Lorenzen (2005).

7.2.5. Hatchery protocols to maintain genetic diversity

Because all releases have the potential to alter the genetic structure of the target population, steps should be taken to minimise these changes. Principles for managing broodstock to keep such risks at very low levels are discussed in Allendorf and Ryman (1987). Because this is such an important issue, we outline below the theoretical basis for their recommended broodstock management practices.

The expected proportion of the original heterozygosity remaining (E(h)) after a bottleneck broodstock size of N for one generation is:

$$E_{(h)} = 1 - \frac{1}{2N} \qquad (3)$$

Two parents will, therefore, have 75% of the original heterozygosity. This might seem adequate, but heterozygosity is only one measure of genetic variation and is relatively insensitive to bottleneck effects. The number of alleles is a far more sensitive measure. The maximum number of alleles a single parent pair can have at any one locus is four, whereas hypervariable loci such as microsatellites may have dozens of alleles in the natural population.

For a locus with n alleles, the expected number of alleles remaining (E(n)) after a bottleneck of size N is:

$$E(n) = n - \sum (1 - p_j)^{2N} \qquad (4)$$

where p_j are the allele frequencies (Denniston, 1978). For example, with 10 equally frequent alleles, only 3.4 alleles will be retained, on average, from two parents. Thus, allelic diversity in cultured progeny can be lost rapidly, unless great care is taken to manage broodstock.

These loss estimates are based on broodstock taken directly from the wild. If the broodstock come from animals that have already been in the hatchery for several generations, losses of alleles will be much more pronounced. In such cases, the proportion of heterozygosity remaining after t generations is:

$$E_{(h)(t)} = \left(1 - \frac{1}{2N}\right)^t \qquad (5)$$

So, for example, after three generations with just one parent pair per generation, only 42% of the original heterozygosity of alleles will remain. Losses of genetic variation in cultured animals are commonplace and have been recorded for several marine invertebrates, including shrimp (Sbordoni et al., 1986; Xu et al., 2001), pearl oysters (Durand et al., 1993), Pacific oysters (Gosling, 1982; Hedgecock and Sly, 1990), giant clams (Benzie and Williams, 1996) and abalone (Smith and Conroy, 1992; Mgaya et al., 1995). Note that mtDNA is expected to be a more sensitive indicator of loss of variation among hatchery-reared animals than nuclear DNA markers, because of its four-fold smaller effective population size.

Using small numbers of broodstock for multiple generations in the hatchery also leads to rapid inbreeding and the likely onset of inbreeding depression and poor performance. Also, broodstock maintained in a hatchery for several generations may become selected for good performance under hatchery conditions but produce offspring with an increased incidence of the types of morphological and behavioural deficits described in Chapters 2 and 3.

A potential pitfall for hatchery operations is that even when large numbers of broodstock are used, there can be highly selective spawning, resulting in a much reduced effective population size. In equations (3) to (5), N is the effective population size (N_e), not the number of broodstock. The effective population size will also be affected by sex ratio where it differs from 1:1, such that:

$$N_e = 4N_f N_m / (N_f + N_m) \qquad (6)$$

where N_f and N_m are numbers of females and males, respectively. Thus, a mass-spawning population of 99 females and 1 male has an effective size of 4, the same as 2 females and 2 males. Where multiple males and females are used, then equalising the reproductive contributions of the prospective parents maximises N_e. However, even after taking care to balance gametic contributions at the fertilisation stage, parental contribution can be quite uneven, leading to significant reductions in effective population size (Boudry et al., 2002).

To minimise the effects of hatchery releases on the genotypes of populations that are the targets of restocking and stock enhancement programmes, the considerations outlined previously must be translated into the following basic protocols.

1. Broodstock should be taken directly from the target population. Even then, care may be needed to avoid using closely-related individuals if there are reasons to believe that a large proportion of the broodstock may have been derived from one cohort. Relatedness can be estimated from molecular markers such as microsatellites (Goodnight and Queller, 1999; Wilson and Ferguson, 2002). Use of broodstock taken directly from the wild will negate the genetic losses resulting from broodstock that have been in hatcheries for several generations.

2. Use as many parents as possible in the production of young. It might be tempting to harness the enormous fecundity of many marine invertebrates and culture large numbers of progeny from just one or two females and males. This temptation must be strenuously avoided so that the target population does not get flooded with a limited range of genotypes. Care should also be taken to balance parental contributions to maximise effective population size. This includes using equal numbers of males and females, fertilising about the same number of eggs for each female and using single-pair matings where possible. If hatchery production is limited to only a small number of parents at any one time, then sequential spawnings should be planned so that, over time, a diverse range of genotypes is released into the target population.

7.2.6. Genetic impacts of restocking and stock enhancement programmes

Unfortunately, the genetic impacts of hatchery releases have not often been studied for restocking and stock enhancement programmes involving marine invertebrates. However, some insights are available on the range of effects that might be expected from the extensive studies on salmonids (Utter, 1998), albeit that they deal with species that home to natal rivers and therefore have pronounced genetic population heterogeneity. There is evidence that hatchery-reared fish often have a reduced reproductive capacity in the wild (Fleming *et al.*, 1997; Reisenbichler and Rubin, 1999; McLean *et al.*, 2003), and hybridisation of native fish with released cultured individuals is leading to reduced productivity and viability of the resulting population (Gharret and Smoker, 1991; Reisenbichler and Rubin, 1999).

Some of the assessments of genetic impacts have produced variable results. Microsatellite-based studies of brown trout (Hansen, 2002) have shown that after a decade of extensive stocking in one population, domesticated trout contributed only 6% to contemporary samples rather than the 64% anticipated from the numbers released, but in another population, domesticated trout contributed ~70%. In a different study of the same species, levels of gene flow from a population derived from a hatchery to the local wild population were estimated to be ~80% per generation, with both the cultured and naturally produced trout having similar reproductive success in the wild (Palm *et al.*, 2003).

These studies, and a very large number of others on salmonids, support the thesis that if managers decide that releases of hatchery-reared juveniles are needed to rebuild or enhance populations, the protocols outlined in Section 7.2.5 must be applied. The risk that restocking and stock enhancement programmes might reduce the effective population size of the supplemented population is considered further by Tringali and Bert (1998).

7.3. DISEASE RISKS

The possibility that hatchery-reared juveniles could affect target populations adversely is not restricted to changes in gene frequencies. There is also potential for released cultured animals to carry pathogens and parasites and to transfer these organisms to conspecifics and other species. Much of the knowledge about such risks has been learned from the way diseases have spread from non-indigenous species used for aquaculture in coastal waters to wild populations (Sindermann, 1993; Stickney, 2002). Two main categories of risk have emerged in the use of non-indigenous species for marine aquaculture. The first involves transfer of known pathogens or parasites associated with introduced animals to native species. This category also includes situations where infected, resistant carriers transmit a disease to non-resistant native species. The second is that organisms that have not been reported to cause problems for the introduced animal are pathogenic to other species. There are also other risks, but these relate mainly to the fact that introduced animals themselves can be susceptible to diseases that are already enzootic in the habitats into which they are placed.

In the relatively short history of marine aquaculture, there have been many transfers of pathogens and parasites among marine invertebrates because of introductions and translocations. Some of the most infamous examples are viral diseases in shrimp wherever they are cultured (Sindermann, 1993; Baldock *et al.*, 1999; Subasinghe *et al.*, 2001; Stickney, 2002, and references therein); viruses, bacteria and protozoa in oysters in Europe (Sindermann, 1993); viral diseases among pearl oysters in French Polynesia (Eldredge, 1994 and references therein); and sabellid polychaete parasites of abalone in California (McBride, 1998). In many cases, complex transfer networks have been created by the global production and trade in cultured species of marine invertebrates, which have resulted in rapid distribution of diseases among and within countries.

Although the transfer networks for species cultured for restocking and stock enhancement projects are not as complex or widespread as those established for aquaculture of introduced species, and the diseases caused by local pathogens are often not as virulent as those that occur in animals introduced for aquaculture (Sindermann, 1993), many of the same problems exist. In particular, there is scope for creating conditions suitable for the spread of diseases in hatcheries and for transfer of diseases from places where animals are caught and/or cultured to conspecifics, and related and non-related species, in the habitats where they are released. As mentioned in Chapter 3, these problems can and do occur. For example, withering syndrome (WS) has devastated populations of abalone in California (Richards, 2000), and white spot syndrome virus (WSSV) has reduced the scale of shrimp harvests and shrimp stock enhancement programmes in China.

Nevertheless, it is interesting to note that the incidence of disease in cultured or wild populations of several of the species/groups reviewed in Chapters 2 and 3 is relatively low (e.g., for giant clams, topshell, sea cucumbers, queen conch and lobsters). Importantly, however, in virtually all these cases, the level of hatchery production of juveniles has been low. Conversely, for those species in which diseases have been a regular and severe problem (i.e., shrimp, oysters and abalone), hatchery production has been orders of magnitude greater, and there has been extensive introduction and translocation of species and large-scale commercial aquaculture. This is not surprising because the intensive rearing facilities used in large-scale aquaculture can promote the incidence of disease in at least two ways: they amplify pathogens that are already enzootic and they increase their virulence (Sindermann, 1993).

Restocking and stock enhancement programmes that rely on intensive, large-scale hatchery production of juveniles, particularly when the same species are produced by commercial aquaculture, have the potential to cause more harm than good unless great care is taken to minimise the inherent risks of promoting and transferring diseases (Blankenship and Leber, 1995; Munro and Bell, 1997). It should also be noted that the potential for damage is not limited to wild conspecifics and that diseases can be more pathogenic in atypical hosts (Langdon, 1989; Sinderman, 1993). The ecosystem-based management approach to fisheries (see Section 7.4) now actively being promoted by local, regional and global fisheries agencies, recommends strongly that careful measures be taken to ensure that cultured juveniles are released in a way that will not unreasonably increase the risk of diseases. Even so, it is well recognized that it will never be possible to eliminate all disease risk associated with the propagation and translocation of marine invertebrates (Sindermann, 1993). The persistence of risk is illustrated well by the failure of diligent quarantine protocols and inspection measures to detect the transfer of a *Perkinsus* s/p. protozoan in bay scallops transferred from the United States to Canada (Chapter 3, Section 3.1.6). Nevertheless, many practical recommendations have now been made for aquaculture, restocking and stock enhancement practices that should reduce these risks to acceptable levels (Gibbons and Blogoslawski, 1989; Langdon, 1989; Norton et al., 1993b; Sinderman, 1993; Humphrey, 1995; Subasinghe et al., 2001) and, therefore, give managers confidence that the release of cultured juveniles will deliver net benefits where it has been deemed to add value to other forms of management. These recommendations encourage national management and regulatory authorities to:

1. Promote the use of native species in aquaculture to reduce the need to rely on introduced species, thereby restricting the transfer networks for diseases.

2. Develop and maintain maps of target host species showing the presence and abundance of each disease species as the basis for prohibition of transfers of broodstock and cultured animals from infected to non-infected areas. These maps should be done at the level of the self-replenishing populations that form the basis of management (see Chapter 6, Section 6.1). In keeping with the need to maintain genetic diversity (see Section 7.2), an added proviso is that broodstock and juveniles should not be transferred between these populations even in the absence of disease risk.

3. Identify pathogen-free populations to be used as broodstock.

4. Maintain proper sanitation, high-quality water and a constant environment in hatcheries to reduce stress.

5. Test broodstock and progeny for known diseases in quarantine. This should be done at least twice during the quarantine period and should be capable of detecting known and possible pathogens at their lowest prevalence with 95% confidence limits.

6. Grant health certification to juveniles before release in the wild.

7. Ensure that occurrence of any disease receives immediate attention, because once a new pathogen is established, it is difficult to stop it spreading further. This will involve destruction of diseased animals, sterilization of facilities and other measures to prevent further transfer of pathogens.

8. Use of the smallest-sized animals that is practical in release programmes, because the risk of contracting diseases is lower for the smaller size classes in several species. This should be taken into consideration when developing optimal release strategies but represents a potential dilemma, because the chances of surviving other causes of mortality are greater at larger sizes.

It is encouraging to see that strategies to reduce the risk of spreading diseases during restocking and stock enhancement initiatives have already been developed for some species (see Chapter 2, Section 2.1.6, 2.3.6 and Chapter 3, Section 3.5.6). However, there is no room for complacency. The message is clear: restocking and stock enhancement will be counterproductive if diseases are introduced to wild populations through lack of responsible protocols or careless application of those already in place. Particular care is needed when the target species, or related species, are being farmed commercially, because aquaculture increases the prevalence and virulence of diseases. Stringent regulations must be tempered, however, by the realization that the risks of spreading diseases can never be eliminated completely, and so the inability to remove all risk is not a valid reason to abandon restocking or stock enhancement where these management measures promise to be beneficial.

7.4. OTHER ENVIRONMENTAL IMPACTS

Changes to the genetic diversity of target stocks, and the risks of introducing diseases to conspecifics and other species in the ecosystem, are not the only potential impacts of restocking and stock enhancement. There is also concern that changes in relative abundance of target species, brought about by successful large-scale releases or associated manipulations, may be at the expense of other valuable components in the ecosystem (FAO, 1995b; Kearney and Andrew, 1995; Hilborn, 1998; Travis *et al.*, 1998; Leber, 1999, 2002; Caddy and Defeo, 2003; Molony *et al.*, 2003). The primary concerns are that: (1) increased abundances of target species may have negative impacts on other species through interactions such as competition for food and space or predation; (2) alterations to habitats made to favour target species may disadvantage other species and (3) removal of competitors or predators may have undesirable effects on animal and plant communities. This review suggests that any such impacts are only likely to occur for stock enhancement, not restocking. By definition, a population requiring restocking has such a low abundance that addition of cultured juveniles will not increase the numbers to the point where they were higher than historical levels. Thus, unprecedented interactions with other species in the ecosystem should not occur while the population is rebuilt. Similarly, restocking is unlikely to involve any decisions to manipulate habitats or other species unless managers deem that this is necessary to redress changes that have occurred since the demise of the target species.

On the other hand, stock enhancement in its various forms has the potential to cause impacts. For example, where several species depend on the same supplies of food and shelter, releases of target species that usurp the vacant carrying capacity can be expected to disadvantage potential competitors. However, this should not occur if releases for stock enhancement are designed as outlined in Chapter 6, Section 6.4 (i.e., to optimise production of the target species by leaving much of the carrying capacity vacant for subsequent year classes to recruit naturally, or to be added if necessary, to provide the desired level of harvests each year). Clearly, estimates of available carrying capacity should also include assessments of how much of the resource is currently being used by competitors or needs to be allocated to other valuable species. Thus, a well-designed stock enhancement programme should not fill the vacant habitat rapidly with the target species and deny access to competitors. Rather, there should always be scope for settlement of other species the following year. However, if releases of the target species are effective year after year and competitors settle in low numbers, increases in abundance relative to competitors can be expected and may have some undesirable effects. For example, valuable sea urchins may be eliminated by successful stock enhancement of abalone or vice versa.

Even where releases are designed to make room for other species, fine-scale tuning of abundances of target species will be difficult to achieve, and releases are likely to result in coarser effects, such as an over-supply of animals in some years. It remains to be seen whether any unintended dominance of the target species will be reversible. However, this might be possible: the dominance will have been created by managers and could perhaps be counteracted if society wishes to change the way the resources of coastal ecosystems are allocated to support different species (e.g., by ceasing releases and increasing harvests). In practice, this may be easier said than done, especially if the target species itself modifies habitats so that they are no longer available for use by other species or displaces other functionally important species that play a major role in maintaining productive habitat. Bartley and Kapuscinski (in press) make the case that stock enhancement must always guard against reducing the resilience of the ecosystem.

Much consideration needs to be given to this issue, because there are cases in the marine invertebrate literature of undesirable outcomes when abundance of a key species is reduced greatly. Although it had nothing to do with competition resulting from stock enhancement, the die-off of 99% of the population of the sea urchin *Diadema antillarum* in Jamaica is a pertinent example. After this mass-mortality, there was a striking phase-shift from a coral-dominated to an algal-dominated system (Hughes, 1994). In the absence of herbivory, there were negligible opportunities for corals to settle successfully.

On the other hand, it is also important that managers, and society, do not interpret all changes in species composition of communities and the structural complexity of habitats after the restoration or enhancement of a species as being negative. A pertinent example is the restoration of sea otter populations in Alaska after they were almost driven to extinction by commercial exploitation in the 18th and 19th centuries. In the absence of otters, populations of sea urchins *Strongylocentrotus* spp. became so dense that they overgrazed the macro-algae, removing much of the structural complexity from inshore sub-littoral habitats. After the restoration of otter populations at several locations through effective protection and a transplantation programme, the numbers of sea urchins were reduced to the point where a large biomass of kelp re-established (Duggins, 1980; Estes and Duggins, 1995 and references therein).

Much of the concern about potential negative effects from stock enhancement undoubtedly stems from the undesirable consequences of introducing alien species (McNeely, 2001; McNeely *et al.*, 2001; Wittenberg and Cock, 2001; Baskin, 2002; Caddy and Defeo, 2003; but see also Ruiz *et al.*, 1999). However, changing abundance of indigenous species through stock enhancement is unlikely to have the same potential to change ecosystems as the introduction of alien species, which in some cases have established themselves to the exclusion of the preferred native biota.

An instructive way of assessing the potential impact of manipulating abundances of naturally occurring species through stock enhancement would be to compare the effects to previous natural variations in abundance of the target species. In other words, how might the effects of releases differ from the large-scale interannual variations in natural settlement and recruitment success typical of many marine invertebrates? In most cases, such variations pass without comment, except for concerns in years of low abundance, and are a regular factor in estimating yields and setting quotas. For example, the settlement success of spiny lobsters can vary widely among years, resulting in annual harvest fluctuations of up to 50% (see Chapter 3, Section 3.6.2) without apparent concern about the effects on other species. When extraordinarily large numbers of marine invertebrates do settle naturally, as has occurred for hard clams on the east coast of the United States on several occasions (MacKenzie *et al.*, 2001), managers can take various actions to reduce abundances and dampen the effect on other species in the ecosystem. Thus, although there is some risk in increasing the abundance of target species artificially, particularly for habitat-modifying species, such effects are likely to fall within the historical range of variation in abundance and may well be reversible.

Another feature of the stock enhancement of marine invertebrates to be considered when assessing potential impacts of releases is that many of the target species feed low in the food chain. Increased abundances of such species can, therefore, be expected to have minimal effects on other conspicuous species through predation. This is not true for the predatory crustaceans (shrimp, lobsters and spiny lobsters), although only shrimp have been released in sufficient numbers to potentially cause an impact. There have been no studies to determine whether increased populations of shrimp in China and Japan have affected abundances of other harvested species, or the ecosystem, and no apparent objections to the releases. Rather, it can be argued that enhancing the harvests from wild stocks of shrimp is potentially more benign to the ecosystem than the methods used to increase production of shrimp through farming. This argument is of course weakened considerably by the fact that mass-releases of shrimp for stock enhancement in China and Japan have so far depended heavily on surplus production of juveniles from aquaculture (Chapter 3, Section 3.5.2).

Addition of habitat to promote increased settlement of target species, or provide additional shelter and food for a range of life history stages, will usually be at the expense of soft substrata and the epifauna and infauna that occur there. However, such impacts will often be minor because areas of soft substrata frequently dominate the inshore zone, and managers are likely to be persuaded that minimal losses to such areas are more than offset by the benefits of additional appropriate hard substrata, which is often in short supply. This generalization does not always hold, and special care should be taken to avoid reducing aquatic vegetation on soft substrata, because these plant communities can provide nursery habitat for commercially important marine invertebrates (Bell and Pollard, 1989 and references therein).

Provision of additional hard substrata also has the potential benefit of mitigating any effects of increases in abundance by target species, because it should also provide additional resources for competitors.

One feature of some stock enhancement programmes does have considerable potential to cause undesirable effects on the ecosystem: the culling of predators. This practice could have a direct impact 'up the food chain' and indirect effects at lower levels as well. Predatory starfish have been removed in large numbers from enhanced scallop fisheries in Japan (Chapter 3, Section 3.1.9), abundances of starfish and crabs have been actively reduced in smaller-scale restocking initiatives for some other bivalves (Chapter 3, Section 3.2.4), and octopus have been culled where abalone have been released in California (Chapter 3, Section 3.3.4). Removal of predators should be conducted only after the full implications have been investigated (Caddy and Defeo, 2003; FAO, 2003b).

In summary, changes to the status quo of ecosystems could eventuate from stock enhancement. However, given that coastal ecosystems used for fishing are already altered substantially from historical conditions (Jackson *et al.*, 2001), and additions of juveniles will often be restoring abundances of target species toward former levels, the impacts should not usually be a cause for concern provided they do not degrade the diversity of the ecosystem further. Indeed, increased manipulations of coastal ecosystems can be expected as technology and market forces shape the supply and demand for seafood. Managers of coastal resources now need to integrate the requirements of more stakeholders to supply a diverse range of products and services. Ideally, this needs to be done in a way that maintains the integrity of the ecosystem while providing opportunities to increase production through restored habitats and populations of fish and invertebrates. Preliminary guidelines for such an ecosystem approach to fisheries have been issued by FAO (2003b).

Chapter 8

Conclusions

8.1. ACHIEVEMENTS

This review makes it clear that considerable efforts have already been made to investigate and apply methods for restocking and stock enhancement of marine invertebrates. In particular, the widespread availability of technology for propagating and rearing juveniles in recent decades has led to a diverse array of initiatives variously designed to restore fisheries for valuable species, and increase productivity. These initiatives are at different stages of development, depending on the biological knowledge of the species and the availability of resources, and range from small-scale preliminary trials to well-established large-scale operations. Most releases have been aimed at increasing yields in stock enhancement programmes, which surpass restocking efforts to restore severely depleted fisheries both in number of initiatives and in the quantity of juveniles released. The distinction between restocking and stock enhancement has not always been evident, and the objectives of releasing cultured juveniles have not always been clear. In several cases, it has become apparent that restoration of spawning biomass is more important than short-term increases in yields, and so the emphasis has switched from stock enhancement to restocking.

Despite the relatively large numbers of stock enhancement initiatives for invertebrate fisheries, there have been few successes at the industrial scale. The programmes for scallops in Japan and the south island of New Zealand and for shrimp in China and Japan are the only initiatives to boast great increases in commercial catches after release of cultured juveniles. These interventions stand out because they are the only ones to have released

0065-2881/05 $35.00
DOI: 10.1016/S0065-2881(05)49008-2

hundreds of millions, or billions, of juveniles. It is tempting to jump to the conclusion, therefore, that effective stock enhancement will depend on truly massive releases of juveniles. However, as will be discussed in Section 8.3, this may not necessarily be the case. It is more important to identify the scale of releases needed to match the target population size desired by managers than to strive for huge numbers. It is also evident that the quantity of juveniles needed for effective stock enhancement can be reduced through adaptive management. This has been achieved for scallops but not yet for shrimp. Releases of shrimp in China and Japan have been based largely on surplus production from aquaculture, on the assumption that 'more is better'. Such releases have sometimes resulted in density-dependent mortality.

Another key message from stock enhancement of scallops and shrimp is that success depends not only on the technology and resources needed to release very large numbers of juveniles but also on incentives to make the necessary investments. This has occurred for scallops, where fishermen in Japan and New Zealand have been encouraged to bear the costs of stock assessments and releases of juveniles in return for reaping the benefits, but not yet for shrimp. Conversely, releases of shrimp have been scaled back substantially in China in recent years. This reduction is partly related to the reduced supply of juveniles stemming from a variety of factors, including disease, but it may also reflect the fact that fishermen are not yet convinced that the costs of mass releases previously made by the government will be lower than the value of increased yields.

A somewhat surprising conclusion of this review is that other fairly large-scale stock enhancement initiatives (e.g., release of tens of millions of sea urchins and abalone in Japan) have not succeeded in increasing national production to former levels, although some of these releases are credited with arresting declining harvests. It is difficult to determine to what extent failure of these releases to return catches to previous levels was related to: (1) production of insufficient animals to support increased yields; (2) failure of management to regulate the levels of fishing needed to obtain the greatest benefit from the released animals or (3) release of juveniles at a size and in a way that exposed them unnecessarily to the very high levels of predation that beset most young marine invertebrates in the wild. The considerable investments required to produce and to release tens of millions of animals demand answers to these questions (see Section 8.3 for a suggested way forward).

Smaller releases of juvenile marine invertebrates for stock enhancement, at scales of millions or less than millions, have rarely had significant economic benefit. Nevertheless, many of these initiatives have provided useful information on the rates of survival and recapture expected from larger scale operations. It is clear, however, that scaling-up the numbers of juveniles released will not always be possible or practical. In some cases, the biology of the species is incompatible with mass-production (e.g., the cannibalistic behaviour of lobsters). In other cases, the costs involved may be beyond the capability of management agencies and fishermen. Does this mean that the investments made to investigate stock enhancement for many species will be lost? Not necessarily. As mentioned previously, there has been growing awareness that the initial aspirations of many stock enhancement programmes–greater numbers of shellfish to grow-on in the wild for subsequent capture–are not the primary need of the fishery. Rather, the fishery requires more spawning adults to improve natural recruitment and catches in the future–i.e., the population needs restocking. Thus, the hatchery technology and optimal release strategies developed for many species will be useful to managers where it can be demonstrated that spawning biomass is likely to be restored most effectively by release of cultured juveniles (see Section 8.3). Indeed, this review demonstrates that many of the future assessments of the potential for managing fisheries with the use of cultured juveniles should be re-oriented from stock enhancement to restocking. This applies to lobsters in Europe and queen conch in the Caribbean and in part to abalone and sea urchins, depending on the species and location. On the other hand, the primary potential application of releasing cultured juveniles remains stock enhancement for shrimp, scallops, some other bivalves and spiny lobsters. Provision of additional habitat to increase the survival of settling juveniles is expected to provide the greatest gains for some bivalves, and for spiny lobsters, in the short to medium term.

Perhaps the greatest benefit of the diverse array of initiatives reviewed here is the rich set of lessons, from past investments, for future restocking and stock enhancement efforts (Chapter 5). Two compelling messages emerge from these lessons: (1) that there is no generic technology available for restocking or stock enhancement; and (2) that the successful application of restocking or stock enhancement in one place does not mean that the methods can be transferred to closely related species or even to the same species elsewhere. There are no shortcuts to reliable information on the processes that influence replenishment of the fishery and the best methods for producing and releasing fit cultured juveniles. Rather, the search for a generic technology should be replaced by adoption of the responsible approach to restocking and stock enhancement described throughout this review. Each situation demands its own set of studies.

8.2. THE RESPONSIBLE APPROACH

This review has confirmed that many of the potential problems associated with the release of cultured marine animals foreshadowed by Blankenship and Leber (1995) can and do occur. It also demonstrates that there is now considerable awareness of the need for responsible practice.

Some of the most serious problems have involved the lack of sufficient care in maintenance of wild gene frequencies among the accumulated cohorts of released juveniles. The most pronounced effects seem to be for the shrimp, *Penaeus chinensis*, in China, although it remains to be seen whether the apparent variation in gene frequencies for this species caused by stock enhancement affects the fitness of the shrimp or the sustainability of catches. It is pleasing to see that general and species-specific protocols for maintaining natural gene frequencies during release programmes are in place for many marine invertebrates. There are now few technical reasons to inhibit responsible practice in this area; governments simply need to ensure that the relevant protocols are implemented. Diseases have also been introduced to the wild through the hatchery production of juveniles (e.g., for abalone). Protocols have been developed to reduce the risk of biological damage to wild stocks that can occur through release of diseased hatchery-reared juveniles, although they are not in place for all initiatives nor are they always implemented where they do exist. Greatly increased public awareness about the issues of biological diversity and the need for an ecosystem approach to fisheries management can be expected to focus further attention on the application of responsible methods for producing genetically diverse, healthy juveniles for release. Public opinion can also be expected to influence the types of species considered for stock enhancement, with the emphases on minimising adverse effects on other valuable species in the ecosystem.

There has been encouraging progress in three components of the responsible approach for several restocking and stock enhancement initiatives (see Chapter 4 for details). These are: (1) understanding the factors that affect survival of juveniles released in the wild; (2) releasing juveniles in ways that optimise their survival; and (3) developing methods for identifying released individuals to measure success.

Predation has emerged as the single greatest obstacle to be overcome in ensuring that released juveniles survive to contribute to subsequent generations in restocking programmes, or to future harvests for stock enhancement initiatives. A diverse range of strategies has been devised to reduce predation, including: releasing juveniles into purpose-built habitats; removing predators; protecting juveniles in cages and pens in intermediate culture; rearing juveniles in ways to overcome morphological and behavioural deficits otherwise associated with hatchery production; saturating predators with high densities of small juveniles and releasing animals at a size and (low) density where they escape much of the mortality. However, predation remains a fatal problem for many restocking and stock enhancement initiatives, particularly those without the capacity to produce very large numbers of juveniles at low cost.

A creative range of systems for marking marine invertebrates has been developed to provide managers with robust methods for measuring the success of releases. Few experimental studies on the behaviour and survival of cultured animals released in the wild have been impeded by the lack of a suitable way of identifying the animals. This has not always been true, however, when it comes to distinguishing cultured animals from wild ones in mass-releases. Although many of the molluscs have proved easy to mark by inducing changes to the colour of the shell during culture, practical methods have yet to be developed for crustaceans and echinoderms. Until effective tagging systems are available for marking juveniles of these species *en masse*, which will probably rely on genetic approaches, measuring the success of stock enhancement for such species will depend on more laborious, and sometimes more equivocal, methods. These include length–frequency analysis of juveniles released out of synchrony with natural recruitment and 'Before versus After, Control versus Impact (BACI)' sampling designs (Chapter 7, Section 7.1).

The development of tagging systems for measuring the success of restocking programmes for marine invertebrates is in a different category. The only way to measure the contribution of the cultured released animals, and their progeny, to restoration of spawning biomass is to mark released individuals with a heritable, genetic tag. The technology for this is currently under active investigation. Care is needed here not to alter the genome to confer disadvantages to the progeny of released animals and wild individuals with which they may breed or in a way that will be unacceptable to consumers.

There are four components of the responsible approach for which achievements have only been modest and where much attention is needed in the immediate future if restocking and stock enhancement of marine invertebrates are to be used appropriately. These relate to: (1) the need for integration of restocking and stock enhancement into fisheries management plans; (2) defining measures of success; (3) identifying economic and policy objectives and (4) using adaptive management. With the exception of scallop stock enhancement in New Zealand and Japan, well-integrated plans for maximizing the benefits of investments by identifying the objectives of the releases and refining the approach to consistently achieve or surpass the objectives are not in place.

The first of these components, the need for integrating restocking and stock enhancement into management plans, is critical and must be the priority that dictates the need for all other activities related to the release of juveniles. In short, unless the management plan for a fishery can provide convincing arguments for investing in the culture of juveniles and specify the number of animals that should survive after release, there is no justification for proceeding with restocking or stock enhancement. This does not negate the need for research to develop responsible genetic and disease protocols or optimal release strategies. However, until the need for releases can be justified, such research should serve only to provide managers with the best possible information on rates of survival and the likely effects on the target population(s) and the ecosystem for modelling the benefits of including restocking or stock enhancement in the management regime. This has not been the focus for much of the past research. Instead, the assumption has been that releases of cultured juveniles will be beneficial. This has resulted in striving to ensure that releases are as effective and benign as possible in the hope that such efforts will eventually result in success. The approach we advocate for assessing the need for restocking or stock enhancement is described in more detail in the following section.

8.3. THE WAY FORWARD

This review emphasises that investments in hatchery technology are likely to fail unless managers know the status of the target stocks and define the objectives of the releases. Basically, managers need to determine whether the primary goals of any interventions they make using cultured juveniles are to rebuild spawning biomass to levels where reasonable harvests can be made again (restocking) or to improve the productivity of a reasonably 'healthy' fishery (stock enhancement). Defining the 'fishery' is critical to this process.

There is growing awareness that many exploited species of marine invertebrates do not occur as a 'unit stock'. Rather, a species is often subdivided into several, largely self-recruiting populations or composed of loosely linked units in a meta-population. Although this may not have many consequences for other forms of fisheries management provided the biology (growth rate, size or age at first maturity, fecundity etc.) of the species is fairly uniform across its distribution, this is not the case for restocking and stock enhancement. Success of these interventions must be measured against the status of the functional population unit into which juveniles are released. Therefore, the basis for assessing the need for restocking, and measuring its success, must be sound estimates of the number of spawning adults for each self-replenishing population in the fishery. Cultured juveniles should only be released in those populations where modelling has shown that other forms of management cannot be used to increase the number of spawners to the desired level in acceptable time frames (see Chapter 6, Section 6.1). Release of juveniles in the absence of knowledge of the spatial structure of the stocks runs the severe risk of squandering investments in hatchery production, because juveniles may be released where they are not needed. There are also the consequences of losing local adaptations through introduction of individuals derived from one population into another one with markedly different gene frequencies.

Similar constraints apply to stock enhancement, where measures need to be taken to maximise the production from each self-replenishing population. To obtain the maximum benefit from stock enhancement, managers need to calculate the carrying capacity of the ecosystem supporting each population and identify the levels of annual replenishment needed to optimise production, whether the juveniles are supplied through natural recruitment and/or from culture operations (see Chapter 6, Section 6.4). Without reliable information on the spatial structure of stocks, indiscriminant releases of juveniles run the risk of over-supplying areas already receiving the desired number of juveniles naturally, at the expense of areas in need of additional juveniles.

This model for basing the use of restocking and stock enhancement on a sound understanding of the distribution and status of each self-replenishing population within the fishery requires investment in identifying the spatial and genetic population structure of the target species and also assessments, or indicators, of adult and juvenile abundance. The approach we advocate has three particularly appealing features. First, it integrates the potential use of restocking and stock enhancement with plans to manage the spawning biomass, and the number of juveniles, to prevent recruitment over-fishing and optimise production. This principle has been espoused often, but implemented rarely. Second, it promises to be particularly suitable for many sessile or sedentary marine invertebrates because fisheries for these species are often made up of multiple self-replenishing populations. In addition, the size of a given population is likely to be small enough for hatchery releases to make a difference in many cases. Third, it provides the basis for identifying the beneficiaries of investments in hatchery releases (i.e., the fishermen operating regularly within the distribution of the self-replenishing population being targeted). Negotiations can then be held with these stakeholders to assess their willingness to contribute to the costs of stock enhancement in return for preferential rights to use the improved resource.

We acknowledge that identifying the spatial structure of stocks will incur additional costs. However, these costs are minor compared with the losses that are likely to occur if release programmes proceed without foundational information on the potential benefits of releases and on the scale of any releases required to achieve the desired increases. In practice, it may be relatively easy to identify some of the self-replenishing populations within a fishery. For example, coastal embayments with current regimes that retain larvae would be expected to promote self-recruitment for shallow-water species with a short larval duration.

8.4. FUTURE RESEARCH

We agree with Leber (1999, 2002, 2004), Molony *et al.* (2003) and Lorenzen (2005) that the science of restocking and stock enhancement is at a relatively early stage, and we fully endorse the call for systematic and thorough research to allow managers to make informed and logical decisions about the potential benefits of these interventions. In our view, two categories of research are needed to provide managers with the information they need to decide whether restocking and stock enhancement will be useful tools. The first involves identifying the spatial structure of the stock to determine whether it is homogeneous or composed of multiple, largely self-replenishing populations. Attention needs to be given here to using population genetic methods that are sensitive enough to detect meaningful differences and to the fact that gene flow within the distribution of the species does not necessarily mean that there are no major barriers to reproduction within the 'stock'. The best population genetics approach for determining the spatial structure of marine invertebrate fisheries is to combine mitochondrial DNA and nuclear DNA analyses, relying mainly on microsatellite analysis in the case of nuclear DNA. Where the results of these analyses are equivocal, managers should look for additional evidence from parasites, morphometrics and prevailing currents systems to be confident that they have a sound description of the spatial structure of the populations.

Once managers know how many self-replenishing populations comprise the fishery, they then need to know the status of each population and identify the changes to the size/age structure and abundances of size/age classes required to improve production. This information does not differ from the analyses undertaken routinely by fisheries biologists, except that it may well need to be done at the finer scale of self-replenishing populations. As the history of fisheries science attests, assessing the status of marine populations and identifying how to maximise production on a sustainable basis is rarely easy. Such assessments also need to be done at varying levels of sophistication depending on the value of the resource and the financial and scientific capacity of the country. Assessments can be expected to vary from precise and accurate estimations of age structure, fecundity, recruitment, mortality and carrying capacity to coarse indicators of the status of the stock and relative abundance of adults and juveniles. Regardless of the situation, managers will need sufficient information to judge whether the population is likely to respond as desired with or without the release of cultured juveniles. Obviously, the better the information, the better the predictive models will be for rebuilding spawning biomass with and without restocking, and improving productivity with and without stock enhancement.

The second category of research involves identifying how to produce and release juveniles at low cost so that they survive in high proportions. Reliable information on these important aspects is pivotal to the modelling described previously. The lower the cost of juveniles, and the better the rates of survival after release, the greater the probability that releases of cultured juveniles may have a role to play in restoring spawning biomass and improving productivity.

Much of the research on restocking and stock enhancement to date has already contributed to this second category. However, across the 11 species/groups covered here, five key areas for further research in the second category have emerged.

1. Scaling-up cost-effective production of juveniles to provide enough individuals to achieve the desired impact. Although the numbers of juveniles required will be determined by modelling, and will vary among populations and species, a key outcome of this review is that large numbers of juveniles (millions to hundreds of millions) are likely to be needed for many restocking and stock enhancement programmes. In the case of hatchery production, this will mean reducing the cost of nursery systems needed to rear genetically diverse, environmentally fit juveniles to the best release size. For species that can be collected effectively as postlarvae or spat, it will depend on identifying 'hotspots' where the settling juveniles are consistently abundant.

2. Identifying the best sizes, densities, habitats, times and methods to release juveniles so that survival is predictable and optimal. The effects of density dependence on growth and survival of wild and released juveniles is particularly important here to allay concerns that hatchery releases may simply replace wild individuals.

3. Locating the primary settlement areas for species where suitable nursery habitat is frequently in short supply and identifying how to construct and install more habitat at low cost to enhance natural recruitment.

4. Developing low-cost, benign genetic tags. In the first instance, these will be needed for several crustacean and echinoderm species to obtain survival estimates for modelling. In the longer term, where modelling has demonstrated that the use of cultured juveniles should be advantageous, genetic tags will be essential for measuring the contribution of released animals and their progeny to restored spawning biomass in restocking programmes, and to increased harvests in stock enhancement initiatives.

5. Identifying the number of individuals to place in aggregations of released or translocated animals to optimise reproductive success in restocking programmes, and the types of locations to place the aggregations to maximise settlement and survival of their progeny.

8.5. SUMMARY REMARKS

Managers and resource owners are actively looking for ways to restore marine capture fisheries. Restocking and stock enhancement are intuitively appealing; hatchery technology is now widely available, and there have been some success stories. However, managers and resource owners should not adopt the premise that, given sufficient research, these technologies can be made to work to solve their problems. Rather, objective questions need to be asked about whether investments in release programmes will add value to other forms of management.

To identify whether restocking and stock enhancement do have a contribution to make, it is necessary to model whether measures such as the release of cultured juveniles or provision of additional habitats for juveniles are likely to deliver the expected benefits before the interventions are made. These analyses will need to be done at the spatial scale of self-recruiting populations and should integrate information about the status and dynamics of the population with sound estimates of the cost of producing genetically diverse juveniles that survive well when released.

In many cases, further research will be needed on methods for propagating, releasing and monitoring cultured juveniles. However, the need for such research should not be taken to indicate that large-scale releases might be made once the problems of mass-production and overcoming predation after release have been conquered. Rather, research is necessary to provide robust estimates of the cost and survival of cultured juveniles for modelling the potential benefits of hatchery releases. Ultimately, restocking and stock enhancement should only be used where comprehensive analysis of alternative models for management of self-recruiting populations demonstrates that releases of juveniles will deliver greater benefits than other interventions.

The approach of applying restocking and stock enhancement at the spatial scale of self-recruiting populations not only provides a sound rationale and basis for measuring success, it also helps to overcome the conceptual blockage provided by the daunting scale of coastal waters. Release of juveniles into discrete populations is easier to envisage and organize than release into the 'open sea'. In this sense, restocking and stock enhancement initiatives for marine invertebrates will be much easier than for marine finfish, because the sedentary or sessile habits and inshore distributions of many invertebrates generally promote a higher incidence of smaller-scale, self-replenishing populations. The relatively low position of marine invertebrates on the food chain compared with many finfish also has the advantage of minimising the effects of stock enhancement on other species in the ecosystem. We conclude, therefore, that restocking and stock enhancement promise to have their greatest applications among the marine invertebrates.

ACKNOWLEDGEMENTS

We are grateful to the many people who helped bring this review to fruition. We especially thank Guat Khim Chew and Shu Shu Foo for their tireless efforts in assisting us to produce all components of the manuscript through several drafts. We also thank Stephanie Pallay, Bee Leng Chew, Stewart Grant, Toni Cannard and Jan Vaughan for their help in assembling the literature; Mei Norizam, Len Garces and Garrick Tan for drawing some of the figures; and Ann-Lizbeth Agnalt, Bill Arnold, Brian Beal, Megan Davis, Rob Day, Kim Drummond, Peter Gardiner, Ed Gomez, Mick Haywood, Sylvain Huchette, Mike Heasman, Rob Kenyon, Chan Lee, Steve Purcell, Scoresby Shepherd, Alan Stoner and David Tarbath for their useful comments on sections of the draft manuscript and/or for providing information for parts of the text. The photographs were generously made available by the various people who took them, including P. Crocos, C. Berg, D. Dennis, E. Farestveit, M. Francis, C. Hair, C. Hesse, R. Kuiter, M. Kurata, J. Lucas, M. McCoy, A. Mercier, R. Naylor, S. Purcell, D. Rothaus, A. Stoner, S. Talbot and C. Woods, or generously provided by M. Davis, D. Grimes, CSIRO Marine Research, the National Oceanic and Atmospheric Administration of the United States of America and the Virginia Institute of Marine Science. Finally, we would like to thank Alan Southward and Craig Young for their encouragement to write on this topic and for their guidance along the way. This is the WorldFish Center Contribution no. 1742.

References

Ablan, M. C. A., Macaranas, J. M. and Gomez, E. D. (1993). Genetic structure of the giant clam *Tridacna derasa* from five areas in Asia and the Pacific. *In* "Proceedings of the International Workshop on Genetics in Aquaculture and Fisheries Management, 31 August–4 September 1992" (D. Penman, N. Roongratri and B. McAndrew, eds), pp. 57–62. University of Stirling, Scotland.

Acosta, B. A. (1994). Bibliography of the conch genus *Strombus* (Gastropoda: Strombidae). *In* "Queen Conch Biology, Fisheries and Mariculture" (R. Appeldoorn and B. Rodriguez, eds), pp. 321–356. Fundación Científica Los Roques, Caracas.

Adamkewicz, L. and Castagna, M. (1988). Genetics of shell color and pattern in the bay scallop *Argopecten irradians*. *Journal of Heredity* **79**, 14–17.

Adams, T. (1992). Resource aspects of the Fiji beche-de-mer industry. *South Pacific Commission Beche-de-mer Information Bulletin* **4**, 13–16.

Addison, J. T. (1997). Lobster stock assessment: Report from a workshop; I. *Marine and Freshwater Research* **48**, 941–944.

Addison, J. T. and Bannister, R. C. A. (1994). Re-stocking and stock enhancement of clawed lobster stocks: A review. *Crustaceana* **67**, 131–155.

Agatsuma, Y. (1987). On the releasing of the cultured seeds of sea urchin, *Strongylocentrotus intermedius* (A. Agassiz), at Shikabe in the coast of Funka Bay. Discrimination between releasing group and aboriginal group by rings formed in genital plate. *Hokusuishi Geppo* **44**, 118–126 (in Japanese).

Agatsuma, Y. (1991). *Strongylocentrotus intermedius* (A. Agassiz). *In* "Fishes and Marine Invertebrates of Hokkaido: Biology and Fisheries" (K. Nagasawa and M. Torisawa, eds), pp. 324–329. Kita-Nihon Kaiyo Center Co. Ltd, Sapporo (in Japanese).

Agatsuma, Y. (1998). Aquaculture of the sea urchin (*Strongylocentrotus nudus*) transplanted from coralline flats in Hokkaido, Japan. *Journal of Shellfish Research* **17**, 1541–1547.

Agatsuma, Y. (1999a). Gonadal growth of the sea urchin, *Strongylocentrotus nudus*, from trophically poor coralline flats and fed excess kelp, *Laminaria religiosa*. *Suisan Zoshoku* **47**, 325–330.

Agatsuma, Y. (1999b). Marine afforestation off the Japan Sea coast in Hokkaido. *In* "The Ecological Mechanism of "Isoyake" and Marine Afforestation" (K. Taniguchi and K. Kouseikaku, eds), pp. 84–97. Blackwell, Tokyo (in Japanese).

Agatsuma, Y. (2001a). Ecology of *Strongylocentrotus intermedius*. *In* "Edible Sea Urchins: Biology and Ecology" (J. M. Lawrence, ed.), pp. 333–346. Elsevier, Amsterdam.

0065-2881/05 $35.00
DOI: 10.1016/S0065-2881(05)49014-8

Agatsuma, Y. (2001b). Ecology of *Strongylocentrotus nudus*. *In* "Edible Sea Urchins: Biology and Ecology" (J. M. Lawrence, ed.), pp. 347–361. Elsevier, Amsterdam.

Agatsuma, Y. (2001c). Ecology of *Hemicentrotus pulcherrimus*, *Pseudocentrotus depressus*, and *Anthocidaris crassispina* in southern Japan. *In* "Edible Sea Urchins: Biology and Ecology" (J. M. Lawrence, ed.), pp. 363–374. Elsevier, Amsterdam.

Agatsuma, Y. and Kawai, T. (1997). Seasonal migration of the sea urchin *Strongylocentrotus nudus* in Oshoro Bay of southwestern Hokkaido, Japan. *Nippon Suisan Gakkaishi* **63**, 557–562 (in Japanese with English abstract).

Agatsuma, Y., Sakai, Y. and Matsuda, T. (eds) (1995). "Manual for Transplantation of Sea Urchin Seeds *Strongylocentrotus intermedius*", p. 81. Hokkaido Central Fisheries Experimental Station, Otaru, Hokkaido (in Japanese).

Agatsuma, Y., Sakai, Y., Andrew, N. and Saiko, Y. (2003). Enhancement of Japan's sea urchin fisheries. *In* "Sea Urchins: Fisheries and Ecology. Proceeding of the International Conference on Sea Urchin Fisheries and Aquaculture" (J. M. Lawrence, ed.), pp. 18–36. CRC Press, Boca Raton, Florida.

Agnalt, A.-L., van der Meeren, G., Jørstad, K. E., Næss, H., Farestveit, E., Nøstvold, E., Svåsand, T., Korsøen, E. and Ydstebø, L. (1999). Stock enhancement of European lobster (*Homarus gammarus*): A large-scale experiment off southwestern Norway. *In* "Stock Enhancement and Sea Ranching" (B. R. Howell, E. Moksness and T. Svåsand, eds), pp. 401–419. Fishing News Books, Blackwell, Oxford.

Agnalt, A.-L., Jørstad, K. E., Kristiansen, T., Nostvold, E., Farestveit, E., Næss, H., Paulsen, O. I. and Svåsand, T. (2004). Enhancing the European lobster (*Homarus gammarus*) stock at Kvitsøy Islands: Perspectives of rebuilding Norwegian stocks. *In* "Stock Enhancement and Sea Ranching: Developments, Pitfalls and Opportunities" (K. M. Leber, S. Kitada, L. Blankenship and T. Svåsand, eds), pp. 415–426. Blackwell, Oxford.

Ahn, I.-Y., Malouf, R. and Lopez, G. (1993). Enhanced larval settlement of the hard clam *Mercenaria mercenaria* by the gem clam *Gemma gemma*. *Marine Ecology Progress Series* **99**, 51–59.

Aiken, D. E. and Waddy, S. L. (1985). Production of seed stock for lobster culture. *Aquaculture* **44**, 103–114.

Aiken, K., Kong, G. A., Smikle, S., Mahon, R. and Appeldoorn, R. (1997). Status of the fisheries and regulations regarding queen conch in Jamaica in 1996. *Proceedings of the Gulf and Caribbean Fisheries Institute* **49**, 485–498.

Allendorf, F. W. and Ryman, N. (1987). Genetic management of hatchery stocks. *In* "Population Genetics and Fishery Management" (N. Ryman and F. Utter, eds), pp. 141–159. Washington Sea Grant Program, University of Washington Press, Seattle.

Altstatt, J. M., Ambrose, R. F., Engle, J. M., Haaker, P. L., Lafferty, K. D. and Raimondi, P. T. (1996). Recent declines of black abalone, *Haliotis cracherodii*, on the mainland coast of central California. *Marine Ecology Progress Series* **142**, 185–192.

Amaratunga, T. and Misra, R. K. (1989). Identification of soft-shell clam (*Mya arenaria* Linnaeus, 1758) stocks in eastern Canada based on multivariate morphometric analysis. *Journal of Shellfish Research* **8**, 391–397.

Amos, M. J. (1997). Management policy for the trochus fishery in the Pacific. *In* "Trochus: Status, Hatchery Practice and Nutrition" (C. H. Lee and P. W. Lynch, eds), pp. 164–169. ACIAR Proceedings No. 79, Australian Centre for International Agricultural Research, Canberra.

Amos, M. J. and Purcell, S. W. (2003). Evaluation of strategies for intermediate culture of *Trochus niloticus* (Gastropoda) in sea cages for restocking. *Aquaculture* **218**, 235–249.

Andrew, N. L. and O' Neill, A. L. (2000). Large-scale patterns in habitat structure on subtidal rocky reefs in New South Wales. *Marine and Freshwater Research* **51**, 255–263.

Andrew, N. L., Worthington, D. G., Brett, P. A., Bentley, N., Chick, R. C. and Blount, C. (1998). "Interactions between the Abalone Fishery and Sea Urchins in New South Wales. Final Report Fisheries Research and Development Corporation", p. 64. New South Wales Fisheries, Sydney, Australia.

Andrew, N. L., Agatsuma, Y., Ballesteros, E., Bazhin, A. G., Creaser, E. P., Barnes, D. K. A., Botsford, L. W., Bradbury, A., Campbell, A., Dixon, J. D., Einarsson, S., Gerring, P., Hebert, K., Hunter, M., Hur, S. B., Johnson, C. R., Juinio–0Meñez, M. A., Kalvass, P., Miller, R. J., Moreno, C. A., Palleiro, J. S., Rivas, D., Robinson, S. M. L., Schroeter, S. C., Steneck, R. S., Vadas, R. I., Woodby, D. A. and Xiaoqi, Z. (2002). Status and management of world sea urchin fisheries. *Oceanography and Marine Biology: An Annual Review* **40**, 343–425.

Andrews, J. D. (1996). History of *Perkinsus marinus*, a pathogen of oysters in Chesapeake Bay 1950–1984. *Journal of Shellfish Research* **15**, 13–16.

Anon. (1984). On the natural seeds collection, intermediate culture and release of the sea urchin, *Strongylocentrotus intermedius*. *Hokusuishi Geppo* **41**, 270–315 (in Japanese).

Anon. (1992). "Manual for Artificial Seed Production of the Sea Urchin", p. 29. Hokkaido Institute of Mariculture, Shikabe, Hokkaido (in Japanese).

Anon. (2000). "Americans on the Wrong Side—The Lobster *Homarus americanus* Captured in Norwegian Waters", p. 20. Copenhagen, Denmark.

Annala, J., Sullivan, K. S., O'Brien, C. J., Smith, N. W. M. and Varian, S. J. A. (2002). "Report from the Fisheries Assessment Plenary, May 2002: Stock Assessments and Yield Estimated", p. 639. New Zealand Ministry of Fisheries, Wellington.

Aoyama, S. (1989). The Mutsu Bay scallop fisheries: Scallop culture, stock enhancement and resource management. *In* "Marine Invertebrate Fisheries: Their Assessment and Management" (J. F. Caddy, ed.), pp. 525–539. John Wiley and Sons, New York.

Appeldoorn, R. (1985). Growth, mortality and dispersion of juvenile laboratory-reared conchs, *Strombus gigas* and *S. costatus,* released at an offshore site. *Bulletin of Marine Science* **37**, 785–793.

Appeldoorn, R. (1988). Fishing pressure and reproductive potential in strombid conchs: Is there a critical stock density for reproduction. *Memorias de la Sociedad de Ciencias Naturales "La Salle"* **48** (suppl. 3), 275–288.

Appeldoorn, R. and Ballantine, D. (1983). Field release of cultured queen conch in Puerto Rico: Implications for stock restoration. *Proceedings of the Gulf and Caribbean Fisheries Institute* **35**, 89–98.

Appeldoorn, R. and Rodriguez, B. (eds) (1994). "*Strombus gigas* Queen Conch Biology Fisheries and Mariculture", p. 356. Fundación Científica Los Roques, Caracas.

Appeldoorn, R., Ballantine, D. and Chanley, P. (1983). Observations on the growth and survival of laboratory-reared juvenile conchs, *Strombus gigas* and *S. costatus.* *Journal of Shellfish Research* **3**, 82.

Apte, S. and Gardner, P. A. (2002). Population genetic subdivision in the New Zealand greenshell mussel (*Perna canaliculus*) inferred from single-strand conformation polymorphism analysis of mitochondrial DNA. *Molecular Ecology* **11**, 1617–1628.

Arbuckle, M. and Drummond, K. (2000). Evolution of self-governance within a harvesting system governed by individual transferable quota. *In* "Use of Property Rights in Fisheries Management" (R. Shotton, ed.), pp. 370–382. FAO Fisheries Technical Paper 404/2, FAO, Rome.

Arbuckle, M. and Metzger, M. (2000). "Food for Thought. A Brief History of the Future of Fisheries' Management", p. 26. Challenger Scallop Enhancement Company Limited, New Zealand.

Arce, A. M., Aquilar-Dávila, W., Sosa-Cordero, E. and Caddy, J. F. (1997). Artificial shelters (casitas) as habitats for juvenile spiny lobsters *Panulirus argus* in the Mexican Caribbean. *Marine Ecology Progress Series* **158**, 217–224.

Ariyama, H., Uratani, F., Ohyama, H., Sano, M. and Yamochi, S. (1994). Survival, growth, and tag retention of the kuruma prawn *Penaeus japonicus* and the greasy back prawn *Metapenaeus ensis* injected with gold bit tags. *Fisheries Science* **60**, 785–786.

Arnason, A. N. (1972). Parameter estimates from mark-recapture experiments on two populations subject to migration and death. *Researches on Population Ecology* **3**, 97–113.

Arnason, A. N. (1973). The estimation of population size, migration rates and survival in a stratified population. *Researches on Population Ecology* **15**, 1–8.

Arndt, A. and Smith, M. J. (1998). Genetic diversity and population structure in two species of sea cucumber: Differing patterns according to mode of development. *Molecular Ecology* **7**, 1053–1064.

Arnold, S. (1984). The effects of prey size, predator size, and sediment composition on the rate of predation of the blue crab, *Callinectes sapidus* Rathbun, on the hard clam, *Mercenaria mercenaria* (Linné). *Journal of Experimental Marine Biology and Ecology* **80**, 207–219.

Arnold, S. (2001). Bivalve enhancement and restoration strategies in Florida, USA. *Hydrobiologia* **465**, 7–19.

Arnold, S., Bert, T., Marelli, D. C., Cruz-Lopez, H. and Gill, P. A. (1996). Genotype-specific growth of hard clams (genus *Mercenaria*) in a hybrid zone: Variation among habitats. *Marine Biology* **125**, 129–139.

Arnold, S., Marelli, D. C., Parker, M., Hoffman, P., Frischer, M. and Scarpa, J. (2002). Enhancing hard clam (*Mercenaria* spp.) population density in the Indian River Lagoon, Florida: A comparison of strategies to maintain the commercial fishery. *Journal of Shellfish Research* **21**, 659–672.

Aswani, S. (1997). Troubled water in south-western New Georgia Solomon Islands: Is codification of the commons a viable avenue for resource use regulation? *South Pacific Commission Traditional Marine Resource Management and Knowledge Information Bulletin* **8**, 2–16.

Aviles, O. J. G. G. (2000). Seeding competent cultured larvae of the blue abalone *Haliotis fulgens* into some wild stocks off the Island of Cedros, Baja California, Mexico. 4th International Abalone Symposium., Cape Town, South Africa, February, 2000, *Journal of Shellfish Research* **19**, 514.

Avise, J. C. (1994). "Molecular Markers, Natural History and Evolution", p. 511. Chapman and Hall, New York.

Bahlo, M. and Griffiths, R. C. (2000). Inference from gene trees in a subdivided population. *Theoretical Population Biology* **57**, 79–95.

Baldock, C., Callinan, R. and Loneragan, N. (1999). Environmental impact of the establishment of exotic prawn pathogens in Australia. Report to the Australian Quarantine and Inspection Service (Ausvet) Animal Health Services, Box 3180, South Brisbane, Qld 4101, Australia.

Baldwin, J. D., Bass, A. L., Bowen, B. W. and Clark, W. H. (1998). Molecular phylogeny and biogeography of the marine shrimp *Penaeus*. *Molecular Phylogenetics and Evolution* **10**, 399–407.

Ball, B., Linnane, A., Munday, B., Browne, R. and Mercer, J. P. (2001). The effect of cover on *in situ* predation in early benthic phase European lobster *Homarus gammarus*. *Journal of the Marine Biological Association of the United Kingdom* **81**, 639–642.

Ballantine, D. L. and Appeldoorn, R. (1983). Queen conch culture and future prospects in Puerto Rico. *Proceedings of the Gulf and Caribbean Fisheries Institute* **35**, 57–63.

Bannister, R. C. A. (1991). Stock enhancement. *ICES Journal of Marine Science Symposium* **192**, 191–192.

Bannister, R. C. A. and Addison, J. T. (1998). Enhancing lobster stock: A review of recent European methods, results, and future prospects. *Bulletin of Marine Science* **62**, 369–387.

Bannister, R. C. A. and Howard, A. E. (1991). A large-scale experiment to enhance a stock of lobster (*Homarus gammarus* L.) on the English east coast. *ICES Marine Science Symposia* **192**, 99–107.

Bannister, R. C. A., Addison, J. T. and Lovewell, S. R. J. (1994). Growth, movement, recapture rate and survival of hatchery-reared lobsters *Homarus gammarus* (Linnaeus, 1785) released into the wild on the English east coast. *Crustaceana* **67**, 155–172.

Barbeau, M. A., Scheibling, R. E., Hatcher, B. G., Taylor, L. H. and Hennigar, A. W. (1994). Survival analysis of tethered juvenile sea scallops *Placopecten magellanicus* in field experiments: Effects of predators, scallop size and density, site and season. *Marine Ecology Progress Series* **115**, 243–256.

Barbeau, M. A., Hatcher, B. G., Scheibling, R. E., Hennigar, A. W., Taylor, L. H. and Risk, A. C. (1996). Dynamics of juvenile sea scallop (*Placopecten magellanicus*) and their predators in bottom seeding trials in Lunenburg Bay, Nova Scotia. *Canadian Journal of Fisheries and Aquatic Sciences* **53**, 2494–2512.

Barber, P. H., Palumbi, S. R., Erdmann, M. V. and Moosa, M. K. (2000). A marine Wallace's line? *Nature* **406**, 692–693.

Barber, P. H., Palumbi, S. R., Erdmann, M. V. and Moosa, M. K. (2002). Sharp genetic breaks among populations of *Haptosquilla pulchella* (Stomatopoda) indicate limits to larval transport: Patterns, causes, and consequences. *Molecular Ecology* **11**, 659–674.

Barshaw, D. E. (1989). Growth and survival of early juvenile American lobsters, *Homarus americanus*, on a diet of plankton. *Fishery Bulletin, US* **87**, 366–370.

Barshaw, D. E. and Lavalli, K. (1988). Predation upon postlarval lobsters *Homarus americanus* by cunners *Tautoglabrus adspersus* and mud crabs *Neopanope sayi* on three different substrates: Eelgrass, mud and rocks. *Marine Ecology Progress Series* **48**, 119–123.

Bartley, D. M. and Kapuscinski, A. R. (in press). What makes fisheries enhancements responsible? 4th World Fisheries Congress Proceedings, Vancouver, Canada.

Bartley, D. M., Born, A. and Immink, A. (2004). Stock enhancement and sea ranching in developing countries. *In* "Stock Enhancement and Sea Ranching: Developments, Pitfalls and Opportunities" (K. M. Leber, S. Kitada, L. Blankenship and T. Svåsand, eds), pp. 48–58. Blackwell, Oxford.

Baskin, Y. (2002). "A Plague of Rats and Rubber Vines: The Growing Threat of Species Invasions", p. 377. Island Press, Washington.

Battaglene, S. C. (1999). Culture of tropical sea cucumbers for the purposes of stock restoration and enhancement. *NAGA, The ICLARM Quarterly* **22**(4), 4–11.

Battaglene, S. C. and Bell, J. D. (1999). Potential of the tropical Indo-Pacific sea cucumber, *Holothuria scabra*, for stock enhancement. *In* "Stock Enhancement and Sea Ranching" (B. R. Howell, E. Moksness and T. Svåsand, eds), pp. 478–490. Fishing News Books, Blackwell, Oxford.

Battaglene, S. C. and Bell, J. D. (2004). The restocking of sea cucumbers in the Pacific Islands. *In* "Marine Ranching" (D. M. Bartley and K. M. Leber, eds), pp. 109–132. Fishery Technical Paper No 429. FAO, Rome.

Battaglene, S. C., Seymour, E. J. and Ramofafia, C. (1999). Survival and growth of cultured juvenile sea cucumbers *Holothuria scabra*. *Aquaculture* **178**, 293–322.

Beal, B. F. (1991). The fate of hatchery-reared juveniles of *Mya arenaria* L. in the field: How predation and competition are affected by initial clam size and stocking density. *Journal of Shellfish Research* **10**, 292–293.

Beal, B. F. (1993a). Effects of initial clam size and type of protective mesh netting on the survival and growth of hatchery-reared individuals of *Mya arenaria* in eastern Maine. *Journal of Shellfish Research* **12**, 138–139.

Beal, B. F. (1993b). Overwintering hatchery-reared individuals of *Mya arenaria*: A field test of site, clam size and intraspecific density. *Journal of Shellfish Research* **12**, 153.

Beal, B. F. and Chapman, S. R. (2001). Methods for mass rearing stages I–IV larvae of the American lobster, *Homarus americanus* H. Milne-Edwards, 1837, in static systems. *Journal of Shellfish Research* **20**, 337–346.

Beal, B. F. and Kraus, M. G. (1991). The importance of initial size and density on the survival and growth of hatchery-reared individuals of *Mya arenaria* L. *Journal of Shellfish Research* **10**, 288–289.

Beal, B. F., Chapman, S. R., Irvine, C. and Bayer, R. C. (1998). Lobster (*Homarus americanus*) culture in Maine: A community-based, fishermen-sponsored public stock enhancement programme. *In* "Proceedings of a Workshop on Lobster Stock Enhancement Held in the Magdalen Islands (Quebec) from October 20th to 31st 1997" (L. Gendron, ed.), pp. 47–54. Canadian Industry Report of Fisheries and Aquatic Sciences, Number 244.

Beal, B. F., Mercer, J. P. and O'Conghaile, A. (2002). Survival and growth of hatchery-reared individuals of the European lobster, *Homarus gammarus* (L.), in field-based nursery cages on the Irish west coast. *Aquaculture* **210**, 137–157.

Beamish, R. J. and Noakes, D. L. J. (2004). Global warming, aquaculture and commercial fisheries. *In* "Stock Enhancement and Sea Ranching: Developments, Pitfalls and Opportunities" (K. M. Leber, S. Kitada, L. Blankenship and T. Svåsand, eds), pp. 25–47. Blackwell, Oxford.

Beard, T. W. and Wickins, J. F. (1992). "Techniques for the production of juvenile lobsters, *Homarus gammarus* L. Fisheries Research Technical Report 92", p. 22. MAFF Directorate Fisheries Research, Lowestoft, England.

Bearden, C. M. and McKenzie, M. D. (1972). Results of a pilot shrimp tagging project using internal anchor tags. *Transactions of the American Fisheries Society* **101**, 358–362.

Beattie, J. H. (1992). Geoduck enhancement in Washington State. *Bulletin of the Aquacultural Association of Canada* **92**, 18–24.

Beattie, H. and Blake, B. (1999). Development of culture methods for the geoduck clam in the USA (Washington State) and Canada (British Columbia). *Magazine of the World Aquaculture Society* **30**, 50–53.

Beaumont, A. R. (1982). Geographic variation in allele frequencies at three loci in *Chlamys opercularis* from Norway to the Brittany Coast. *Journal of the Marine Biological Association United Kingdom* **62**, 243–261.

Beaumont, A. (2000). Genetic considerations in transfer and introductions of scallops. *Aquaculture International* **8**, 493–512.

Beaumont, A. R. and Zouros, E. (1991). Genetics of scallops. *In* "Scallops: Biology, Ecology and Aquaculture" (S. E. Shumway, ed.), pp. 585–623. Elsevier, New York.

Beaumont, A. R., Morvan, C., Huelvan, S., Lucas, A. and Ansell, A. D. (1993). Genetics of indigenous and transplanted populations of *Pecten maximus*: No evidence for the existence of separate stocks. *Journal of Experimental Marine Biology and Ecology* **169**, 77–88.

Beerli, P. and Felsenstein, J. (1999). Maximum-likelihood estimation of migration rates and effective population numbers in two populations using a coalescent approach. *Genetics* **152**, 763–773.

Beerli, P. and Felsenstein, J. (2001). Maximum likelihood estimation of a migration matrix and effective population sizes in subpopulations by using a coalescent approach. *Proceedings of the National Academy of Sciences USA* **98**, 4563–4568.

Bell, J. D. (1999a). Aquaculture: A development opportunity for Pacific Islands. *Development Bulletin* **49**, 49–52.

Bell, J. D. (1999b). Restocking giant clams: Progress, problems and potential. *In* "Stock Enhancement and Sea Ranching" (B. R. Howell, E. Moksness and T. Svasand, eds), pp. 437–452. Fishing News Books, Blackwell, Oxford.

Bell, J. D. (1999c). Transfer of technology on marine ranching to small island states. *In* "Marine Ranching: Global Perspectives with Emphasis on the Japanese Experience" (unedited), pp. 53–65. FAO Fisheries Circular 943, FAO, Rome.

Bell, J. D. (2004). Management of restocking and stock enhancement programs: The need for different approaches. *In* "Stock Enhancement and Sea Ranching: Developments, Pitfalls and Opportunities" (K. M. Leber, S. Kitada, L. Blankenship and T. Svåsand, eds), pp. 213–224. Blackwell, Oxford.

Bell, J. D. and Pollard, D. A. (1989). Ecology of fish assemblages and fisheries associated with seagrasses. *In* "Biology of Seagrasses: A Treatise on the Biology of Seagrass with Special Reference to the Australian Region" (A. W. D. Larkum, A. J. McComb and S. A. Shepherd, eds), pp. 565–609. Elsevier, Amsterdam.

Bell, J. D., Hart, A. M., Foyle, T. P., Gervis, M. and Lane, I. (1997a). Can aquaculture help restore and sustain production of giant clams? *In* "Developing and Sustaining World Fisheries Resources: The State of Science and Management" (D. A. Hancock, D. C. Smith, A. Grant and J. P. Beumer, eds), pp. 509–513. 2nd World Fisheries Congress Proceedings, Brisbane, Australia.

Bell, J. D., Lane, I., Gervis, M., Soule, S. and Tafea, H. (1997b). Village-based farming of the giant clam, *Tridacna gigas* (L.), for the aquarium market: Initial trials in the Solomon Islands. *Aquaculture Research* **28**, 121–128.

Benton, R. C. and Lightner, D. (1972). Spray-marking juvenile shrimp with granular fluorescent pigment. *Contributions in Marine Science* **16**, 62–69.

Benzie, J. A. H. (1993a). Review of the Population Genetics of Giant Clams. *In* "Genetic Aspects of Conservation and Cultivation of Giant Clams" (P. Munro, ed.), pp. 1–6. ICLARM Conference Proceedings No. 39, Manila, Philippines.

Benzie, J. A. H. (1993b). Conservation of wild stocks: Policies for the preservation of biodiversity. *In* "Genetic Aspects of Conservation and Cultivation of Giant Clams" (P. Munro, ed.), pp. 13–15. ICLARM Conference Proceedings No. 39, Manila, Philippines.

Benzie, J. A. H. (2000). Population genetic structure in penaeid prawns. *Aquaculture Research* **31**, 95–119.

Benzie, J. A. H. and Williams, S. T. (1995). Gene flow among giant clam (*Tridacna gigas*) populations in Pacific does not parallel ocean circulation. *Marine Biology* **123**, 781–787.

Benzie, J. A. H. and Williams, S. T. (1996). Limitations in the genetic variation in hatchery-produced batches of the giant clam, *Tridacna gigas. Aquaculture* **139**, 225–241.

Benzie, J. A. H. and Williams, S. T. (1997). Genetic structure of giant clam (*Tridacna maxima*) populations in the West Pacific is not consistent with dispersal by present-day ocean currents. *Evolution* **51**, 768–783.

Benzie, J. A. H., Ballment, E. and Frusher, S. (1993). Genetic structure of *Penaeus monodon*. Australia: concordant results from mtDNA and allozymes. *Aquaculture* **111**, 89–93.

Berg, C. J. (1976). Growth of the queen conch, *Strombus gigas,* with a discussion of the practicality of mariculture. *Marine Biology* **34**, 191–199.

Berg, C. J., Mitton, J. B. and Orr, K. S. (1984). Genetic analyses of the queen conch, *Strombus gigas.* 1. Preliminary implications for fisheries management. *Proceedings of the Gulf and Caribbean Fisheries Institute* **37**, 112–118.

Bert, T. and Arnold, S. (1995). An empirical test of predictions of two competing models for the maintenance and fate of hybrid zones: Both models are supported in a hard-clam hybrid zone. *Evolution* **49**, 276–289.

Bert, T., Hesselman, D. M., Arnold, S., Moore, W. S., Cruz-Lopez, H. and Marelli, D. C. (1993). High frequency of gonadal neoplasia in a hard clam (*Mercenaria* spp.) hybrid zone. *Marine Biology* **117**, 97–104.

Bertelsen, R. D. and Matthews, T. R. (2001). Fecundity dynamics of female spiny lobster (*Panulirus argus*) in a south Florida fishery and Dry Tortugas National Park lobster sanctuary. *Marine and Freshwater Research* **52**, 1559–1565.

Beverton, R. J. H. and Holt, S. J. (1957). On the Dynamics of Exploited Fish Populations. *Fisheries Investigations* **19**, 1–533.

Bianchini, M. L., Chessa, L., Greco, S. and Ragonese, R. (1996). An Italian enhancement programme for slipper lobster *Scyllarides latus.* "2nd World Fisheries Congress, Brisbane, Australia Book of Abstracts", p. 91.

Bisker, R. and Castagna, M. (1989). Biological control of crab predation on hard clams *Mercenaria mercenaria* (Linnaeus) by the toadfish *Opsanus tau* (Linnaeus) in tray cultures. *Journal of Shellfish Research* **8**, 33–36.

Blankenship, H. L. and Leber, K. M. (1995). A responsible approach to marine stock enhancement. *American Fisheries Society Symposium* **15**, 167–175.

Blaxter, J. H. S. (2000). The enhancement of marine fish stocks. *Advances in Marine Biology* **38**, 1–54.

Boettcher, A., Dyer, A., Casey, C. and Targett, N. M. (1997). Hydrogen peroxide induced metamorphosis of queen conch, *Strombus gigas*: Tests at the commercial scale. *Aquaculture* **148**, 247–258.

Bohnsack, J. A. (1989). Are high densities of fishes at artificial reefs the result of habitat limitation or behavioural preference? *Bulletin of Marine Science* **44**, 631–645.

Bohnsack, J. A., Ecklund, A. M. and Szmant, A. M. (1997). Artificial reef research: Is there more than the attraction-production issue? *Fisheries* **22**, 14–16.

Booth, J. D. (1979). Settlement of the rock lobster, *Jasus edwardsii* (Decapoda: Palinuridae), at Castlepoint, New Zealand. *New Zealand Journal of Marine and Freshwater Research* **13**, 395–406.

Booth, J. D. (1994). *Jasus edwardsii* larval recruitment off the east coast of New Zealand. *Crustaceana* **66**, 295–317.

Booth, J. D. and Kittaka, J. (2000). Spiny lobster growout. *In* "Spiny Lobsters: Fisheries and Culture (Second Edition)" (B. F. Phillips and J. Kittaka, eds), pp. 556–585. Fishing News Books, Oxford.

Booth, J. D. and Tarring, S. C. (1986). Settlement of the red rock lobster *Jasus edwardsii* near Gisborne New Zealand. *New Zealand Journal of Marine and Freshwater Research* **20**, 291–297.

Booth, J. D., Street, R. J. and Smith, P. J. (1990). Systematic status of the rock lobsters *Jasus edwardsii* from New Zealand and *J. novaehollandiae* from Australia. *New Zealand Journal of Marine and Freshwater Research* **24**, 239–249.

Booth, J., Davies, P. and Zame (1999). Commercial scale collections of young rock lobsters for aquaculture. *World Aquaculture* **99**, 84.

Borsa, P. and Benzie, J. A. H. (1993). Genetic relationships among the topshells *Trochus* and *Tectus* (Prosobranchia: Trochidae) from the Great Barrier Reef. *Journal of Molluscan Studies* **59**, 275–284.

Borthen, J., Agnalt, A.-L. and van der Meeren, G. (1999). A bio-economic evaluation of a stock-enhancement project of European lobster: The simulation model LOBST.ECO with some preliminary results. *In* "Stock Enhancement and Sea Ranching" (B. R. Howell, E. Moksness and T. Svåsand, eds), pp. 583–596. Fishing News Books, Blackwell, Oxford.

Botsford, L. W. (2001). Physical influences on recruitment to California Current invertebrate populations on multiple scales. *ICES Journal of Marine Science* **58**, 1081–1091.

Boudry, P., Collet, B., Cornette, F., Hervouet, V. and Bonhomme, F. (2002). High variance in reproductive success of the Pacific oyster (*Crassostrea gigas*, Thunberg) revealed by microsatellite-based parentage analysis of multifactorial crosses. *Aquaculture* **204**, 283–296.

Bower, S. M. (2000). Infectious diseases of abalone (*Haliotis* spp.) and risks associated with transplantation. *In* "Workshop on Rebuilding Abalone Stocks in British Columbia" (A. Campbell, ed.). *Canadian Special Publication in Fisheries and Aquatic Sciences* **130**, 111–122.

Bower, S.M, McGladdery, S. E. and Price, I. M. (1994). Synopsis of infectious diseases and parasites of commercially exploited shellfish. *Annual Reviews of Fish Diseases* **4**, 1–199.

Braley, R. D. (ed.) (1992). "The Giant Clam: Hatchery and Nursery Culture Manual", p. 144. ACIAR Monograph 15, Australian Centre for International Agricultural Research, Canberra.

Braley, R. D. (1996). The importance of aquaculture and establishment of reserves for restocking of giant clams on overharvested reefs in the Indo-Pacific region. *In* "The Role of Aquaculture in World Fisheries" (T. G. Heggberget, ed.), pp. 136–147. Oxford University Press & IBH Publishing Co, New Delhi.

Bravington, M. V. and Ward, R. D. (2004). Microsatellite DNA markers: Evaluating their potential for estimating the proportion of hatchery-reared offspring in a stock enhancement program. *Molecular Ecology* **13**, 1287–1297.

Breen, P. A. (1986). Management of the British Columbia fishery for northern abalone (*Haliotis kamtschatkana*). *Canadian Special Publication in Fisheries and Aquatic Sciences* **92**, 300–312.

Breitburg, D. L. (1999). Are three dimensional structure and healthy oyster populations the keys to an ecologically interesting and important fish community? *In* "Oyster Reef Habitat Restoration: A Synopsis and Synthesis of Approaches" (M. W. Luckenbach, R. Mann and J. A. Wesson, eds), pp. 239–250. Virginia Institute of Marine Science, School of Marine Science, College of William and Mary, VIMS Press, Gloucester Point, Virginia.

Briones-Fourzán, P., Lozano-Álvarez, E. and Eggleston, D. B. (2000). The use of artificial shelters (*Casitas*) in research and harvesting of Caribbean spiny lobsters in Mexico. Chapter 23. *In* "Spiny Lobsters: Fisheries and Culture (Second Edition)" (B. F. Phillips and J. Kittaka, eds), pp. 420–446. Fishing News Books, Blackwell, Oxford.

Brooker, A. L., Benzie, J. A. H., Blair, D. and Versini, J.-J. (2000). Population structure of the giant tiger prawn *Penaeus monodon* in Australian waters, determined using microsatellite markers. *Marine Biology* **136**, 149–157.

Brown, L. D. (1991). Genetic variation and population structure in the blacklip abalone, *Haliotis rubra. Australian Journal of Marine and Freshwater Research* **42**, 77–90.

Brown, L. D. and Murray, N. D. (1992). Genetic relations within the genus *Haliotis. In* "Abalone of the World: Biology, Fisheries and Culture" (S. A. Shepherd, M. J. Tegner and S. A. Guzmán del Próo, eds), pp. 19–23. Blackwell, Oxford.

Brown, W. M., George, M. and Wilson, A. C. (1979). Rapid evolution of animal mitochondrial DNA. *Proceedings of the National Academy of Sciences USA* **76**, 1967–1971.

Browne, R. and Mercer, J. P. (1999). A review of innovations concerning Ireland's lobster (*Homarus gammarus*) fishery (1992 to 1998). *Journal of Shellfish Research* **18**, 710.

Browne, R., Mercer, J. P. and Duncan, M. J. (2001). A historical overview of the Republic of Ireland's lobster (*Homarus gammarus* L.) fishery, with reference to European and North American (*Homarus americanus* Milne-Edwards) landings. *Hydrobiologia* **465**, 49–62.

Brownell, W. N., Berg, C. J. and Haines, K. C. (1977). Fisheries and aquaculture of the conch, *Strombus gigas*, in the Caribbean. *In* "FAO Fisheries Report 200" pp. 59–69. FAO, Rome.

Bull, M. (1989). New Zealand scallop enhancement project—Costs and benefits. *In* "Proceedings of the Australian Scallop Workshop" (M. C. L. Dredge, W. F. Zacharin and L. M. Joll, eds), pp. 154–165. Tasmanian Government Printer, Hobart, Australia.

Bull, M. (1994). Enhancement and management of New Zealand's southern scallop fishery. *In* "Proceedings of the 9th International Pectinid Workshop, Nanamio, British Colombia, Canada, April 22–27, 1993" (N. F. Bourne, B. L. Bunting and L. D. Townsend, eds), Vol. 2, pp. 131–136.

Buroker, N. E. (1983). Population genetics of the American oyster *Crassostrea virginica* along the Atlantic coast and the Gulf of Mexico. *Marine Biology* **75**, 99–112.

Burton, C. A. (1999). The role of lobster hatcheries in ranching, restoration and remediation programmes. *Journal of Shellfish Research* **18**, 711.

Burton, C. A. and Adamson, K. (2002). "A Preliminary Study of the Costs of Operating a Lobster Hatchery in Orkney and the Development of an Economic

Model for Future Hatchery Programmes". Sea Fish Industry Authority, Edinburgh, United Kingdom.

Burton, R. S. and Tegner, M. J. (2000). Enhancement of red abalone *Haliotis rufescens* stocks at San Miguel Island: Reassessing a success story. *Marine Ecology Progress Series* **202**, 303–308.

Butler, M. J. and Herrnkind, W. F. (1997). A test of recruitment limitation and the potential for artificial enhancement of spiny lobster (*Panulirus argus*) populations in Florida. *Canadian Journal of Fisheries and Aquatic Sciences* **54**, 452–463.

Butler, M. J., Herrnkind, W. F. and Hunt, J. H. (1997). Factors affecting the recruitment of juvenile Caribbean spiny lobsters dwelling in macroalgae. *Bulletin of Marine Science* **61**, 3–19.

Caddy, J. F. (1986). Modelling stock-recruitment processes in Crustacea: Some practical and theoretical perspectives. *Canadian Journal of Fisheries and Aquatic Sciences* **43**, 2330–2344.

Caddy, J. F. (1989). A perspective on the population dynamics and assessment of scallop fisheries, with special reference to the sea scallop *Placopecten magellanicus* (Gmelin). *In* "Marine Invertebrate Fisheries: Their Assessment and Management" (J. F. Caddy, ed.), pp. 559–589. John Wiley and Sons, New York.

Caddy, J. F. and Defeo, O. (2003). "Enhancing or Restoring the Productivity of Natural Populations of Shellfish and Other Marine Invertebrate Resources", p. 159. FAO Fisheries Technical Paper No. 448, FAO, Rome.

Caetano-Anollés, G. and Gresshoff, P. M. (eds) (1997). "DNA Markers: Protocols, Applications and Overviews". Wiley, New York.

Calumpong, H. P. (ed.) (1992). "The Giant Clam: An Ocean Culture Manual", p. 65. ACIAR Monograph No. 16, Australian Centre for International Agricultural Research, Canberra.

Campbell, A. (1983). Growth of tagged American lobsters, *Homarus americanus*, in the Bay of Fundy. *Canadian Journal of Fisheries and Aquatic Sciences* **40**, 1667–1675.

Campbell, A. (ed.) (2000a). "Workshop on Rebuilding Abalone Stocks in British Columbia". *Canadian Special Publication in Fisheries and Aquatic Sciences* **130**.

Campbell, A. (2000b). Review of northern abalone, *Haliotis kamtschatkana,* stock status in British Columbia. *In* "Workshop on Rebuilding Abalone Stocks in British Columbia" (A. Campbell, ed.) *Canadian Special Publication in Fisheries and Aquatic Sciences* **130**, 41–50.

Campbell, A. and Pezzack, D. S. (1986). Relative egg production and abundance of berried lobsters, *Homarus americanus*, in the Bay of Fundy and off south-western Nova Scotia. *Canadian Journal of Fisheries and Aquatic Sciences* **43**, 2190–2196.

Campbell, A. and Robinson, D. G. (1983). Reproductive potential of three lobster (*Homarus americanus*) stocks in the Canadian Maritimes. *Canadian Journal of Fisheries and Aquatic Sciences* **40**, 1958–1967.

Campton, D. E., Berg, C. J., Robison, L. M. and Glazer, R. A. (1992). Genetic patchiness among populations of queen conch *Strombus gigas* in the Florida Keys and Bimini. *Fishery Bulletin, US* **90**, 250–259.

Caputi, N. and Brown, R. S. (1993). The effect of environment on puerulus settlement of the western rock lobster (*Panulirus cygnus*) in Australia. *Fisheries Oceanography* **2**, 1–10.

Caputi, N., Brown, R. S. and Phillips, B. F. (1995a). Prediction of catches of the western rock lobster (*Panulirus cygnus*) based on indices of puerulus and juvenile abundance. *ICES Marine Science Symposia* **199**, 287–293.

Caputi, N., Chubb, C. F. and Brown, R. S. (1995b). Relationship between spawning stock, environment, recruitment and fishing effort for the western rock lobster, *Panulirus cygnus,* fishery in Western Australia. *Crustaceana* **68**, 213–226.

Caputi, N., Penn, J. W., Joll, L. M. and Chubb, C. F. (1998). Stock–recruitment–environment relationships for invertebrate species of Western Australia. *In* "Proceedings of the North Pacific Symposium on Invertebrate Stock Assessment and Management" (G. S. Jamieson and A. Campbell, eds). *Canadian Special Publications in Fisheries and Aquatic Sciences* **125**, 247–255.

Carlton, J. T. and Mann, R. (1996). Transfers and world-wide introductions. *In* "The Eastern Oyster, *Crassostrea virginica*" (V. S. Kennedy, R. E. I. Newell and A. F. Eble, eds), pp. 691–706. Maryland Sea Grant Program, College Park, Maryland.

Carvalho, G. R. and Hauser, L. (1995). Molecular genetics and the stock concept in fisheries. *In* "Molecular Genetics in Fisheries" (G. R. Carvalho and T. J. Pitcher, eds), pp. 55–79. Chapman and Hall, London.

Castell, L. L. (1996). "Ecology of Wild and Cultured Juvenile *Trochus niloticus* Relevant to the Use of Juveniles for Population Enhancement", Ph.D. Thesis, James Cook University, Australia.

Castell, L. L. (1997). Population studies of juvenile *Trochus niloticus* on a reef flat on the North-Eastern Queensland Coast, Australia. *Marine and Freshwater Research* **48**, 211–217.

Castell, L. L. and Sweatman, H. P. A. (1997). Predator–prey interactions among some intertidal gastropods on the Great Barrier Reef. *Journal of Zoology London* **241**, 145–159.

Castell, L. L., Naviti, W. and Nguyen, F. (1996). Detectability of cryptic juvenile *Trochus niloticus* Linnaeus in stock enhancement experiments. *Aquaculture* **144**, 91–101.

Catchpole, E. A., Freeman, S. N., Morgan, B. J. T. and Nash, W. J. (2001). Abalone I: Analysing mark-recapture-recovery data, incorporating growth and delayed recovery. *Biometrics* **57**, 469–477.

Chakalall, B. and Cochrane, K. L. (1997). The queen conch fishery in the Caribbean—An approach to responsible fisheries management. *Proceedings of the Gulf and Caribbean Fisheries Institute* **49**, 531–554.

Chang, E. S. and Conklin, D. E. (1983). Lobster (*Homarus*) hatchery techniques. *In* "CRC Handbook of Mariculture, Vol. 1. Crustacean Aquaculture" (J. P. McVey, ed.), pp. 271–275. CRC Press, Boca Raton, Florida.

Chang, J. W., Baik, K. K., Hwang, Y. T. and Rho, Y. I. (1985). Studies on the release effects of abalone in the eastern waters of Korea. *Bulletin of Fisheries Research Development Agency* **36**, 61–68 (In Korean, with English abstract, figures and tables).

Chapman, R. W., Sedberry, G. R., McGovern, J. C. and Wiley, B. A. (1999). The genetic consequences of reproductive variance: Studies of species with different longevities. *American Fisheries Society Symposium* **23**, 169–181.

Charles, A. (2001). "Sustainable Fisheries Systems", p. 370. Blackwell, Oxford.

Chen, J. (2003). Overview of sea cucumber farming and sea ranching practices in China. *Secretariat of the Pacific Community Beche-de-mer Information Bulletin* **18**, 18–23.

Chen, J. (2004). Present status and prospects of sea cucumber industry in China. *In* "Advances in Sea Cucumber Aquaculture and Management" (A. Lovatelli, C. Conand, S. Purcell, S. Uthicke, J.-F. Hamel and A. Mercier, eds), pp. 25–38. FAO Fisheries Technical Paper. No. 463. FAO, Rome.

Chen, L.-C. (1990). "Aquaculture in Taiwan", p. 273. Fishing News Books, Oxford.

Cheneson, R. (1997). Status of the trochus resource in French Polynesia. *In* "Workshop on Trochus Resource Assessment, Management and Development" (T. D. N. Integrated Fisheries Management Project, ed.), pp. 35–37. South Pacific Commission, Noumea, New Caledonia.

Chew, K. K. (1989). Manila clam biology and fishery development in western North America. *In* "Clam Mariculture in North America" (J. J. Manzi and M. Castagna, eds), pp. 243–261. Developments in Aquaculture and Fisheries Science 19, Elsevier, Amsterdam.

Childress, M. and Herrnkind, W. F. (1994). The behavior of juvenile Caribbean spiny lobsters in Florida Bay: Seasonality, ontogeny and sociality. *Bulletin of Marine Science* **54**, 819–827.

Chivas, A. R., Garcia, A., van der Kaars, S., Couapel, M. J. J., Holt, S., Reeves, J. M., Wheeler, D. J., Switzer, A. D., Murray-Wallace, C. V., Banerjee, D., Price, D. M., Wang, S. X., Pearson, G., Edgar, N. T., Beaufort, L., De Deckker, P., Lawson, E. and Cecil, C. B. (2001). Sea-level and environmental changes since the last interglacial in the Gulf of Carpentaria, Australia: An overview. *Quaternary International* **83–85**, 19–46.

Choquet, R., Pradel, R., Gimenez, O., Reboulet, A. M. and Lebreton, J. D. (2002). M-SURGE Version 1.0: Multistate Survival Generalized Estimation. A general software for multistate capture–recapture models. CEFE/CNRS, Montpellier, France. http://www.cefe.cnrs-mop.fr/wwwbiom/M-SURGE.htm.

Choquet, R., Reboulet, A. M., Pradel, R., Gimenez, O. and Lebreton, J. D. (2003). User's manual for M-SURGE 1.0. Mimeographed document. CEFE/CNRS, Montpellier, France. ftp://ftp.cefe.cnrs-mop.fr/biom/Soft-CR.

Clarke, P. J., Komatsu, T., Bell, J. D., Lasi, F., Oengpepa, C. P. and Leqata, J. (2003). Combined culture of *Trochus niloticus* and giant clams (Tridacnidae): Benefits for restocking and farming. *Aquaculture* **215**, 123–144.

Cliche, G., Vigneau, S. and Giguere, M. (1997). Status of a commercial sea scallop enhancement project in Iles-de-la-Madeleine (Quebec, Canada). *Aquaculture International* **5**, 259–266.

Cobb, J. S. and Wang, D. (1985). Fishery biology of lobsters and crayfish. *In* "The Biology of Crustacea. Volume 10. Economic Aspects: Fisheries and Culture" (A. J. Provenzano, ed.), pp. 168–247. Academic Press, San Diego, California.

Cochard, J. C. and Devauchelle, N. (1993). Spawning, fecundity and larval survival and growth in relation to controlled conditioning in native and transplanted populations of *Pecten maximus* (L.): Evidence for the existence of separate stocks. *Journal of Experimental Marine Biology and Ecology* **169**, 41–56.

Cochrane, K. L., Chakalall, B. and Munro, B. (2004). The whole could be greater than the sum of the parts: The potential benefits of cooperative management of the Caribbean spiny lobster. *In* "Management of Shared Fish Stocks" (A. I. L. Payne, C. M. O'Brien and S. I. Rogers, eds), pp. 223–239. Blackwell, Oxford.

Collie, J., Richardson, K. and Steele, J. H. (2004). Regime shifts: Can theory illuminate the mechanisms? *Progress in Oceanography* **60**, 281–302.

Colquhoun, J. R. (2001). Habitat preferences of juvenile trochus in Western Australia: Implications for stock enhancement and assessment. *Secretariat of the Pacific Community Trochus Information Bulletin* **7**, 14–20.

Commito, J. A. and Boncavage, E. M. (1989). Suspension-feeders and co-existing infauna: An enhancement counter example. *Journal of Experimental Marine Biology and Ecology* **125**, 33–42.

Conan, G. Y. (1986). Summary of Session 5: Recruitment enhancement. International Workshop on Lobster Recruitment. *Canadian Journal of Fisheries and Aquatic Sciences* **43**, 2384–2388.

Conand, C. (1989). "Les Holothuries Aspidochirotes du Lagon de Nouvelle-Caledonie: Biologie, Ecologie et Exploitation", p. 393. Etudes et Theses ORS-TOM, Paris.

Conand, C. (1990). "The Fishery Resources of Pacific Island Countries: *Holothurians.* Part 2", p. 143. FAO Fisheries Technical Paper 272.2, FAO, Rome.

Conand, C. (2004). Present status of world sea cucumber resources and utilization: An international overview. *In* "Advances in Sea Cucumber Aquaculture and Management" (A. Lovatelli, C. Conand, S. Purcell, S. Uthicke, J. F. Hamel and A. Mercier, eds), pp. 13–24. FAO Fisheries Technical Paper No. 463. FAO, Rome.

Conand, C. and Byrne, M. (1993). A review of recent developments in the world sea cucumber fisheries. *Marine Fisheries Review* **55**(4), 1–13.

Conand, C. and Sloan, N. A. (1989). World fisheries for Echinoderms. *In* "Marine Invertebrate Fisheries: Their Assessment and Management" (J. F. Caddy, ed.), pp. 647–663. Wiley, Chichester.

Condie, S. A., Loneragan, N. R. and Die, D. J. (1999). Modelling the recruitment of tiger prawns *Penaeus esculentus* and *P. semisulcatus* to the nursery grounds in the Gulf of Carpentaria, northern Australia: Implications for assessing stock-recruitment relationships. *Marine Ecology Progress Series* **178**, 55–68.

Conod, N., Bartlett, J. P., Evans, B. S. and Elliott, N. G. (2002). Comparison of mitochondrial and nuclear DNA analyses of population structure in blacklip abalone, *Haliotis rubra* Leach. *Marine and Freshwater Research* **53**, 711–718.

Cook, P. A. and Sweijd, N. A. (1999a). The potential for abalone ranching and enhancement in South Africa. *In* "Stock Enhancement and Sea Ranching" (B. R. Howell, E. Moksness and T. Svåsand, eds), pp. 453–467. Fishing News Books, Blackwell, Oxford.

Cook, P. A. and Sweijd, N. A. (1999b). Some genetic considerations of shellfish ranching: A case-study of the abalone, *Haliotis midae,* in South Africa. *Journal of Shellfish Research* **18**, 712.

Copland, J. W. and Lucas, J. S. (eds) (1988). "Giant Clams in Asia and the Pacific", p. 274. ACIAR Monograph No. 9, Australian Centre for International Agricultural Research, Canberra.

Cowan, D. F. (1999). Method for assessing relative abundance, size distribution, and growth of recently settled and early juvenile lobsters (*Homarus americanus*) in the lower intertidal zone. *Journal of Crustacean Biology* **19**, 738–751.

Cowen, R. K., Lwiza, K. M. M., Sponaugle, S., Paris, C. B. and Olson, D. B. (2000). Connectivity of marine populations: Open or closed? *Science* **287**, 857–859.

Crear, B. J., Thomas, C. W., Hart, P. R. and Carter, C. G. (2000). Growth of juvenile southern rock lobsters, *Jasus edwardsii,* is influenced by diet and temperature, whilst survival is influenced by diet and tank environment. *Aquaculture* **190**, 169–182.

Crear, B., Hart, T., Thomas, C. and Barclay, M. (2002). Evaluation of commercial shrimp grow-out pellets as diets for juvenile southern rock lobster, *Jasus edwardsii:* Influence on growth, survival, colour and biochemical composition. *Journal of Applied Aquaculture* **12**, 43–57.

Creswell, R. L. (1994). An historical overview of queen conch mariculture. *In* "Queen Conch Biology, Fisheries and Mariculture" (R. Appeldoorn and B. Rodriguez, eds), pp. 223–230. Fundación Científica Los Roques, Caracas.

Crocos, P., Arnold, S., Sellars, M., Burford, M., Barnard, R. and McCulloch, R. (2003). Technology development for the high-density production of juvenile *Penaeus esculentus*. In "Developing Techniques for Enhancing Prawn Fisheries, with a Focus on Brown Tiger Prawns (*Penaeus esculentus*) in Exmouth Gulf" (N. Loneragan, ed.), Final Report Australian Fisheries Research and Development Corporation Project 1999/222, p. 276. CSIRO, Cleveland, Australia.

Cross, T. F. (2000). Genetic implications of translocation and stocking of fish species, with particular reference to Western Australia. *Aquaculture Research* **31**, 83–94.

Crowe, T. P., Amos, M. J. and Lee, C. L. (1997). The potential of reseeding with juveniles as a tool for the management of trochus fisheries. In "Trochus: Status, Hatchery Practice and Nutrition" (C. L. Lee and P. W. Lynch, eds), pp. 170–177. ACIAR Proceedings No. 79, Australian Centre for International Agricultural Research, Canberra.

Crowe, T. P., Dobson, G. and Lee, C. (2001). A novel method for tagging and recapturing animals in complex habitats and its use in research into stock enhancement of *Trochus niloticus*. *Aquaculture* **194**, 383–391.

Crowe, T. P., Lee, C. L., Mc Guinness, K. A., Amos, M. J., Dangeubun, J., Dwiono, S. A. P., Makatipu, P. C., Manuputly, J., N'guyen, F., Pakoa, K. and Tetelepta, J. (2002). Experimental evaluation of the use of hatchery-reared juveniles to enhance stocks of the topshell *Trochus niloticus* in Australia, Indonesia and Vanuatu. *Aquaculture* **206**, 175–197.

Cruz, R. and Phillips, B. F. (2000). The artificial shelters (*Pesqueros*) used for the spiny lobster (*Panulirus argus*) fisheries in Cuba. In "Spiny Lobsters: Fisheries and Culture (Second Edition)" (B. F. Phillips and J. Kittaka, eds), pp. 400–419. Fishing News Books, Oxford.

Cruz, R., DeLeon, M. E. and Puga, R. (1995). Prediction of commercial catches of the spiny lobster *Panulirus argus* in the Gulf of Batabano, Cuba. *Crustaceana* **68**, 238–244.

Crozier, W. W. and Moffett, I. J. J. (1995). Application of low frequency genetic marking at GPI-3* and MDH-B1,2* to assess supplementary stocking of Atlantic salmon, *Salmo salar* L., in a Northern Irish stream. *Fisheries Management and Ecology* **2**, 27–36.

Dall, W., Hill, B. J., Rothlisberg, P. C. and Staples, D. J. (1990). The biology of the Penaeidae. *Advances in Marine Biology* **27**, 1–489.

Dalton, A. (1994). Mariculture of the queen conch (*Strombus gigas* L.): Development of growout and nursery techniques. In "Queen Conch Biology, Fisheries and Mariculture" (R. Appeldoorn and B. RodriguezE, eds), pp. 253–260. Fundación Científica Los Roques, Caracas.

Dance, S. K., Lane, I. and Bell, J. D. (2003). Variation in short-term survival of cultured sandfish (*Holothuria scabra*) released in mangrove-seagrass and coral reef flat habitats in Solomon Islands. *Aquaculture* **220**, 495–505.

Dao, J., Fleury, P. and Barret, J. (1999). Scallop sea bed culture in Europe. In "Stock Enhancement and Sea Ranching" (B. R. Howell, E. Moksness and T. Svåsand, eds), pp. 423–426. Fishing News Books, Blackwell, Oxford.

D'Asaro, C. N. (1965). Organogenesis, development and metamorphosis in the queen conch, *Strombus gigas*, with notes on breeding habits. *Bulletin of Marine Science* **15**, 359–416.

Davenport, J., Ekaratne, S. U. K., Walgama, S. A., Lee, D. and Hills, J. M. (1999). Successful stock enhancement of a lagoon prawn fishery at Rekawa, Sri Lanka using cultured post-larvae of penaeid shrimp. *Aquaculture* **180**, 65–78.

Davis, G. E. (1995). Recruitment of juvenile abalone (*Haliotis* spp.) measured in artificial habitats. *Marine and Freshwater Research* **46**, 549–554.

Davis, G. E. (2000). Refugia-based strategies to restore and sustain abalone (*Haliotis* spp.) populations in Southern California. *In* "Workshop on Rebuilding Abalone Stocks in British Columbia" (A. Campbell, ed.), *Canadian Special Publication in Fisheries and Aquatic Sciences* **130**, 133–138.

Davis, G. E., Haaker, P. L. and Richards, D. V. (1998). The perilous condition of white abalone, *Haliotis sorenseni* Bartsch 1940. *Journal of Shellfish Research* **17**, 871–876.

Davis, J. A. (1986). Boundary layers, flow microenvironments and stream benthos. *In* "Limnology in Australia" (P. de Dekker and W. D. Williams, eds), pp. 303–312. Junk, Dordrecht, Netherlands.

Davis, M. (1994). Mariculture techniques for queen conch (*Strombus gigas* L.): Egg mass to juvenile stages. *In* "Queen Conch Biology, Fisheries and Mariculture" (R. Appeldoorn and B. Rodriguez, eds), pp. 231–252. Fundación Científica Los Roques, Caracas.

Davis, M. (2000). The combined effects of temperature and salinity on growth, development, and survival for tropical gastropod veligers of *Strombus gigas*. *Journal of Shellfish Research* **19**, 883–889.

Davis, M. (2001). The effects of natural foods on the growth and development of queen conch larvae (*Strombus gigas*). *Proceedings of the Gulf and Caribbean Fisheries Institute* **52**, 757–766.

Davis, M. and Dalton, A. (1991). New large-scale culturing techniques for *Strombus gigas* post larvae, in the Turks and Caicos Islands. *Proceedings of the Gulf and Caribbean Fisheries Institute* **40**, 257–266.

Davis, M. and Stoner, A. W. (1994). Trophic cues induce metamorphosis of queen conch larvae (*Strombus gigas* Linnaeus). *Journal of Experimental Marine Biology and Ecology* **180**, 83–102.

Davis, M., Mitchell, B. A. and Brown, J. L. (1984). Breeding behavior of the queen conch, *Strombus gigas*. *Journal of Shellfish Research* **4**, 17–21.

Davis, M., Hesse, C. and Hodgkins, G. (1987). Commercial hatchery produced queen conch, *Strombus gigas,* seed for the research and grow-out market. *Proceedings of the Gulf and Caribbean Fisheries Institute* **38**, 326–335.

Davis, M., Heyman, W. D., Harvey, W. and Withstandley, C. A. (1990). A comparison of two inducers, KCL and *Laurencia* extracts, and techniques for the commercial scale induction of metamorphosis in queen conch, *Strombus gigas*, larvae. *Journal of Shellfish Research* **9**, 67–73.

Davis, M., Dalton, A. and Higgs, P. (1992). Recent developments in conch mariculture in the Turks and Caicos Islands. *Proceedings of the Gulf and Caribbean Fisheries Institute* **42**, 397–402.

Day, E. and Branch, G. M. (2000). Evidence for a positive relationship between juvenile abalone *Haliotis midae* and the sea urchin *Parechinus angulosus* in the south-western Cape, South Africa. *South Africa Journal of Marine Science* **22**, 145–156.

De Waal, S. W. P. and Cook, P. (2001a). Use of a spreadsheet model to investigate the dynamics and the economics of a seeded abalone population. *Journal of Shellfish Research* **19**, 863–866.

De Waal, S. W. P. and Cook, P. (2001b). Quantifying the physical and biological attributes of successful ocean seeding sites for farm reared juvenile abalone (*Haliotis midae*). *Journal of Shellfish Research* **19**, 857–861.

De Waal, S. W. P., Branch, G. M. and Navarro, R. (2003). Interpreting evidence of dispersal by *Haliotis midae* juveniles seeded in the wild. *Aquaculture* **221**, 299–310.

Debenham, P., Brzezinski, M., Foltz, K. and Gaines, S. (2000). Genetic structure of populations of the red sea urchin, *Strongylocentrotus franciscanus*. *Journal of Experimental Marine Biology and Ecology* **253**, 49–62.

Delgado, G. A., Glazer, R. A., Stewart, N. J., McCarthy, K. J. and Kidney, J. A. (2000). Modification of behavior and morphology in hatchery-reared queen conch (*Strombus gigas*, L.): A preliminary report. *Proceedings of the Gulf and Caribbean Fisheries Institute* **51**, 80–86.

Del Río-Portilla, M. A. and González-Aviles, J. G. (2000). Population genetics of the yellow abalone, *Haliotis corrugata*, in Cedros and San Benito Islands: A preliminary survey. *Journal of Shellfish Research* **20**, 765–770.

Dennis, D. M., Skewes, T. D. and Pitcher, C. R. (1997). Use of habitats and growth of juvenile ornate rock lobsters *Panulirus ornatus* (Fabricius, 1798) in Torres Strait, Australia. *Marine and Freshwater Research* **48**, 663–670.

Denniston, C. D. (1978). Small population size and genetic diversity. *In* "Endangered Birds: Management Techniques for Preserving Endangered Species" (S. A. Temple, ed.) University of Wisconsin Press, Madison.

Dexter, D. M. (1972). Moulting and growth in laboratory reared phyllosomas of the California spiny lobster, *Panulirus interruptus*. *California Fish and Game* **58**, 107–115.

Dillon, R. T. and Manzi, J. J. (1987). Hard clam, *Mercenaria mercenaria*, broodstocks: Genetic drift and loss of rare alleles without reduction in heterozygosity. *Aquaculture* **60**, 99–105.

Dillon, R. T. and Manzi, J. J. (1989). Genetics and shell morphology in a hybrid zone between the hard clams *Mercenaria mercenaria* and *M. campechiensis*. *Marine Biology* **100**, 217–222.

Doherty, P. (1999). Recruitment limitation is the theoretical basis for stock enhancement in marine populations. *In* "Stock Enhancement and Sea Ranching" (B. R. Howell, E. Moksness and T. Svåsand, eds), pp. 9–21. Fishing News Books, Blackwell, Oxford.

Doherty, P. J., Planes, S. and Mather, P. (1995). Gene flow and larval duration in seven species of fish from the Great Barrier Reef. *Ecology* **76**, 2373–2391.

Doumenge, F. (1973). Developing the exploitation of *Trochus niloticus* stock on the Tahiti reefs. *South Pacific Commission Fisheries Newsletter* **10**, 35–36.

Drummond, K. (2004). The role of stock enhancement in the management framework for New Zealand's southern scallop fishery. *In* "Stock Enhancement and Sea Ranching: Developments, Pitfalls and Opportunities" (K. M. Leber, S. Kitada, L. Blankenship and T. Svåsand, eds), pp. 397–412. Blackwell, Oxford.

Duggins, D. (1980). Kelp beds and sea otters: An experimental approach. *Ecology* **61**(3), 447–453.

Durand, P., Wada, K. T. and Blanc, F. (1993). Genetic variations in wild and hatchery stocks of the black pearl oyster *Pinctada margaritifera* from Japan. *Aquaculture* **110**, 27–40.

Ebert, T. A. (2001). Growth and survival of post-settlement sea urchins. *In* "Edible Sea Urchins: Biology and Ecology" (J. M. Lawrence, ed.), pp. 79–102. Elsevier Science, Amsterdam.

Ebert, T. B. and Ebert, E. E. (1988). An innovative technique for seeding abalone and preliminary results of laboratory and field trials. *California Fish and Game* **74**, 68–81.

Edmands, S., Moberg, P. E. and Burton, R. S. (1996). Allozyme and mitochondrial DNA evidence of population subdivision in the purple sea urchin *Strongylocentrotus purpuratus. Marine Biology* **126**, 443–450.

Eeckhaut, I., Parmentier, E., Becker, P., Gomez da Silva, S. and Jangoux, M. (2004). Parasites and biotic diseases in field and cultivated sea cucumbers. *In* "Advances in Sea Cucumber Aquaculture and Management" (A. Lovatelli, C. Conand, S. Purcell, S. Uthicke, J. F. Hamel and A. Mercier, eds), pp. 311–326. FAO Fisheries Technical Paper No. 463. FAO, Rome.

Eggleston, D. B. (1999). Application of landscape ecological principles to oyster reef habitat restoration. *In* "Oyster Reef Habitat Restoration: A Synopsis and Synthesis of Approaches" (M. W. Luckenbach, R. Mann and J. A. Wesson, eds), pp. 213–228. Virginia Institute of Marine Science, School of Marine Science, College of William and Mary, VIMS Press, Gloucester Point, Virginia.

Eggleston, D. B., Lipcius, R. N. and Miller, D. L. (1992). Artificial shelters and survival of juvenile Caribbean spiny lobster *Panulirus argus*: Spatial, habitat, and lobster size effects. *Fishery Bulletin, US* **90**, 691–702.

Eldredge, L. G. (1994). "Introductions of Commercially Significant Aquatic Organisms to the Pacific Islands: Perspectives in Aquatic Exotic Species Management in the Pacific Islands. Inshore Fisheries Research Project Technical Document No 7", p. 127. South Pacific Commission, Noumea, New Caledonia.

Eldredge, P. J., Eversole, A. G. and Whetstone, J. M. (1979). Comparative survival and growth rate of hard clams *Mercenaria mercenaria*, planted in trays subtidally and intertidally at varying densities in a South Carolina estuary. *Proceedings of the National Shellfish Association* **69**, 30–39.

Elliott, N. G., Bartlett, J., Evans, B., Officer, R. and Haddon, M. (2002). "Application of Molecular Genetics to the Australian Abalone Fisheries: Forensic Protocols for Species Identification and Blacklip Stock Structure", p. 132. Final Report Fisheries Research and Development Corporation Project 1999/164. CSIRO, Hobart, Australia.

Elner, R. W. and Jamieson, G. S. (1979). Predation of sea scallops, *Placopecten magellanicus*, by the rock crab, *Cancer irroratus*, and the American lobster, *Homarus americanus. Journal of the Fisheries Research Board Canada* **36**, 537–543.

Elston, R. A. and Burge, R. (1984). Pathology and certification of the Japanese scallop, *Pactionpecten yessoensis*: A case history. *Journal of Shellfish Research* **4**, 110.

Emery, L. and Wydoski, R. (1987). Marking and tagging of aquatic animals: An indexed bibliography. *US Fish and Wildlife Service Resource Publication* **165**, 1–57.

Emmett, B. and Jamieson, G. S. (1989). An experimental transplant of northern abalone, *Haliotis kamtschatkana*, in Barkley Sound, British Columbia. *Fishery Bulletin, US* **87**, 95–104.

Ennis, G. P. (1972). Growth per molt of tagged lobsters (*Homarus americanus*) in Bonavista Bay, Newfoundland. *Journal of the Fisheries Research Board of Canada* **29**, 143–148.

Ennis, G. P. (1986). Sphyrion tag loss from the American lobster *Homarus americanus. Transactions of the American Fisheries Society* **115**, 914–917.

Estes, J. A. and Duggins, D. (1995). Sea otters and kelp forests in Alaska: Generality and variation in a community ecological paradigm. *Ecological Monographs* **65**, 75–100.

Estrella, B. T. and Morrisey, T. D. (1997). Seasonal movement of offshore American lobster, *Homarus americanus*, tagged along the eastern shore of Cape Cod, Massachusetts. *Fishery Bulletin, US* **95**, 466–476.

Etaix-Bonnin, R. and Fao, B. (1997). Country statement—New Caledonia. "Workshop on Trochus Resource Assessment, Management and Development. Integrated Fisheries Management Project" (unedited), pp. 43–46. Technical Document No. 13, South Pacific Commission, Noumea, New Caledonia.

Evans, B., White, W. R. and Elliott, G. N. (2000). Characterization of microsatellite loci in the Australian blacklip abalone (*Haliotis rubra* Leach). *Molecular Ecology* **9**, 1183–1184.

Evans, L. H., Jones, J. B. and Brock, J. A. (2000). Diseases of spiny lobsters. *In* "Spiny Lobsters: Fisheries and Culture (Second Edition)" (B. F. Phillips and J. Kittaka, eds), pp. 586–600. Fishing News Books, Oxford.

Evans, B., Sweijd, N., Bowie, R. C. K., Cook, P. and Elliott, N. G. (2004). Population genetic structure of the Perlemoen, *Haliotis midae* in South Africa: Evidence of range expansion and founder events. *Marine Ecology Progress Series* **270**, 163–172.

FAO (1995a). "Code of Conduct for Responsible Fisheries", p. 41. FAO, Rome.

FAO (1995b). "Precautionary Approach to Fisheries. Part 1: Guidelines on the Precautionary Approach to Capture Fisheries and Species Introductions", p. 52. FAO Fisheries Technical Paper 350/1. FAO, Rome.

FAO (2002). "Yearbook of Fishery Statistics. 1999 Catches and Landings". FAO, Rome (online database).

FAO (2003a). FishStat Plus [http://www.fao.org/fi/statist/FISOFT/FISHPLUS.asp].

FAO (2003b). "The Ecosystem Approach to Fisheries", p. 112. FAO Technical Guidelines for Responsible Fisheries No. 4, Suppl. 2, FAO, Rome.

Farmer, A. S. D. (1981). Historical review of the Kuwait shrimp culture project. *Kuwait Bulletin of Marine Science* **2**, 3–9.

Farmer, A. S. D. (1981). A review of crustacean marking methods with particular reference to penaeid shrimp. *Kuwait Bulletin of Marine Science* **2**, 167–183.

Ferguson, M. M. and Danzmann, R. G. (1998). Role of genetic markers in fisheries and aquaculture: Useful tools or stamp collecting? *Canadian Journal of Fisheries and Aquatic Sciences* **55**, 1553–1563.

Ferguson, A. F., Prodöhl, P. A., Hughes, M., Heath, P., Hynes, R., Taggart, J.B, Jørstad, K., Svasand, T., Agnalt, A.-L., Farestveit, E., Naess, H., Wennevik, V., Mercer, J. P., Ball, B., Kelly, E., Browne, R., Linnane, A., Brown, D., Maddock, T., Triantaphyllidis, C., Kouvatsi, A., Apostolidis, A., Triantafyllidis, A. and Katsares, V. (2002). Genetic diversity in the European lobster (*Homarus gammarus*): Population structure and impacts of stock enhancement. http://www.qub.ac.uk/bb/prodohl/GEL/gel.html.

Ferrer, L. T. and Alcolado, P. M. (1994). Panoramica actual del *Strombus gigas* en Cuba. *In* "Queen Conch Biology, Fisheries and Mariculture" (R. S. Appeldoorn and B. Rodriguez, eds), pp. 73–78. Fundación Científica Los Roques, Caracas.

Fevolden, S. V. (1989). Genetic differentiation of the Iceland scallop *Chlamys islandica* (Pectinidae) in the Northern Atlantic ocean. *Marine Ecology Progress Series* **51**, 77–85.

Finley, C. A., Wendell, F. and Friedman, C. S. (2001). Geographic distribution of *Candidatus Xenohaliotis californiensis* in northern California abalone. "Aquaculture 2001: Book of Abstracts", p. 225. World Aquaculture Society, Baton Rouge, Louisiana.

Fisher, W. S., Nilsen, E. H., Steenbergen, J. F. and Lightner, D. V. (1978). Microbial diseases of cultured lobsters: A review. *Aquaculture* **14**, 115–140.

Fisschler, K. J. and Walburg, C. H. (1962). Blue crab movement in coastal South Carolina, 1958–59. *Transactions of the American Fisheries Society* **91**, 275–278.

Flassch, J. P. and Aveline, C. (1984). "Production de jeunes ormeaux à la station expérimentale d'Argenton". Centre National pour l'Exploitation des Océans, Rapports Scientifiques et Techniques 50.

Fleming, A. E. (1995). Growth, intake, feed conversion efficiency and chemosensory preference of the Australian abalone *Haliotis rubra*. *Aquaculture* **132**, 297–311.

Fleming, I. A., Lamberg, A. and Jonsson, B. (1997). Effects of early experience on the reproductive performance of Atlantic salmon. *Behavioral Ecology* **8**, 470–480.

Fletcher, W. J., Fielder, D. R. and Brown, I. W. (1989). Comparison of freeze- and heat-branding techniques to mark the coconut crab *Birgus latro* (Crustacea, Anomura). *Journal of Experimental Marine Biology and Ecology* **127**, 245–251.

Fleury, P. G., Mingant, C. and Castillo, A. (1996). A preliminary study of behaviour and vitality of reseeded juvenile great scallops of three sizes in three seasons. *Aquaculture International* **4**, 325–337.

Fleury, P. G., Doa, J. C., Mikolajunas, J. P., Minchin, D., Norman, M. and Strand, O. (1997). Concerted action on scallop seabed cultivation in Europe. Final Report, specific community programme for research, technological development and demonstration in the field of agriculture and agro-forestry, inclusive fisheries AIR2-CT93–1647, pp. 52.

Foale, S. (1997). Ownership and management of traditional trochus fisheries at West Ngella, Solomon Islands. *In* "Developing and Sustaining World Fisheries Resources: The State of Science and Management" (D. A. Hancock, D. C. Smith, A. Grant and J. P. Beumer, eds), pp. 266–272. 2nd World Fisheries Congress Proceedings, Brisbane, CSIRO, Melbourne, Australia.

Fogarty, M. J. (1995). Populations, fisheries and management. *In* "Biology of the Lobster, *Homarus americanus*" (J. R. Factor, ed.), pp. 111–138. Academic Press, San Diego, California.

Fogarty, M. J. and Idoine, J. S. (1986). Recruitment dynamics in an American lobster (*Homarus americanus*) population. *Canadian Journal of Fisheries and Aquatic Sciences* **43**, 2368–2376.

Fogarty, M. J., Sissenwine, M. P. and Cohen, E. B. (1991). Recruitment variability and the dynamics of exploited marine populations. *Trends in Ecology and Evolution* **6**, 241–246.

Ford, S. E. (2001). Pests, parasites, diseases, and defense mechanisms of the hard clam, *Mercenaria mercenaria*. *In* "Biology of the Hard Clam" (J. N. Kraeuter and M. Castagna, eds), pp. 591–621. Developments in Aquaculture and Fisheries Science, 31. Elsevier, Amsterdam.

Ford, S. E. and Haskin, H. H. (1982). History and epizootiology of *Haplosporidium nelsoni* (MSX), an oyster pathogen, in Delaware Bay, 1957–1980. *Journal of Invertebrate Pathology* **40**, 118–141.

Foyle, T. P., Bell, J. D., Gervis, M. and Lane, I. (1997). Survival and growth of juvenile fluted giant clams, *Tridacna squamosa*, in large-scale village grow-out trials in the Solomon Islands. *Aquaculture Research* **148**, 85–104.

Friedman, C. S., Andree, K. B., Beauchamp, K., Moore, J. D., Robbins, T. T., Shields, J. D. and Hedrick, R. P. (2000). *Candidatus Xenohaliotis californiensis*, a newly described pathogen of abalone, *Haliotis* spp., along the west coast of North America. *International Journal of Systematic and Evolutionary Microbiology* **50**, 847–855.

Fry, B. (1981). Natural stable carbon isotope tag traces Texas shrimp migrations. *Fishery Bulletin, US* **79**, 337–345.

Fry, B., Mumford, P. L. and Robblee, M. B. (1999). Stable isotope studies of pink shrimp (*Farfantepenaeus duorarum* Burkenroad) migrations on the southwestern Florida shelf. *Bulletin of Marine Science* **65**, 419–430.

Fujio, Y. and von Brand, E. (1991). Differences in degree of homozygosity between seed and sown populations of the Japanese scallop *Patinopecten yessoensis*. *Nippon Suisan Gakkaishi* **57**(1), 45–50.

Fujita, N. and Mori, K. (1990). Effects of environmental instability on the growth of the Japanese scallop *Patinopecten yessoensis* in Abashiri sowing-culture grounds. *In* "Marine Farming and Enhancement. Proceedings of the Fifteenth U.S.-Japan Meeting on Aquaculture, Kyoto, Japan, October 22–23, 1986" (A. K. Sparks, ed.), pp. 81–89. US Department of Commerce, NOAA Technical Report NMFS 85.

Fushimi, H. (1999). How to detect the effect in releasing operation of hatchery raised Kuruma prawn postlarvae?—Case study of the operation in the Hamana Lake. *Bulletin of the Tohoku National Fisheries Research Institute* **62**, 1–12.

Gaffney, P. M., Rubin, V. P., Hedgecock, D., Powers, D. A., Morris, G. and Hereford, L. (1996). Genetic effects of artificial propagation—Signals from wild and hatchery populations of red abalone in California. *Aquaculture* **143**, 257–266.

Garcia, S. (1983). The stock–recruitment relationship in shrimps: Reality or artifacts and misinterpretations? *Oceanography Tropicana* **18**, 25–48.

Geiger, D. L. (2000). Distribution and biogeography of the recent Haliotidae (Gastropoda: Vetigastropoda) world-wide. *Bolletino Malacologico* **35**, 57–120.

Geiger, D. L. and Herrmann, R. (1998). A black abalone with "withering foot disease" from the San Diego area. *Festivus* **30**, 101–102.

Gharrett, A. J. and Smoker, W. W. (1991). Two generations of hybrids between even- and odd-year pink salmon (*Oncorhynchus gorbuscha*): A test for out-breeding depression. *Canadian Journal of Fisheries and Aquatic Sciences* **48**, 1744–1749.

Gibbons, M. C. and Blogoslawski, W. J. (1989). Predators, pests, parasites, and diseases. *In* "Clam Mariculture in North America" (J. J. Manzi and M. Castagna, eds), pp. 167–200. Developments in Aquaculture and Fisheries Science 19, Elsevier, Amsterdam.

Giese, A. C. (1966). On the biochemical constitution of some echinoderms. *In* "Physiology of Echinodermata" (R. A. Boolootian, ed.), pp. 547–576. Interscience, New York.

Gillett, R. (1986). A summary of the Tokelau trochus transplant project. *South Pacific Commission Fisheries Newsletter* **38**, 36–43.

Gillett, R. (1993). Pacific Islands trochus introductions. *South Pacific Commission Trochus Information Bulletin* **2**, 13–16.

Glaister, J. (1988). Tag types and uses—Penaeid prawns. *In* "Tagging—Solution or Problem? Australian Society for Fish Biology Tagging Workshop, 21–22 July 1988" (D. A. Hancock, ed.), pp. 62–64. Bureau of Rural Resources Proceedings No. 5, Australian Government Printing Office, Canberra.

Glazer, R. A. and Berg, C. J. (1994). Queen conch research in Florida: An overview. *In* "Queen Conch Biology, Fisheries and Mariculture" (R. S. Appeldoorn and B. Rodriguez, eds), pp. 79–95. Fundación Científica Los Roques, Caracas.

Glazer, R. A., McCarthy, K. J., Anderson, L. and Kidney, J. A. (1997a). Recent advances in the culture of the queen conch in Florida. *Proceedings of the Gulf and Caribbean Fisheries Institute* **49**, 510–522.

Glazer, R. A., McCarthy, K. J., Jones, R. L. and Anderson, L. A. (1997b). The use of underwater metal detectors to recover outplants of the mobile marine gastropod

Strombus gigas L. *Proceedings of the Gulf and Caribbean Fisheries Institute* **49**, 503–509.

Godin, D. M., Carr, W. H., Hagino, G., Segura, F., Sweeney, J. N. and Blankenship, L. (1996). Evaluation of a fluorescent elastomer internal tag in juvenile and adult shrimp *Penaeus vannamei*. *Aquaculture* **139**, 243–248.

Goldstein, D. B. and Schlötterer, C. (eds) (1999). "Microsatellites: Evolution and Applications". Oxford University Press, Oxford.

Goodnight, K. F. and Queller, D. C. (1999). Computer software for performing likelihood tests of pedigree relationship using genetic markers. *Molecular Ecology* **8**, 1231–1234.

Gopurenko, D. and Hughes, J. M. (2002). Regional patterns of genetic structure among Australian populations of the mud crab, *Scylla serrata* (Crustacea: Decapoda): Evidence from mitochondrial DNA. *Marine and Freshwater Research* **53**, 849–857.

Gosling, E. M. (1982). Genetic variability in hatchery-produced Pacific oysters (*Crassostrea gigas* Thunberg). *Aquaculture* **26**, 273–287.

Govan, H. (1995). *Cymatium muricinum* and other ranellid gastropods: Major predators of cultured tridacnid clams. ICLARM Technical Report 49, Manila, Philippines.

Govind, C. K. and Kent, K. S. (1982). Transformation of fast fibres to slow prevented by lack of activity in developing lobster muscle. *Nature* **298**, 755–757.

Gracia, A. (1991). Spawning stock–recruitment relationships of white shrimp in the southwestern Gulf of Mexico. *Transactions of the American Fisheries Society* **120**, 519–527.

Gracia, A. (1996). White shrimp (*Penaeus setiferus*) recruitment overfishing. *Marine and Freshwater Research* **47**, 59–65.

Grice, A. M. and Bell, J. D. (1997). Enhanced growth of the giant clam, *Tridacna derasa* (Roding, 1798), can be maintained by reducing the frequency of ammonium supplements. *Journal of Shellfish Research* **16**, 523–525.

Grice, A. M. and Bell, J. D. (1999). Application of dissolved inorganic nutrients to enhance growth of giant clams (*Tridacna maxima*): Effects of size-class, stocking density, and nutrient concentration. *Aquaculture* **170**, 17–28.

Griffin, D. A., Wilkin, J. L., Chubb, C. F., Pearce, A. F. and Caputi, N. (2001). Ocean currents and the larval phase of Australian western rock lobster, *Panulirus cygnus*. *Marine and Freshwater Research* **52**, 1187–1199.

Grimes, C. B. (1998). Marine stock enhancement: Sound management or technoarrogance? *Fisheries* **23**(9), 18–23.

Grove, R. S., Sonu, C. J. and Nakamura, M. (1991). Design and engineering of manufactured habitats for fisheries enhancement. *In* "Artificial Habitats for Marine and Freshwater Fisheries" (W. Seaman, Jr and L. M. Sprague, eds), pp. 109–152. Academic Press, San Diego.

Grove, R. S., Nakamura, M., Kakimoto, H. and Sonu, C. J. (1994). Aquatic habitat technology innovation in Japan. *Bulletin of Marine Science* **55**, 276–294.

Guichard, F. and Bouchet, E. (1998). Topographic heterogeneity, hydrodynamics, and benthic community structure: A scale-dependent cascade. *Marine Ecology Progress Series* **171**, 59–70.

Gulka, G., Chang, P. W. and Marti, K. A. (1983). Prokaryotic infection associated with a mass mortality of the sea scallops, *Placopecten magellanicus*. *Journal of Fish Diseases* **6**, 355–364.

Gulland, J. A. and Rothschild, B. R. (1984). "Penaeid Shrimps—Their Biology and Management", p. 308. Fishing News Books, Oxford.

Guo, X., Guo, X. and Lin, Z. (1996). A study on intertidal abalone cage culture technique. *Transactions of Oceanology and Limnology* **1996**(3), 54–58. (In Chinese with English abstract.)

Guo, X., Zhang, X., Luan, Z. and Guan, X. (1998). Studies on technique for farming abalone with piling stones covered with net in the intertidal zone. *Shandong Fisheries* **15**(2), 7–9. (In Chinese with English abstract.)

Guo, X., Ford, S. E. and Zhang, F. (1999). Molluscan aquaculture in China. *Journal of Shellfish Research* **18**, 19–31.

Haaker, P. L., Parker, D. O., Togstad, H., Richards, D. V., Davis, G. E. and Friedman, C. S. (1992). Mass mortality and withering syndrome in black abalone, *Haliotis cracherodii*, in California. *In* "Abalone of the World: Biology, Fisheries and Culture" (S. A. Shepherd, M. J. Tegner and S. A. Guzmán del Próo, eds), pp. 214–224. Blackwell, Oxford.

Haakonsen, H. O. and Anoruo, A. O. (1994). Tagging and migration of the American lobster *Homarus americanus*. *Reviews in Fisheries Science* **2**, 79–93.

Hagen, N. (1996). Echinoculture: From fishery enhancement to closed cycle cultivation. *World Aquaculture* **27**, 6–19.

Hahn, K. O. (1989). "A Handbook of the Culture of Abalone and Other Gastropods", p. 438. CRC Press Inc, Boca Raton, Florida.

Hair, C., Bell, J. and Doherty, P. (2002). The use of wild-caught juveniles in coastal aquaculture and its application to coral reef fishes. *In* "Responsible Marine Aquaculture" (R. R. Stickney and J. P. McVey, eds), pp. 327–353. CAB International, New York.

Hall, S. J. (1999). "The Effects of Fishing on Marine Ecosystems and Communities", p. 274. Blackwell, Oxford.

Hamel, J. F., Conand, C., Pawson, D. and Mercier, A. (2001). The sea cucumber *Holothuria scabra* (Holothuroidea: Echinodermata): Its biology and exploitation as beche-de-mer. *Advances in Marine Biology* **41**, 129–223.

Hamm, D. E. and Burton, R. S. (2000). Population genetics of black abalone, *Haliotis cracherodii*, along the central California coast. *Journal of Experimental Marine Biology and Ecology* **254**, 235–247.

Hammond, L. S. (1982). Analysis of grain-size selection by deposit-feeding holothurians and echinoids (Echinodermata) from a shallow reef lagoon, Discovery Bay, Jamaica. *Marine Ecology Progress Series* **8**, 25–36.

Hansen, M. M. (2002). Estimating the long-term effects of stocking domesticated trout into wild brown trout (*Salmo trutta*) populations: An approach using microsatellite DNA analysis of historical and contemporary samples. *Molecular Ecology* **11**, 1003–1015.

Hansen, M. M., Hynes, R. A., Loeschcke, V. and Rasmusen, G. (1995). Assessment of the stocked or wild origin of anadromous brown trout (*Salmo trutta* L.) in a Danish river system, using mitochondrial DNA RFLP analysis. *Molecular Ecology* **4**, 189–198.

Hanson, J. M. and Lanteigne, M. (2000). Evaluation of Atlantic cod predation on American lobster in the southern Gulf of St. Lawrence, with comments on other potential fish predators. *Transactions of the American Fisheries Society* **129**, 13–29.

Hara, M. and Fujio, Y. (1992). Geographic distribution of isozyme genes in natural abalone. *Bulletin of the Tohoku National Fisheries Research Institute* **54**, 115–124.

Hara, M. and Kikuchi, S. (1992). Genetic variability and population structure in the abalone *Haliotis discus hannai*. *Bulletin of the Tohoku National Fisheries Research Institute* **54**, 107–114.

Harding, G. C., Kenchington, E. L., Bird, C. J., Pezzack, D. S. and Landry, D. C. (1997). Genetic relationships amongst subpopulations of the American lobster (*Homarus americanus*) as revealed by random, amplified polymorphic DNA. *Canadian Journal of Fisheries and Aquatic Sciences* **54**, 1762–1771.

Hare, M. P. and Avise, J. C. (1998). Population structure in the American oyster as inferred by nuclear gene genealogies. *Molecular Biology and Evolution* **15**, 119–128.

Hare, M. P., Karl, S. A. and Avise, J. C. (1996). Anonymous nuclear DNA markers in the American oyster and their implications for the heterozygote deficiency phenomenon in marine bivalves. *Molecular Biology and Evolution* **13**, 334–345.

Hargis, W. J., Jr. (1999). The evolution of the Chesapeake oyster reef system during the holocene epoch. *In* "Oyster Reef Habitat Restoration: A Synopsis and Synthesis of Approaches" (M. W. Luckenbach, R. Mann and J. A. Wesson, eds), pp. 5–24. Virginia Institute of Marine Science, School of Marine Science, College of William and Mary, VIMS Press, Gloucester Point, Virginia.

Hargis, W. J., Jr and Haven, D. S. (1999). Chesapeake Bay Oyster Reefs, their importance, destruction and guidelines for restoring them. *In* "Oyster Reef Habitat Restoration: A Synopsis and Synthesis of Approaches" (M. W. Luckenbach, R. Mann and J. A. Wesson, eds), pp. 329–358. Virginia Institute of Marine Science, School of Marine Science, College of William and Mary, VIMS Press, Gloucester Point, Virginia.

Hart, A. M., Bell, J. D. and Foyle, T. P. (1998). Growth and survival of the giant clams *Tridacna derasa*, *T. maxima* and *T. crocea* at village farms in the Solomon Islands. *Aquaculture* **165**, 203–220.

Hatcher, B. G., Scheibling, R. E., Barbeau, M. A., Hennigar, A. W., Taylor, L. H. and Windust, A. J. (1996). Dispersion and mortality of a population of sea scallop (*Placopecten magellanicus*) seeded in a tidal channel. *Canadian Journal of Aquatic Sciences* **53**, 38–54.

Haugum, G. A., Strand, O. E. and Minchin, D. (1999). "Are Cultivated Scallops Wimps", pp. 92–93. *In* "Proceedings of the 12th International Pectinid Workshop", Institute of Marine Research, Department of Aquaculture, Bergen, Norway.

Hawtof, D. B., McCarthy, K. J. and Glazer, R. A. (1998). Distribution and abundance of queen conch, *Strombus gigas,* larvae in the Florida Current: Implications for recruitment to the Florida Keys. *Proceedings of the Gulf and Caribbean Fisheries Institute* **50**, 94–103.

Hayakawa, Y., Booth, J. D., Nishida, S., Sekiguchi, H., Saisho, T. and Kittaka, J. (1990). Daily settlement of puerulus stage of the red rock lobster *Jasus edwardsii* at Castle Point, New Zealand. *Nippon Suisan Gakkaishi* **56**, 1703–1716.

Hayashi, I. and Yamakawa, H. (1988). Population fluctuations of three sympatric species of *Haliotis* (Mollusca: Gastropoda) in artificial habitats at Kominato, central Japan. *Aquaculture* **73**, 67–84.

Haywood, E. L., III, Soniat, T. M. and Broadhurst, R. C., III (1999). Alternatives to clam and oyster shell as clutch for eastern oysters. *In* "Oyster Reef Habitat Restoration: A Synopsis and Synthesis of Approaches" (M. W. Luckenbach, R. Mann and J. A. Wesson, eds), pp. 295–304. Virginia Institute of Marine Science, School of Marine Science, College of William and Mary, VIMS Press, Gloucester Point, Virginia.

Haywood, M. D. E., Vance, D. J. and Loneragan, N. R. (1995). Seagrass and algal beds as nursery habitats for tiger prawns (*Penaeus semisulcatus* and *P. esculentus*) in a tropical Australian estuary. *Marine Biology* **122**, 213–223.

Haywood, M. D. E., Heales, D. S., Kenyon, R. A., Loneragan, N. R. and Vance, D. J. (1998). Predation of juvenile tiger prawns in a tropical Australian estuary. *Marine Ecology Progress Series* **162**, 201–214.

Heales, D. S., Vance, D. J. and Loneragan, N. R. (1996). Field observations of moult cycle, feeding behaviour, and diet of small juvenile tiger prawns *Penaeus semisulcatus* in a tropical seagrass bed in the Embley River, Australia. *Marine Ecology Progress Series* **145**, 43–51.

Heasman, M. (2002). Enhancement of populations of abalone in New South Wales using hatchery-produced seed. [http://www.fisheries.nsw.ogv.au/sci/projects/enhancement_abalone.htm]

Heasman, M., Chick, R., Savva, N., Worthington, D., Brand, C., Gibson, P. and Diemar, J. (2003). "Enhancement of Populations of Abalone in New South Wales using Hatchery-Produced Seed", p. 262. Final Report Australian Fisheries Research and Development Corporation Project 1998/219. New South Wales Fisheries, Australia.

Hedgecock, D. (1977). Biochemical genetic markers for broodstock identification in aquaculture. *Proceedings of the World Mariculture Society* **8**, 523–531.

Hedgecock, D. (1994a). Temporal and spatial genetic structure of marine animal populations in the California Current. *California Cooperative Oceanic Fisheries Investigations Reports* **35**, 73–81.

Hedgecock, D. (1994b). Does variance in reproductive success limit effective population sizes of marine organisms? *In* "Genetics and Evolution of Marine Organisms" (A. R. Beaumont, ed.), pp. 122–134. Chapman and Hall, London.

Hedgecock, D. and Sly, F. (1990). Genetic drift and effective population sizes of hatchery propagated stocks of the Pacific oyster, *Crassostrea gigas*. *Aquaculture* **88**, 21–38.

Hedgecock, D., Shleser, R. A. and Nelson, K. (1976). Applications of biochemical genetics to aquaculture. *Journal of the Fisheries Research Board of Canada* **33**, 1108–1119.

Hedgecock, D., Nelson, K., Simsons, J. and Shleser, R. A. (1977). Genetic similarity of European and American species of the lobster *Homarus*. *Biological Bulletin* **152**, 41–50.

Heppell, S. S. and Crowder, L. B. (1998). Prognostic evaluation of enhancement programs using population models and life history analysis. *Bulletin of Marine Science* **62**, 495–507.

Hendrickson, S., Barnes, D. and Dequillfeldt, C. (1988). Hard clam transplants in New York State. *Journal of Shellfish Research* **7**, 199.

Herrnkind, W. F. and Butler, M. J. (1986). Factors regulating postlarval settlement and juvenile microhabitat use by spiny lobsters *Panulirus argus*. *Marine Ecology Progress Series* **34**, 23–30.

Herrnkind, W. F. and Butler, M. J. (1994). Settlement of spiny lobster *Panulirus argus* (Latreille, 1804), in Florida: Pattern without predictability? *Crustaceana* **67**, 46–64.

Herrnkind, W. F., Butler, M. J., Hunt, J. H. and Childress, M. (1997). Role of physical refugia: Implications from a mass sponge die off in a lobster nursery in Florida. *Marine and Freshwater Research* **48**, 759–769.

Herrnkind, W. F., Butler, M. J. and Hunt, J. H. (1999). A case for shelter replacement in a disturbed spiny lobster nursery in Florida: Why basic research had to come first. *American Fisheries Society Symposium* **22**, 421–437.

Heslinga, G. A. (1981a). Growth and maturity of *Trochus niloticus* in the laboratory. *Proceedings 4th International Coral Reef Symposium* **1**, 39–45.

Heslinga, G. A. (1981b). Larval development, settlement and metamorphosis of the tropical gastropod *Trochus niloticus*. *Malacologia* **20**, 349–357.

Heslinga, G. A. and Hillman, A. (1981). Hatchery culture of the commercial top snail *Trochus niloticus* in Palau, Caroline Islands. *Aquaculture* **22**, 35–41.

Heslinga, G. A., Orak, O. and Ngiramengior, M. (1984). Coral reef sanctuaries for trochus shells. *Marine Fisheries Review* **46**, 73–80.

Heslinga, G. A., Watson, T. C. and Isamu, T. (1990). "Giant Clam Farming", p. 179. Pacific Fisheries Development Foundation (NMFS/NOAA), Honolulu, Hawaii.

Heyman, W. D., Dobberteen, L. A., Urry, L. A. and Heyman, A. M. (1989). Pilot hatchery for the queen conch, *Strombus gigas,* shows potential for inexpensive and appropriate technology for larval aquaculture in the Bahamas. *Aquaculture* **77**, 277–285.

Hilbish, J. (2001). Genetics of hard clams, *Mercenaria mercenaria*. *In* "Biology of the Hard Clam" (J. N. Kraeuter and M. Castagna, eds), pp. 261–277. Developments in Aquaculture and Fisheries Science, 31, Elsevier, Amsterdam and London.

Hilborn, R. (1997). Lobster stock assessment: Report from a workshop; II. *Marine and Freshwater Research* **48**, 945–947.

Hilborn, R. (1998). The economic performance of marine stock enhancement projects. *Bulletin of Marine Science* **62**, 661–674.

Hilborn, R. (1999). Confessions of a reformed hatchery basher. *Fisheries* **24**, 30–31.

Hilborn, R. (2004). Population management in stock enhancement and sea ranching. *In* "Stock Enhancement and Sea Ranching: Developments, Pitfalls and Opportunities" (K. M. Leber, S. Kitada, L. Blankenship and T. Svasand, eds), pp. 201–210. Blackwell, Oxford.

Hilborn, R., Branch, T., Ernst, B., Magnusson, A., Minte-Vera, C., Scheuerell, M. and Valero, J. (2003). State of the World's Fisheries. *Annual Review of Environment and Resources* **28**, 15.1–15.40.

Hillis, D. M., Moritz, C. and Mable, B. K. (1996). "Molecular Systematics", 2nd Edition. Sinauer Associates, Sunderland, Massachusetts.

Hobday, A. J., Tegner, M. J. and Haaker, P. L. (2001). Over-exploitation of a broadcast spawning marine invertebrate: Decline of the white abalone. *Reviews in Fish Biology and Fisheries* **10**, 493–514.

Hoelzel, A. R. (ed.) (1998). "Molecular Genetic Analysis of Populations: A Practical Approach", 2nd Edition. Oxford University Press, Oxford.

Hoffschir, C. (1990). Introduction of aquaculture–reared juvenile trochus (*Trochus niloticus*) to Lifou, Loyalty Islands, New Caledonia. *South Pacific Commission Fisheries Newsletter* **53**, 32–37.

Holland, A. (1994). The beche-de-mer industry in the Solomon Islands: Recent trends and suggestions for management. *South Pacific Commission Beche-de-mer Information Bulletin* **6**, 2–9.

Holland, L. Z., Giese, A. C. and Phillips, J. H. (1967). Studies on the perivisceral coelomic fluid protein concentration during seasonal and nutritional changes in the purple sea urchin. *Comparative Biochemistry and Physiology* **21**, 361–371.

Holthuis, L. B. (1991). "Marine Lobsters of the World—An Annotated and Illustrated Catalogue of Species of Interest to Fisheries Known to Date", Vol. 13, p. 292. FAO Species Catalogue. FAO Fisheries Synopsis No. 125. FAO, Rome.

Honma, A. (1993). "Aquaculture in Japan", p. 98. Japan FAO Association, Tokyo, Japan.

Honma, N. and Iioka, K. (1980). Experimental studies on the release of the cultured big seeds of abalone. *Report of Yamagata Prefecture Fisheries Experimental Station* **123**, 1–14.

Hooker, S. H. (1998). The demography of paua (*Haliotis iris*) with special reference to juveniles. MSc thesis, University of Auckland, Auckland, New Zealand.

Hooker, S. H., Jeffs, A. G., Creese, R. G. and Sivaguru, K. (1997). Growth of captive *Jasus edwardsii* (Hutton) (Crustacea: Palinuridae) in north-eastern New Zealand. *Marine and Freshwater Research* **48**, 903–909.

Horiguchi, Y., Kimura, S., Iwaki, K. and Ueno, R. (1994). Pigmentation of the nacre of abalone on artificial diet supplemented with natural pigments. 2. *Bulletin of the Faculty of Bioresource Mie University* **12**, 209–216 (in Japanese, English abstract).

Howard, A. E. (1980). Substrate controls on the size composition of lobster (*Homarus gammarus*) populations. *Journal du Conseil Permanent International pour L'Exploration de la Mer* **39**, 130–133.

Howard, A. E. (1988). Lobster behaviour, population structure and enhancement. *Symposia of the Zoological Society of London* **59**, 355–364.

Huang, T., Cottingham, R. J., Ledbetter, D. and Zoghbi, H. (1992). Genetic mapping of four dinucleotide repeat loci, DXS453, DXS458, DXS454, and DXS424 on the X chromosome using multiplex polymerase chain reaction. *Genomics* **13**, 375–380.

Hudon, C., Fradette, P. and Legendre, P. (1986). La répartition horizontale at verticale des larvas de homard (*Homarus americanus*) autour des îles de la Madeline, Golfe du Saint-Laurent. *Journal of the Fisheries Research Board of Canada* **43**, 2164–2176.

Hughes, T. P. (1994). Catastrophes, phase shifts, and large-scale degradation of a Caribbean coral reef. *Science* **265**, 1547–1551.

Humphrey, C. M. and Crenshaw, J. W. (1989). Clam genetics. *In* "Clam Mariculture in North America" (J. J. Manzi and M. Castagna, eds), Developments in Aquaculture and Fisheries Science 19, pp. 323–356. Elsevier, Amsterdam.

Humphrey, J. D. (1988). Disease risks associated with translocation of shellfish, with special reference to the giant clam *Tridacna gigas*. *In* "Giant Clams in Asia and the Pacific" (J. W. Copeland and J. S. Lucas, eds), pp. 241–244. ACIAR Monograph No. 9. Australian Centre for International Agricultural Research, Canberra.

Humphrey, J. D. (1995). "Perspectives in Aquatic Exotic Species Management in the Pacific Islands. Introductions of Aquatic Animals to the Pacific Islands: Disease Threats and Guidelines for Quarantine", Vol. 2, p. 53. Inshore Fisheries Research Project, South Pacific Commission, New Caledonia.

ICES (2000). Report of the Working Group on Introductions and Transfers of Marine Organisms. Advisory Committee on the Marine Environment. Parnu, Estonia, 27–29 March, 2000, p. 65.

Ikenoue, H. and Kafuku, T. (1992). "Modern Methods of Aquaculture in Japan", p. 272. Elsevier/Kodansha (Developments in Aquaculture and Fisheries Science), Amsterdam/Tokyo.

Illingworth, J., Tong, L. J., Moss, G. A. and Pickering, T. D. (1997). Upwelling tank for culturing rock lobster (*Jasus edwardsii*) phyllosomas. *Marine and Freshwater Research* **48**, 911–914.

Imamura, K. (1999). The organization and development of sea farming in Japan. *In* "Stock Enhancement and Sea Ranching" (B. R. Howell, E. Moskness and T. Svasand, eds), pp. 91–101. Fishing News Books, Blackwell, Oxford.

Inoguchi, N. (1993). Abalone seedling and fishing ground utilization. *Bulletin of the Tohoku Branch of the Japanese Society of Scientific Fisheries* **43**, 22–24 (in Japanese).

Inoue, M. (1976). Abalone. *In* "Suison Zoyoshoku Deeta Bukku (Fisheries Propogation Data Book),"*Suisan Shuppon*, 19–61. Translated from Japanese into English by M. G. Mottet (1978) in: A Review of the Fisheries Biology of Abalones. Washington Department of Fisheries Technical Report 37, Washington Department of Fisheries, Seattle.

Irvine, J. C., Bayer, R., Beal, B., Chapman, S. and Stubbs, D. A. (1991). The efficacy of blue colormorphic American lobsters in determining the feasibility of hatch and release programs. *Abstracts 1991 Annual Meeting of the National Shellfisheries Association*, pp. 296.

Isa, J., Kubo, H. and Murakoshi, M. (1997). Mass seed production and restocking of trochus in Okinawa (unedited). *In* "Workshop on Trochus Resource Assessment, Management and Development. Integrated Fisheries Management Project", pp. 75–99. Technical Document No. 13, South Pacific Commission, Noumea, New Caledonia.

Ishioka, K. (1981). Shrimp marking trials in Japan. *In* "Proceedings of the International Shrimp Releasing, Marking and Recruitment Workshop, 25–29 November 1978, Salmiya, State of Kuwait" (A. S. D. Farmer, ed.), pp. 209–226. Kuwait Bulletin of Marine Science **2**.

Ito, H. (1990). Some aspects of offshore spat collection of Japanese scallop. *In* "Marine Farming and Enhancement. Proceedings of the Fifteenth U.S.-Japan Meeting on Aquaculture, Kyoto, Japan, October 22–23, 1986" (A.K. Sparks, ed.), pp. 35–48. NOAA Technical Report NMFS 85, US Department of Commerce.

Ito, M. (1990). Comparative behaviour and morphology of phyllosoma larvae of palinurid and scyllarid lobsters (Crustacea: Decapoda: Palinuridae and Scyllaridae) with reference to their adaptive significance to diurnal vertical migration. PhD Thesis, University of Western Australia, Perth, Australia.

Ito, M. (1995). An overview of palinurid lobster culture and prospect of aquaculture development and resource management in the South Pacific region. *In* "The International Workshop on Present and Future of Aquaculture Research and Development in the Pacific Island Countries. 20–24 November 1995, Nuku'alofa, Tonga", (unedited) pp. 323–337. Japan International Cooperation Agency, Tokyo.

Ito, M. and Lucas, J. S. (1990). The complete larval development of the scyllarid lobster *Scyllarus demani* Holthuis, 1946 (Decapoda, Scyllaridae) in the laboratory. *Crustaceana* **58**, 144–167.

Ito, S., Kawahara, I., Aoto, I. and Eguchi, T. (1994). Rearing of juvenile sea cucumber, *Stichopus japonicus,* in a diked pond. *Bulletin of Saga Prefectural Sea Farming Center* **3**, 35–37 (in Japanese).

IUCN (1998). "Guidelines for Re-introductions", p. 100. Prepared by the IUCN/SSC Re-introduction Specialist Group. IUCN, Gland, Switzerland and Cambridge, UK.

Iversen, E. S. (1982). Feasibility of increasing Bahamian conch production by mariculture. *Proceedings of the Gulf and Caribbean Fisheries Institute* **35**, 83–88.

Iversen, E. S. and Jory, D. E. (1997). Mariculture and enhancement of wild populations of queen conch (*Strombus gigas*) in the Western Atlantic. *Bulletin of Marine Science* **60**, 929–941.

Iversen, E. S., Jory, D. E. and Bannerot, S. P. (1986). Predation on queen conch, *Strombus gigas*, in the Bahamas. *Bulletin of Marine Science* **39**, 61–75.

Iversen, E. S., Rutherford, E. S., Bannerot, S. P. and Jory, D. E. (1987). Biological data on Berry Islands (Bahamas) queen conchs, *Strombus gigas*, with mariculture and fisheries management implications. *Fishery Bulletin, US* **85**, 299–310.

Iversen, E. S., Bannerot, S. P. and Jory, D. E. (1990). Evidence of survival value related to burying behavior in Queen conch, *Strombus gigas*. *Fishery Bulletin, US* **88**, 383–387.

Jackson, D. J., Williams, K. and Degnan, B. (2000). Analysis of the suitability of Australian formulated diets for the aquaculture of the tropical abalone, *Haliotis asinina* Linnaeus. *Journal of Shellfish Research* **19**, 520.

Jackson, J. B. C., Kirby, M. X., Berger, W. H., Bjorndal, K. A., Botsford, L. W., Bourque, B. J., Bradbury, R. H., Cooke, R., Erlandson, J., Estes, J. A., Hughes, T. P., Kidwell, S., Lange, C. B., Lenihan, H. S., Pandolfi, J. M., Peterson, C. H., Steneck, R. S., Tegner, M. J. and Warner, R. R. (2001). Historical overfishing and the recent collapse of coastal ecosystems. *Science* **293**, 629–638.

James, D. B. (1994). Seed production in sea cucumbers. *Aquaculture International* **1**(9), 15–26.

James, D. B. (1996). Culture of sea-cucumber. *Indian Central Marine Fisheries Research Institute Bulletin* **48**, 120–126.

James, D. B., Rajapandian, M. E., Gopinathan, C. P. and Baskar, B. K. (1994). Breakthrough in induced breeding and rearing of the larvae and juveniles of *Holothuria* (*Metriatyla*) *scabra* Jaeger at Tuticorin. *Indian Central Marine Fisheries Research Institute Bulletin* **46**, 66–70.

James, P. J., Tong, L. J. and Paewai, M. P. (2001). Effect of stocking density and shelter on growth and mortality of early juvenile *Jasus edwardsii* held in captivity. *Marine and Freshwater Research* **52**, 1413–1417.

James-Pirri, M. J. and Cobb, J. S. (1999). Influence of coded wire tags on postlarval lobster (*Homarus americanus*) behavior. *Marine Behavior and Physiology* **32**, 255–259.

Jefferts, K. B., Bergman, P. K. and Fiscus, H. F. (1963). A coded wire identification system for macro-organisms. *Nature, London* **198**, 460–462.

Jeffs, A. and Hooker, S. (2000). Economic feasibility of aquaculture of spiny lobsters *Jasus edwardsii* in temperate waters. *Journal of the World Aquaculture Society* **31**, 30–41.

Jeffs, A. G. and James, P. (2001). Sea-cage culture of the spiny lobster *Jasus edwardsii* in New Zealand. *Marine and Freshwater Research* **52**, 1419–1424.

Jensen, A. (1999). Artificial reefs for shellfish habitat: Results and ideas to date. *Journal of Shellfish Research* **18**, 718.

Jensen, A. C., Collins, K. J., Free, E. K. and Bannister, R. C. A. (1994). Lobster (*Homarus gammarus*) movement on an artificial reef: The potential use of artificial reefs for stock enhancement. *Crustaceana* **67**, 198–211.

Jensen, A. C., Collins, K. J. and Lockwood, A. P. (eds) (1999). "Artificial Reefs in European Seas". Kluwer Academic Publishers, Dordrecht, Netherlands.

Jernakoff, P. (1987). An electromagnetic tracking system for use in shallow waters. *Journal of Experimental Marine Biology and Ecology* **113**, 1–8.

Jernakoff, P. (1988). Electromagnetic tags for rock lobsters. *In* "Tagging—Solution or Problem" (D. A. Hancock, ed.), pp. 60–61. Australian Society for Fish Biology Tagging Workshop, 21–22 July 1988. Bureau of Rural Resources. Australian Government Printing Office, Canberra.

Jiang, L., Wu, W. L. and Huang, P. C. (1995). The mitochondrial DNA of Taiwan abalone *Haliotis diversicolor* Reeve, 1846 (Gastropoda, Archaeogastropoda: Haliotidae). *Molecular Marine Biology and Biotechnology* **4**, 353–364.

Johnson, M. S. (2000). Measuring and interpreting genetic structure to minimize the genetic risks of translocations. *Aquaculture Research* **31**, 133–143.

Johnson, M. S. and Black, R. (1982). Chaotic genetic patchiness in an intertidal limpet, *Siphonaria* sp. *Marine Biology* **70**, 157–164.

Johnson, M. S. and Black, R. (1984). Pattern beneath the chaos: The effect of recruitment on genetic patchiness in an intertidal limpet. *Evolution* **38**, 1371–1383.

Jones, C. M., Linton, L., Horton, D. and Bowman, W. (2001). Effect of density on growth and survival of ornate rock lobster, *Panulirus ornatus* (Fabricius, 1798), in a flow through raceway. *Marine and Freshwater Research* **52**, 1425–1429.

Jones, G. P., Milicich, M. J., Emslie, M. J. and Lunow, C. (1999). Self-recruitment in a coral reef fish population. *Nature* **402**, 802–804.

Jones, R. L. and Stoner, A. W. (1997). The integration of GIS and remote sensing in an ecological study of queen conch, *Strombus gigas*, nursery habitats. *Proceedings of the Gulf and Caribbean Fisheries Institute* **49**, 523–530.

Jørstad, K. E. and Farestveit, E. (1999). Population genetic structure of lobster (*Homarus gammarus*) in Norway, and implications for enhancement and sea-ranching operations. *Aquaculture* **173**, 447–457.

Jørstad, K. E., Paulsen, O. I., Nævdal, G. and Thorkildsen, S. (1994). Genetic studies of cod, *Gadus morhua* L., in Masfjord, western Norway: Comparisons between the local stock and released, artificially reared cod. *Aquaculture and Fisheries Management* **25** (suppl. 1), 77–91.

Jory, D. E. and Iversen, E. S. (1983). Queen conch predators: Not a roadblock to mariculture. *Proceedings of the Gulf and Caribbean Fisheries Institute* **35**, 108–111.

Juinio-Meñez, M. A., Macawaris, N. D. and Bangi, H. G. (1998). Community-based sea urchin (*Tripneustes gratilla*) grow-out culture as a resource management tool. *In* "Proceedings of the North Pacific Symposium on Invertebrate Stock Assessment and Management" (G. S. Jamieson and A. Campbell, eds). *Canadian Special Publication in Fisheries and Aquatic Sciences* **125** 393–399.

Kabata, Z. (1963). Parasites as biological tags. *International Commission for the Northwest Atlantic Fisheries Special Publication* **4**, 31–37.

Kafuku, T. and Ikenoue, H. (eds) (1984). "Modern Methods of Aquaculture in Japan". Kodansha Ltd., Tokyo.

Kanciruk, P. (1980). Ecology of juvenile and adult Palinuridae (spiny lobsters). *In* "The Biology and Management of Lobsters, Vol 2. Ecology and Management" (J. S. Cobb and B. F. Phillips, eds), pp. 59–96. Academic Press, New York.

Kassner, J. and Malouf, R. E. (1982). An evaluation of "spawner transplants" as a management tool in Long Island's hard clam fisheries. *Journal of Shellfish Research* **2**, 165–172.

Katada, M. (1963). Life forms of sea-weeds and succession of their vegetation (review). *Bulletin of Japan Socio Science Fisheries* **29**, 798–808 (in Japanese).

Kawahara, I. (1996). "Artificial Production of Two Juvenile Sea Urchins *Pseudocentrotus depressus* and *Hemicentrotus pulcherrimus*". Saga Prefecture Sea Farming Centre, Daido Insatsu Co. Ltd., Saga (in Japanese).

Kawai, T. and Agatsuma, Y. (1995). Predation on released seed of the sea urchin, *Strogylocentrotus intermedius* in Shiribeshi, Hokkaido, Japan. *Fisheries Science* **62**, 317–318.

Kawamura, K. (1969). Sea urchin fisheries and present status of resource management and enhancement in Hokkaido. *Hokusuishi Geppo* **26**, 608–637 (in Japanese).

Kawamura, K. (1973). Fishery biological studies on a sea urchin, *Strongylocentrotus intermedius*. *Scientific Reports of Hokkaido Fisheries Experimental Station* **16**, 1–54 (in Japanese with English abstract).

Kawamura, K. (1993). "Uni Zouyoushoku to Kakou, Ryutsu", p. 82. Hokkai Suisan Company, Sapporo, Japan (in Japanese).

Kearney, R. and Andrew, N. (1995). Enhancement of Australian fish stocks: Issues confronting researches and managers. In "Recent Advances in Marine Science and Technology '94" (O. Bellwood, J. H. Choat and N. K. Saxena, eds), pp. 591–598. James Cook University, Townsville.

Keenan, C. and MacDonald, C. M. (1988). Genetic tagging. In "Tagging—Solution or Problem?" (D. A. Hancock, ed.), pp. 77–82. Australian Society for Fish Biology Tagging Workshop, 21–22 July 1988. Bureau of Rural Resources Proceedings No. 5, Australian Government Printing Office, Canberra.

Keesing, J. K., Grove-Jones, R. and Tagg, P. (1995). Measuring settlement intensity of abalone: Results of a pilot study. Marine and Freshwater Research 46, 539–543.

Kennedy, V. S. and Sanford, L. P. (1999). The morphology and physical oceanography of unexploited oyster reefs in North America. In "Oyster Reef Habitat Restoration: A Synopsis and Synthesis of Approaches" (M. W. Luckenbach, R. Mann and J. A. Wesson, eds), pp. 25–46. Virginia Institute of Marine Science, School of Marine Science, College of William and Mary, VIMS Press, Gloucester Point, Virginia.

Kennedy, V. S., Newell, R. I. E. and Eble, A. F. (eds) (1996). "The Eastern Oyster: Crassostrea virginica". Maryland Sea Grant, College Park, Maryland.

Kemuyama, A., Takeichi, M. and Uchida, A. (1997). Effects of releasing artificial produced seed of ezo abalone, Haliotis discus hannai in Tarumizu, Yamadacho, Iwate Prefecture. Bulletin of the Iwate Prefecture Fisheries Technology Center 1, 37–45 (in Japanese, English abstract).

Kenyon, R. A., Loneragan, N. R. and Hughes, J. (1995). Habitat type and light affect sheltering behaviour of juvenile tiger prawns (Penaeus esculentus Haswell) and success rates of their fish predators. Journal of Experimental Biology and Ecology 192, 87–105.

Kenyon, R., Haywood, M., Loneragan, N., Manson, F. and Toscas, P. (2003a). Benthic habitats of Exmouth Gulf. In "Developing Techniques for Enhancing Prawn Fisheries, with a Focus on Brown Tiger Prawns (Penaeus esculentus) in Exmouth Gulf" (N. Loneragan, ed.), pp. 164–198. Final Report Australian Fisheries Research and Development Corporation Project 1999/222. CSIRO, Cleveland, Australia.

Kenyon, R., Haywood, M., Loneragan, N. and Manson, F. (2003b). Assessing potential release sites for enhancement of juvenile prawns in Exmouth Gulf: Surveys of juvenile prawns and potential fish predators. In "Developing Techniques for Enhancing Prawn Fisheries, with a Focus on Brown Tiger Prawns (Penaeus esculentus) in Exmouth Gulf" (N. Loneragan, ed.), pp. 200–235. Final Report Australian Fisheries Research and Development Corporation Project 1999/222. CSIRO, Cleveland, Australia.

Kessing, B., Croom, H., Martin, A., McIntosh, C., Owen McMillan, W. and Palumbi, S. (1989). "The Simple Fools Guide to PCR". University of Hawaii, Honolulu.

Kikuchi, S. and Uki, N. (1981). Productivity of benthic grazers, abalone and sea urchin in Laminaria beds. Japanese Society of Fisheries Science, Kouseisha-Kouseikaku, Tokyo. Fisheries Science Series 23, 9–23 (in Japanese).

Kim, D. G. and Chang, D. S. (1992). A study on the artificial formation of seaweed beds on artificial reefs. Bulletin of National Fisheries Research and Development Agency, Korea 46, 7–19 (In Korean, English abstract.).

Kirby, V. L., Villa, R. and Powers, D. A. (1998). Identification of microsatellites in California red abalone, Haliotis rufescens. Journal of Shellfish Research 17, 801–804.

Kitada, S. (1999). Effectiveness of Japan's stock enhancement programmes: Current perspectives. In "Stock Enhancement and Sea Ranching" (B. R. Howell, E. Moskness and T. Svasand, eds), pp. 103–131. Fishing News Books, Blackwell, Oxford.

Kitada, S. and Fujishima, H. (1997). The stocking effectiveness of scallop in Hokkaido. *Nippon Suisan Gakkaishi* **63**, 686–693.

Kitada, S., Taga, Y. and Kishino, H. (1992). Effectiveness of a stock enhancement program evaluated by a two-stage sampling survey of commercial landings. *Canadian Journal of Fisheries and Aquatic Sciences* **49**, 1573–1582.

Kittaka, J. (1981). Large-scale production of shrimp for releasing in Japan and in the United States and the results of the releasing programme at Panama City, Florida. *Kuwait Bulletin of Marine Science* **2**, 149–163.

Kittaka, J. (1988). Culture of the palinurid *Jasus lalandii* from egg stage to puerulus. *Nippon Suisan Gakkaishi* **54**, 87–93.

Kittaka, J. (1994). Culture of phyllosomas of spiny lobster and its application to studies of larval recruitment and aquaculture. *Crustaceana* **66**, 258–294.

Kittaka, J. (1997). Culture of larval spiny lobsters: A review of work done in northern Japan. *Marine and Freshwater Research* **48**, 923–930.

Kittaka, J. (2000). Culture of larval spiny lobsters. In "Spiny Lobsters: Fisheries and Culture (2nd Edition)" (B. F. Phillips and J. Kittaka, eds), pp. 508–532. Fishing News Books, Oxford.

Kittaka, J. and Booth, J. D. (1994). Prospectus for aquaculture. In "Spiny Lobster Management" (B. F. Phillips, J. S. Cobb and J. Kittaka, eds), pp. 365–373. Fishing News Books, Oxford.

Kittaka, J. and Booth, J. D. (2000). Prospectus for aquaculture. In "Spiny Lobsters: Fisheries and Culture (2nd Edition)" (B. F. Phillips and J. Kittaka, eds), pp. 465–473. Fishing News Books, Oxford.

Kittaka, J. and Ikegami, E. (1988). Culture of the palinurid *Palinurus elephas* from egg stage to puerulus. *Nippon Suisan Gakkaishi* **54**, 1149–1154.

Kittaka, J. and Kimura, K. (1989). Culture of the Japanese spiny lobster *Panulirus japonicus* from egg to the juvenile stage. *Nippon Suisan Gakkaishi* **55**, 963–970.

Kittaka, J., Iwai, M. and Yoshimura, M. (1988). Culture of a hybrid of spiny lobster genus *Jasus* from egg stage to puerulus. *Nippon Suisan Gakkaishi* **54**, 413–417.

Kittaka, J., Ono, K. and Booth, J. D. (1997). Complete development of the green rock lobster, *Jasus verreauxi,* from the egg to juvenile. *Bulletin of Marine Science* **61**, 57–71.

Kittaka, J., Kudo, R., Onoda, S., Kanemaru, K. and Mercer, J. P. (2001). Larval culture of the European spiny lobster *Palinurus elephas*. *Marine and Freshwater Research* **52**, 1439–1444.

Kittiwattanawong, K., Nugranad, J. and Srisawat, T. (2001). High genetic divergence of *Tridacna squamosa* living at the west and the east coasts of Thailand. *Phuket Marine Biological Center Special Publication* **25**, 343–347.

Kiyomoto, S. and Yamasaki, M. (1999). Size dependent changes in habitat, distribution and food habit of juvenile disc abalone *Haliotis discus discus* on the coast of Nagasaki Prefecture, southwest Japan. *Bulletin of the Tohoku National Fisheries Research Institute* **62**, 71–81.

Klima, E. F. (1965). Evaluation of biological stains, inks and fluorescent pigments as marks for shrimp. *Special Scientific Report. US Fish and Wildlife Service (Fisheries)* **511**, 8.

Knowlton, N. and Weigt, L. A. (1998). New dates and new rates for divergence across the Isthmus of Panama. *Proceedings of the Royal Society of London B* **265**, 2257–2263.

Knudsen, H. and Tveite, S. (1999). Survival and growth of juvenile lobster *Homarus gammarus* L. raised for stock enhancement within in situ cages. *Aquaculture Research* **30**, 421–425.

Kojima, H. (1974). The habitat of young black abalone, *Haliotis discus discus* Reeve in Kaifu-gun, Tokushima Prefecture. *Michurin Seibutugaku Kenkyu* **10**, 155–160.

Kojima, H. (1981). Mortality of young Japanese black abalone *Haliotis discus discus* after transplantation. *Bulletin of the Japanese Society of Scientific Fisheries* **47**, 151–159 (in Japanese with English abstract, tables and figures).

Kojima, H. (1994). Present status, problems, and perspectives of abalone sea ranching. *Kaiyo Monthly* **290**, 485–489.

Kojima, H. (1995). Evaluation of abalone stock enhancement through the release of hatchery-reared seeds. *Marine and Freshwater Research* **46**, 689–695.

Kojima, H. and Imajima, M. (1982). Burrowing polychaetes in the shells of the abalone *Haliotis diversicolor aquatilis* chiefly the species of *Polydora*. *Bulletin of the Japanese Society of Scientific Fisheries* **48**, 31–35.

Kordos, L. M. and Burton, R. S. (1993). Genetic differentiation of Texas Gulf Coast populations of the blue crab *Callinectes sapidus*. *Marine Biology* **117**, 227–233.

Koshikawa, Y., Hagiwara, K., Lim, B. K. and Sakuri, N. (1997). A new marking method for short necked clam *Ruditapes philippinarum* with rust. *Fisheries Science* **63**, 533–538.

Koslow, J. A. (1992). Fecundity and the stock-recruitment relationship. *Canadian Journal of Fisheries and Aquatic Sciences* **49**, 210–217.

Kraeuter, J. N. (2001). Predators and predation. *In* "Biology of the Hard Clam" (J. N. Kraeuter and M. Castagna, eds), pp. 441–568. Developments in Aquaculture and Fisheries Science, 31. Elsevier, Amsterdam.

Kraeuter, J. N. and Castagna, M. (1980). Effects of large predators on the field culture of the hard clam, *Mercenaria mercenaria*. *Fishery Bulletin, US* **78**, 538–541.

Kraeuter, J. N. and Castagna, M. (1989). Factors affecting the growth and survival of clam seed planted in the natural environment. *In* "Clam Mariculture in North America" (J. J. Manzi and M. Castagna, eds), pp. 149–165. Elsevier, Amsterdam.

Kraeuter, J. N. and Castagna, M. (eds) (2001). "Biology of the Hard Clam". Developments in Aquaculture and Fisheries Science, 31. Elsevier, Amsterdam.

Krause, M. K. (1992). Use of genetic markers to evaluate the success of transplanted bay scallops. *Journal of Shellfish Research* **11**, 199.

Krouse, J. S. and Nutting, G. E. (1990). Evaluation of coded microwire tags inserted in legs of small juvenile American lobsters. *American Fisheries Society Symposium* **7**, 304–310.

Kubo, H., Ooshima, H., Nakama, T., Awakuni, T. and Tamaki, H. (1991). Study on artificial nursery for *Trochus* juveniles I. *Annual Report of Okinawa Prefectural Fisheries Experimental Station*, pp. 120–128. (in Japanese).

Kubo, H., Sawashi, Y. and Kawaguchi, A. (1992). Study on artificial nursery for *Trochus*, 2. *Annual Report of Okinawa Prefectural Fisheries Experimental Station* 117–123 (in Japanese).

Kumaki, Y., Douke, A., Hisada, T., Inoue, T. and Hamanaka, Y. (2001). Some aspects of fishing ground on artificial reefs for abalone (*Nordotis discus discus*) in Yabeta area. *Bulletin of the Kyoto Institute of Oceanic and Fishery Science* **23**, 30–38.

Kurata, H. (1981). Shrimp fry releasing techniques in Japan, with special reference to the artificial tideland. *Kuwait Bulletin of Marine Science*, pp. 117–147 (International Shrimp Releasing, Marking and Recruitment Workshop, Kuwait 1978.)

Kuris, A. M. and Culver, C. S. (1999). An introduced sabellid polychaete pest infesting cultured abalone and its potential spread to other California gastropods. *Invertebrate Biology* **118**, 391–403.

Kwiatkowski, D. J., Henske, E. P., Weimer, K., Ozelius, L., Gusella, J. F. and Haines, J. (1992). Construction of a GT polymorphism map of human 9q. *Genomics* **12**, 229–240.

Lafferty, K. D. (2001). Restoration of the white abalone in southern California: Population assessment, brood stock collection, and development of husbandry technology. Final Report. (NOAA)NA96FD0208. (www.nmfs.noaa.gov/sfweb/sk/saltonstallken/husbandry.htm).

Lake, N. C. H., Jones, M. B. and Paul, J. D. (1987). Crab predation on scallop (*Pecten maximus*) and its implication for scallop cultivation. *Journal of the Marine Biological Association of the United Kingdom* **67**, 55–64.

Landers, D. F., Jr, Keser, M. and Saila, S. B. (2001). Changes in female lobster (*Homarus americanus*) size at maturity and implications for the lobster resource in Long Island Sound, Connecticut. *Marine and Freshwater Research* **52**, 1283–1290.

Langdon, J. S. (1989). Disease risks in fish introduction and translocations. *In* "Introduced and Translocated Fishes and Their Ecological Effects", (D.A. Pollard, ed.), pp. 98–107. Australian Society for Fish Biology Workshop, 24–25 August 1989, Bureau of Rural Resources Proceedings No. 8, Australian Government Printing Service, Canberra.

Largier, J. L. (2003). Considerations in estimating larval dispersal distances from oceanographic data *Ecological Applications* **13** (Supplement), S71–S89.

Latrouite, D. and Lorec, J. (1991). L'expérience française de forçage du recrutement du homard européen (*Homarus gammarus*): Résultats préliminaires. *ICES Marine Science Symposia* **192**, 93–98.

Laughlin, G. R. and Weil, M. (1983). Queen conch mariculture and restoration in the Archipelago de Los Roques. *Proceedings of the Gulf and Caribbean Fisheries Institute* **35**, 64–72.

Laurent, V., Planes, S. and Salvat, B. (2002). High variability of genetic pattern in giant clam (*Tridacna maxima*) populations within French Polynesia. *Biological Journal of the Linnean Society* **77**, 221–231.

Lavalli, K. (1991). Survival and growth of early-juvenile American lobsters *Homarus americanus* through their first season while fed on diets of mesoplankton, microplankton and frozen brine shrimp. *Fishery Bulletin, US* **89**, 61–68.

Lavalli, K. and Barshaw, D. E. (1986). Burrows protect postlarval lobsters *Homarus americanus* from predation by non-burrowing cunner *Tautoglabrus adspesus*, but not from burrowing crab, *Neopanope texani*. *Marine Ecology Progress Series* **32**, 13–16.

Lavery, S., Chan, T. Y., Tam, Y. K., and Chu, K. H. (2004). Phylogenetic relationships and evolutionary history of the shrimp genus *Penaeus* s. l. derived from mitochondrial DNA. *Molecular Phylogenetics and Evolution* **31**, 39–49.

Leard, R. L., Dugas, R. and Berrigan, M. (1999). Resource management programs for the Eastern Oyster, *Crassostrea virginica*, in the U.S. Gulf of Mexico...past, present and future. *In* "Oyster Reef Habitat Restoration: A Synopsis and Synthesis of Approaches" (M. W. Luckenbach, R. Mann and J. A. Wesson, eds), pp. 63–92. Virginia Institute of Marine Science, School of Marine Science, College of William and Mary, VIMS Press, Gloucester Point, Virginia.

Leber, K. M. (1999). Rationale for an experimental approach to stock enhancement. *In* "Stock Enhancement and Sea Ranching" (B. R. Howell, E. Moksness and T. Svåsand, eds), pp. 63–75. Fishing News Books, Blackwell, Oxford.

Leber, K. M. (2002). Advances in marine stock enhancement: Shifting emphasis to theory and accountability. *In* "Responsible Marine Aquaculture" (R. R. Stickney and J. P. McVey, eds), pp. 79–90. CAB International, New York.

Leber, K. M. (2004). Marine stock enhancement in the USA: Status, trends and needs. *In* "Stock Enhancement and Sea Ranching: Developments, Pitfalls and Opportunities" (K. M. Leber, S. Kitada, L. Blankenship and T. Svasand, eds), pp. 11–24. Blackwell, Oxford.

Lebreton, J.-D., Burnham, K. P., Clobert, J. and Anderson, D. R. (1992). Modeling survival and testing biological hypotheses using marked animals: A unified approach with case studies. *Ecological Monographs* **62**, 67–118.

Lee, C. L. (1997). Part 3b ACIAR trochus reseeding research: An improved hatchery method for mass production of juvenile trochus. *South Pacific Commission Trochus Information Bulletin* **5**, 39–40.

Lee, C. L. and Amos, M. (1997). Current status of topshell *Trochus niloticus* hatcheries in Australia, Indonesia and the Pacific—A review. *In* "Trochus: Status, Hatchery Practice and Nutrition" (C. L. Lee and P. W. Lynch, eds), pp. 38–42. ACIAR Proceedings No. 79, Australian Centre for International Agricultural Research, Canberra.

Lee, C. L. and Lynch, P. W. (eds) (1997). "Trochus: Status, Hatchery Practice and Nutrition", p. 185. ACIAR Proceedings No. 79, Australian Centre for International Agricultural Research, Canberra.

Lee, D. O'C. and Wickens, J. F. (1992). "Crustacean Farming", p. 329. Blackwell, Oxford.

Lehnert, S., Ward, R. D., Grewe, P. and Bravington, M. (2003). Development and deployment of microsatellite markers. *In* "Developing Techniques for Enhancing Prawn Fisheries, with a Focus on Brown Tiger Prawns (*Penaeus esculentus*) in Exmouth Gulf" (N. Loneragan, ed.), Final report to FRDC for project 1999/222. CSIRO, Clevela Australia.

Leighton, D. L. (1961). Observations on the effects of diet on shell coloration in the red abalone *Haliotis rufescens* Swainson. *Veliger* **4**, 29–32.

Lellis, W. (1991). Spiny lobster, a mariculture candidate for the Caribbean? *World Aquaculture* **22**, 60–63.

Lesser, J. H. R. (1978). Phyllosoma larvae of *Jasus edwardsii* (Hutton) (Crustacea: Decapoda: Palinuridae) and their distribution off the east coast of the North Island, New Zealand. *New Zealand Journal of Marine and Freshwater Research* **12**, 357–370.

Lessios, H. A., Kessing, B. D. and Robertson, D. R. (1998). Massive gene flow across the world's most potent marine biogeographic barrier. *Proceedings of the Royal Society of London B* **265**, 583–588.

Levin, L. A. (1990). A review of methods for labeling and tracking marine invertebrate larvae. *Ophelia* **32**, 115–144.

Levin, L. A. (1993). Rare-earth tagging methods for the study of larval dispersal by marine invertebrates. *Limnology and Oceanography* **38**, 346–360.

Lewis, A. D., Adams, T. J. H. and Ledua, E. (1988). Fiji's giant clam stocks: A review of their distribution, abundance, exploitation and management. *In* "Giant Clams in Asia and the Pacific" (J. W. Copeland and J. S. Lucas, eds), pp. 78–81. ACIAR Monograph No. 9, Australian Centre for International Agricultural Research, Canberra.

Lewontin, R. C. (1974). "The Genetic Basis of Evolutionary Change". Columbia University Press, New York.

Li, Q., Park, C., Kobayashi, T. and Kijima, A. (2003). Inheritance of microsatellite markers in the Pacific abalone *Haliotis discus hannai*. *Marine Biotechnology* **5**, 331–338.

Liao, I.-C. (1999). How can stock enhancement and sea ranching help sustain and increase coastal fisheries? *In* "Stock Enhancement and Sea Ranching" (B. R. Howell, E. Moksness and T. Svasand, eds), pp. 132–149. Fishing News Books, Blackwell, Oxford.

Lincoln-Smith, M. P., Bell, J. D., Ramohia, P. and Pitt, K. A. (2000). "Testing the Use of Marine Protected Areas to Restore and Manage Tropical Multispecies Invertebrate Fisheries at the Arnavon Islands, Solomon Islands", p. 72. Great Barrier Reef Marine Park Authority Research Publication No. 69, Townsville.

Lindberg, W. J. (1997). Can science resolve the attraction–production issue? *Fisheries* **22**, 10–13.

Linnane, A. and Mercer, J. P. (1998). A comparison of methods for tagging juvenile lobsters (*Homarus gammarus* L.) reared for stock enhancement. *Aquaculture* **163**, 195–202.

Linnane, A., Ball, B., Mercer, J. P., van der Meeren, G., Bannister, R. C. A., Mazzoni, D., Munday, B. and Ringvold, H. (1999). Understanding the factors that influence European lobster recruitment: A trans-European study of cobble fauna. *Journal of Shellfish Research* **18**, 719–720.

Linnane, A., Mazzoni, D. and Mercer, J. P. (2000a). A long-term mesocosm study on the settlement and survival of juvenile European lobster *Homarus gammarus* L. in four natural substrata. *Journal of Experimental Marine Biology and Ecology* **249**, 51–64.

Linnane, A., Ball, B., Munday, B. and Mercer, J. P. (2000b). On the occurrence of juvenile lobster *Homarus gammarus* in intertidal habitat. *Journal of the Marine Biological Association of the United Kingdom* **80**, 375–376.

Linnane, A., Ball, B., Mercer, J. P., Browne, R., van der Meeren, G., Ringvold, H., Bannister, R. C. A., Mazzoni, D. and Munday, B. (2001). Searching for the early benthic phase (EBP) of the European lobster: A trans-European study of cobble fauna. *Hydrobiologia* **465**, 63–72.

Linton, L. (1998). "The potential for tropical rock lobster aquaculture in Queensland", p. 22. Queensland Department of Primary Industries Information Series QI 98020, Agdex 486/11, Brisbane, Australia.

Lipcius, R. N., Stockhausen, W. T., Eggleston, D. B., Marshall, L. S., Jr and Hickey, B. (1997). Hydrodynamic decoupling of recruitment, habitat quality and adult abundance in the Caribbean spiny lobster: Source-sink dynamics. *Marine and Freshwater Research* **48**, 807–815.

Lipcius, R. N., Stockhausen, W. T. and Eggleston, D. B. (2001). Marine reserves for Caribbean spiny lobster: Empirical evaluation and theoretical metapopulation recruitment dynamics. *Marine and Freshwater Research* **52**, 1589–1598.

Lipton, D. W., Lavan, E. F. and Strand, I. E. (1992). Economics of molluscan introductions and transfers: The Chesapeake Bay dilemma. *Journal of Shellfish Research* **11**, 511–519.

Liu, J. Y. (1990). Resource enhancement of Chinese shrimp, *Penaeus orientalis*. *Bulletin of Marine Science* **47**, 124–133.

Liu, H. and Loneragan, N. R. (1997). Size and time of day affect the response of postlarvae and early juvenile grooved tiger prawns *Penaeus semisulcatus* (Decapoda: Penaeidae) to natural and artificial seagrass in the laboratory. *Journal of Experimental Marine Biology and Ecology* **211**, 263–277.

Lokani, P. (1995). Illegal fishing for sea cucumber (beche-de-mer) by Papua New Guinea artisanal fishermen in the Torres Strait protected zone. *In* "Joint FFA/SPC Workshop on the Management of South Pacific Inshore Fisheries (26 June–7 July 1995)", pp. 279–288. South Pacific Commission, Noumea, New Caledonia.

Lokani, P., Polon, P. and Lari, R. (1996). Fisheries and management of beche-de-mer fisheries in Western Province of Papua New Guinea. *South Pacific Commission Beche-de-mer Information Bulletin* **8**, 7–11.

Loneragan, N. R., Kenyon, R. A., Haywood, M. D. E. and Staples, D. J. (1994). Population dynamics of juvenile tiger prawns in seagrass habitats of the north-western Gulf of Carpentaria, Australia. *Marine Biology* **119**, 133–143.

Loneragan, N. R., Kenyon, R. A., Staples, D. J., Poiner, I. R. and Conacher, C. A. (1998). The influence of seagrass type on the distribution and abundance of postlarval and juvenile tiger prawns in the western Gulf of Carpentaria, Australia. *Journal of Experimental Marine Biology and Ecology* **228**, 175–196.

Loneragan, N. R., Die, D. J., Kailis, G. M., Watson, R. and Preston, N. (2001a). "Developing and Assessing Techniques for Enhancing Tropical Australian Prawn Fisheries and the Feasibility of Enhancing the Brown Tiger Prawn (*Penaeus esculentus*) Fishery in Exmouth Gulf". Final Report Australian Fisheries Research and Development Corporation Project 98/222 CSIRO, Cleveland, Australia.

Loneragan, N. R., Heales, D. S., Haywood, M. D. E., Kenyon, R. A., Pendrey, R. C. and Vance, D. J. (2001b). Estimating the carrying capacity of seagrass for juvenile tiger prawns (*Penaeus semisulcatus*): Enclosure experiments in high and low biomass seagrass beds. *Marine Biology* **139**, 343–354.

Loneragan, N. (2003). "Developing techniques for enhancing prawn fisheries, with a focus on brown tiger prawns (*Penaeus esculentus*) in Exmouth Gulf". Final Report Australian Fisheries Research and Development Corporation Project 1999/222. CSIRO, Cleveland, Australia.

Loneragan, N. R., Crocos, P. J., Barnard, R., McCulloch, R., Penn, J. W., Ward, R. D. and Rothlisberg, P. C. (2004). An approach to evaluating the potential for stock enhancement of brown tiger prawns (*Penaeus esculentus* Haswell) in Exmouth Gulf, Western Australia. *In* "Stock Enhancement and Sea Ranching: Developments, Pitfalls and Opportunities" (K. M. Leber, S. Kitada, L. Blankenship and T. Svasand, eds), pp. 444–464. Blackwell, Oxford.

Longhurst, A. R. and Pauly, D. (1987). "Ecology of tropical oceans", p. 407. Academic Press, New York.

Loosanoff, V. L. and Davis, H. C. (1963). Rearing of bivalve molluscs. *Advances in Marine Biology* **1**, 1–136.

Lorenzen, K. (2005). Population dynamics and potential of fisheries stock enhancement: practical theory for assessment and policy analysis. *Philosophical Transactions of the Royal Society of London* **B360**, 171–189.

Lorenzen, K., Amarasinghe, U. S., Bartley, D. M., Bell, J. D., Bilio, M., de Silva, S. S., Garaway, C. J., Hartman, W. D., Kapetsky, J. M., Laleye, P., Moreau, J., Sugunan, V. V. and Swar, D. B. (2001). Strategic review of enhancements and culture-based fisheries. *In* "Aquaculture in the Third Millenium. Technical Proceedings of the Conference on Aquaculture in the Third Millennium, Bangkok, Thailand, 20–25 February 2000" (R. P. Subasinghe, P. Beuno, M. J. Phillips, C. Hough, S. E. McGladdery and J. R. Arthur, eds), pp. 221–237. NACA, Bangkok and FAO, Rome.

Losada-Tosteson, V. and Posada, J. M. (2001). Using tyres as shelters for the protection of juvenile spiny lobsters, *Panulirus argus*, or as a fishing gear for adults. *Marine and Freshwater Research* **52**, 1445–1450.

Lovatelli, A., Conand, C., Purcell, S., Uthicke, S., Hamel, J. F. and Mercier, A. (eds), (2004). "Advances in Sea Cucumber Aquaculture and Management", p. 425. FAO Fisheries Technical Paper No. 463. FAO, Rome.

Lozano-Álvarez, E., Briones-Fourzán, P. and Negrete-Soto, F. (1994). An evaluation of concrete block structures as shelter for juvenile Caribbean spiny lobsters *Panulirus argus*. *Bulletin of Marine Science* **55**, 351–362.

Lozano-Álvarez, E., Briones-Fourzán, P., Negrete-Soto, F. and Barradas-Ortiz, C. (1998). Atributos de la población de langostas *Panulirus argus* en la laguna arrecifal de Puerto Morelos, antes y después de la introducción de *casitas*. *In* "Funcionamiento de Refugios Artificiales Para Langosta y su Impacto en Hábitats de Pastizal Marino" (P. Briones-Fourzán, ed.), pp. 178–212. Informe Final UNAM-CONACYT, 1171-N, México.

Lucas, J. S. (1994). The biology, exploitation, and mariculture of giant clams (Tridacnidae). *Reviews in Fisheries Science* **2**, 181–223.

Luckenbach, M. W., Mann, R. and Wesson, J. A. (eds) (1999). "Oyster Reef Habitat Restoration: A Synopsis and Synthesis of Approaches". Virginia Institute of Marine Science, School of Marine Science, College of William and Mary, VIMS Press, Gloucester Point, Virginia.

Luttikhuizen, P. C., Drent, J. and Baker, A. J. (eds) (2003). Disjunct distribution of highly-diverged mitochondrial lineage clade and population subdivision in a marine bivalve with pelagic larval dispersal. *Molecular Ecology* **12** 2215–2229.

Macaranas, J. M. (1993). Means to identify stocks and strains. *In* "Genetic Aspects of Conservation and Cultivation of Giant Clams" (P. E. Munro, ed.), pp. 25–29. ICLARM Conference Proceedings No. 39, Manila, Philippines.

Macaranas, J. M., Ablan, C. A., Pante, M. J. R., Benzie, J. A. H. and Williams, S. T. (1992). Genetic structure of giant clam (*Tridacna derasa*) populations from reefs in the Indo-Pacific. *Marine Biology* **113**, 231–238.

MacFarlane, S. L. (1998). The evolution of a municipal quahog (hardclam), *Mercenaria mercenaria* management program, a 20-year history: 1975–1995. *Journal of Shellfish Research* **17**, 1015–1036.

MacFarlane, J. W. and Moore, R. (1986). Reproduction of the ornate rock lobster, *Panulirus ornatus* (Fabricius), in Papua New Guinea. *Australian Journal of Marine and Freshwater Research* **37**, 55–65.

MacKenzie, C. L., Jr. (1989). A guide for enhancing estuarine molluscan shellfisheries. *Marine Fisheries Review* **51**(3), 1–47.

MacKenzie, C. L., Jr. (1996). History of oystering in the United States and Canada, featuring the eight greatest oyster estuaries. *Marine Fisheries Review* **58** (4), 1–78.

MacKenzie, C. L., Jr, Taylor, D. L. and Arnold, W. S. (2001). A history of hard clamming. *In* "Biology of the Hard Clam" (J. N. Kraeuter and M. Castagna, eds), pp. 651–674. Developments in Aquaculture and Fisheries Science 31, Elsevier, Amsterdam.

Mackie, L. A. and Ansell, A. D. (1993). Differences in reproductive ecology in natural and transplanted populations of *Pecten maximus*: Evidence for the existence of separate stocks. *Journal of Experimental Marine Biology and Ecology* **169**, 57–75.

Maheswarudu, G., Radhakrishnan, E. V., Pillai, N. N., Arputharaj, M. R., Ramakrishnan, A., Mohan, S. and Vairamani, A. (1998). Tagging experiments on sea-ranched *Penaeus indicus* in the Palk Bay, southeast coast of India. *Indian Journal of Fisheries* **45**, 67–74.

Malouf, R. E. (1989). Clam culture as a resource management tool. *In* "Clam Mariculture in North America" (J. J. Manzi and M. Castagna, eds), pp. 427–447. Developments in Aquaculture and Fisheries Science 19, Elsevier, Amsterdam.

Manchenko, G. P. (1994). "Handbook of Detection of Enzymes on Electrophoretic Gels". CRC Press, Boca Raton, Florida.

Manzi, J. J. (1990). The role of aquaculture in the restoration and enhancement of molluscan fisheries in North America. *In* "Marine Farming and Enhancement" (A. K. Sparks, ed.), pp. 53–56. Proceedings of the Fifteenth US-Japan Meeting on Aquaculture, Kyoto, Japan, October 22–23, 1986. NOAA Technical Report NMFS 85, US Department of Commerce.

Manzi, J. J., Eversole, A. G., Hilbish, J., Dillon, R. T. and Hadley, N. H. (1988). Genetic improvement of hard clam, *Mercenaria* spp., populations for commercial mariculture stock development in South Carolina. *Journal of Shellfish Research* **7**, 125.

Marelli, D. C. and Arnold, W. S. (1996). Growth and mortality of transplanted juvenile hard clams *Mercenaria mercenaria,* in the northern Indian River Lagoon, Florida. *Journal of Shellfish Research* **15**, 709–713.

Marshall, L. S. and Lipcius, R. N. (1999). Juvenile queen conch survival in similar seagrass habitats. *Proceedings of the Gulf and Caribbean Fisheries Institute* **45**, 926–931.

Marullo, F., Emiliani, D. A., Caillouet, C. W. and Clark, S. H. (1976). A vinyl streamer tag for shrimp (*Penaeus* spp.). *Transactions of the American Fisheries Society* **6**, 658–663.

Massin, C. (1982). Food and feeding mechanisms: *Holothuroidea*. *In* "Echinoderm Nutrition" (M. Jangoux and J. M. Lawrence, eds), pp. 493–497. A.A. Balkema, Rotterdam, Netherlands.

Masuda, R. and Tsukamoto, K. (1998). Stock enhancement in Japan: Review and perspective. *Bulletin of Marine Science* **62**, 337–358.

Mathews, C. P., El-Musa, M., Al-Hossaini, M., Samuel, M. and Abdul Ghaffar, A. R. (1988). Infestations of *Epipenaeon elegans* on *Penaeus semisulcatus* and their use as biological tags. *Journal of Crustacean Biology* **8**, 53–62.

Matsuda, H. and Yamakawa, T. (1997). The effect of temperature on growth of the Japanese spiny lobster, *Panulirus japonicus* (V. Siebold) phyllosomas under laboratory conditions. *Marine and Freshwater Research* **48**, 791–796.

Matsuda, H. and Yamakawa, T. (2000). The complete development and morphological changes of *Panulirus longipes* (Decapoda, Palinuridae) reared under laboratory conditions. *Fisheries Science* **66**, 278–293.

Mazón-Sástegui, J. M., Muciño-Diaz, M. and Bazúa-Sicre, L. A. (1996). Cultivo de abulón *Haliotis* spp. *In* "Estudio del potencial pesquero y acuicola de Baja California" (S. M. Casas Valdez and G. Ponce-Diaz, eds), pp. 475–511. CIBNOR, La Paz.

McBride, S. C. (1998). Current status of abalone aquaculture in the Californias. *Journal of Shellfish Research* **17**, 593–600.

McCormick, T. B., Henderson, K., Mill, T. S. and Altick, J. (1994). A review of abalone seeding, possible significance, and a new seeding device. *Bulletin of Marine Science* **55**, 680–693.

McDonald, J. H. B., Verrelli, B. C. and Geyer, L. B. (1996). Lack of geographic variation in anonymous nuclear polymorphisms in the American oyster, *Crassostrea virginica*. *Molecular Biology and Evolution* **13**, 1114–1118.

McHugh, J. J. (1981). Recent advances in hard clam mariculture. *Journal of Shellfish Research* **1**, 51–56.

McKoy, J. L. (1980). "Biology, Exploitation and Management of Giant Clams (Tridacnidae) in the Kingdom of Tonga". Fisheries Bulletin 1. Fisheries Division, Ministry of Agriculture and Forestry and Fisheries, Tonga.

McLean, J. E., Bentzen, P. and Quinn, T. P. (2003). Differential reproductive success of sympatric, naturally spawning hatchery and wild steelhead trout (*Oncorhynchus mykiss*) through the adult stage. *Canadian Journal of Fisheries and Aquatic Sciences* **60**, 433–440.

McNeely, J. A. (ed.) (2001). "The Great Reshuffling: Human Dimensions of Invasive Alien Species" p. 242. IUCN, Cambridge.

McNeely, J. A., Mooney, H. A., Neville, L. E., Schee, P. J. and Waage, J. K. (2001). "Global Strategy on Invasive Alien Species", p. 50. IUCN, Cambridge.

McShane, P. E. (1988). Tagging abalone. In "Tagging—Solution or Problem?" (D. A. Hancock, ed.), pp. 65–67. Australian Society for Fish Biology Tagging Workshop, 21–22 July 1988. Bureau of Rural Resources Proceedings No. 5. Australian Government Printing Office, Canberra.

McShane, P. E. (1991). Density-dependent mortality of recruits of the abalone *Haliotis rubra* (Mollusca: Gastropoda). *Marine Biology* **110**, 385–389.

McShane, P. E. (1995). Recruitment variation in abalone: Its importance to fisheries management. *Marine and Freshwater Research* **46**, 555–570.

McShane, P. E. and Smith, M. G. (1988). Measuring abundance of juvenile abalone, *Haliotis rubra* Leach (Gastropoda: Haliotidae); comparison of a novel method with two other methods. *Australian Journal of Marine and Freshwater Research* **39**, 331–336.

McShane, P. E. and Smith, M. G. (1991). Recruitment variation in sympatric populations of *Haliotis rubra* (Mollusca: Gastropoda) in southeast Australian waters. *Marine Ecology Progress Series* **73**, 203–210.

McShane, P. E. and Naylor, J. R. (1995). Depth can affect post-settlement survival of *Haliotis iris* (Mollusca: Gastropoda). *Journal of Experimental Marine Biology and Ecology* **187**, 1–12.

Meadows, J. R. S., Ward, R. D., Grewe, P. M., Dierens, L. M. and Lehnert, S. A. (2003). Characterisation of 23 tri- and tetra-nucleotide microsatellite loci in the brown tiger prawn, *Penaeus esculentus*. *Molecular Ecology Notes* 454–456.

Melville-Smith, R. and Chubb, C. F. (1997). Comparison of dorsal and ventral tag retention in western rock lobsters, *Panulirus cygnus* (George). *Marine and Freshwater Research* **48**, 577–580.

Melville-Smith, R., Jones, J. B. and Brown, R. S. (1997). Biological tags as moult indicators in *Panulirus cygnus* (George). *Marine and Freshwater Research* **48**, 959–965.

Menzel, W. (1989). The biology, fishery and culture of quahog clams, *Mercenaria*. In "Clam Mariculture in North America" (J. J. Manzi and M. Castagna, eds), pp. 201–242. Elsevier, Amsterdam.

Mercier, A., Battaglene, S. and Hamel, J. F. (1999). Daily burrowing cycle and feeding activity of juvenile sea cucumbers *Holothuria scabra* in response to environmental factors. *Journal of Experimental Marine Biology and Ecology* **239**, 125–156.

Mgaya, Y. D., Gosling, E. M., Mercer, J. P. and Donlon, J. (1995). Genetic variation at three polymorphic loci in wild and hatchery stocks of the abalone, *Haliotis tuberculata* Linnaeus. *Aquaculture* **136**, 71–80.

Mikami, S. and Greenwood, J. G. (1997a). Influence of light regimes on phyllosomal growth and timing of moulting in *Thenus orientalis* (Lund) (Decapoda: Scyllaridae). *Marine and Freshwater Research* **48**, 777–782.

Mikami, S. and Greenwood, J. G. (1997b). Complete development and comparative morphology of larval *Thenus orientalis* and *Thenus* sp. (Decapoda: Scyllaridae) reared in the laboratory. *Journal of Crustacean Biology* **17**, 289–308.

Mills, D. J., Gardner, C. and Ibbott, S. (2004). Behaviour of on-grown juvenile spiny lobsters, *Jasus edwardsii*, after re-seeding to a coastal reef in Tasmania, Australia. *In* "Stock Enhancement and Sea Ranching: Developments, Pitfalls and Opportunities" (K. M. Leber, S. Kitada, L. Blankenship and T. Svåsand, eds), pp. 168–180. Blackwell, Oxford.

Minagawa, M. (1990). Early and middle larval development of *Panulirus penicillatus* (Crustacea, Decapoda, Palinuridae) reared in the laboratory. *Researches on Crustacea [Nihon Kokakurui Gakkai]* **18**, 77–93.

Minchin, D. (1991). Decapod predation and the sowing of the scallop *Pecten maximus* (Linneus, 1758). *In* "An International Compendium of Scallop Biology and Culture" (S. E. Shumway and P. A. Sandifer, eds), pp. 191–197. World Aquaculture Workshops No. 1, World Aquaculture Society, Baton Rouge, Louisiana.

Minello, T. J. and Zimmerman, R. J. (1983). Fish predation on juvenile brown shrimp, *Penaeus aztecus* Ives: The effect of simulated *Spartina* structure on predation rates. *Journal of Experimental Marine Biology and Ecology* **72**, 211–231.

Minello, T. J., Zimmerman, R. J. and Martinez, E. X. (1989). Mortality of young brown shrimp *Penaeus aztecus* in estuarine nurseries. *Transactions of the American Fisheries Society* **118**, 693–708.

Mingoa-Licuanan, S. S. and Gomez, E. D. (1996). Giant clam culture. *In* "Perspectives in Asian Fisheries" (S. S. De Silva, ed.), pp. 281–299. Asian Fisheries Society, Manila.

Mingoa-Licuanan, S. S. and Gomez, E. D. (2002). Giant clam conservation in Southeast Asia. *Tropical Coasts* December 2002, 24–31.

Mitton, J. B., Berg, C. J. and Orr, K. S. (1989). Population structure, larval dispersal, and gene flow in the queen conch, *Strombus gigas*, in the Caribbean. *Biological Bulletin* **177**, 356–362.

Miyajima, T. and Toyota, K. (2002). Development of marking technique by tail-cutting in Kuruma prawn *Penaeus japonicus* and the application to estimate recapture rate. Abstract "Second International Symposium on Stock Enhancement and Sea Ranching, Kobe, Japan 28th January–1st February 2002."

Miyamoto, T., Saito, K., Motoya, S., Nishikawa, N., Monma, H. and Kawamura, K. (1982). Experimental studies on the release of the cultured seeds of abalone, *Haliotis discus hannai* Ino in Oshoro Bay, Hokkaido. *Scientific Report Hokkaido Fisheries Experimental Station* **24**, 59–90 (in Japanese with English abstract).

Moberg, P. E. and Burton, R. S. (2000). Genetic heterogeneity among recruit and adult red urchins, *Strongylocentrotus franciscanus*. *Marine Biology* **136**, 773–784.

Moe, M. A., Jr. (1991). "Lobsters—Florida, Bahamas, the Caribbean", p. 511. Green Turtle Publications, Plantation, Florida.

Mohamed, K. H. and George, M. J. (1968). Results of the tagging experiments on the Indian spiny lobster, *Panulirus homarus* (Linnaeus)—Movement and growth. *Indian Journal of Fisheries* **15**, 15–26.

Moksness, E. (2004). Stock enhancement and sea ranching as an integrated part of coastal zone management in Norway. *In* "Stock Enhancement and Sea Ranching: Developments, Pitfalls and Opportunities" (K. M. Leber, S. Kitada, L. Blankenship and T. Svasand, eds), pp. 3–10. Blackwell, Oxford.

Moksness, E., Stole, R. and van der Meeren, G. (1998). Profitability analysis of sea ranching with Atlantic salmon (*Salmo salar*), Arctic charr (*Salvelinus alpinus*) and European lobster (*Homarus gammarus*) in Norway. *Bulletin of Marine Science* **62**, 689–699.

Molony, B. W., Lenanton, R., Jackson, G. and Norriss, J. (2003). Stock enhance-ment as a fisheries management tool. *Reviews in Fish Biology and Fisheries* **13**, 409–432.

Momma, H. (1972). Studies on the release of the seed disc abalone—I. Behavior immediately after release. *Bulletin of the Japanese Society of Scientific Fisheries* **38**, 671–676 (in Japanese, English abstract).

Momma, H., Kobayashi, K., Kato, T., Sasaki, Y., Sakamoto, T. and Murata, H. (1980). On the artificial propagation method of abalone and its effects on rocky shore. I. Remaining ratio of the artificial seed abalone (*Haliotis discus hannai* Ino) on latticed artificial reefs. *The Aquiculture* **28**(2), 59–65. Translated by M.G. Mottet *In* "Summaries of Japanese Papers on Hatchery Technology and Interme-diate Rearing Facilities for Clams, Scallops and Abalone", p. 26. Washington Department of Fisheries Progress Report 203.

Momma, H., Agatsuma, Y. and Sawazaki, M. (1992). Migration with the passage of time and dispersion in the cultured seeds of the sea urchin. *Bulletin of the Japanese Society of Scientific Fisheries* **58**, 1437–1442 (in Japanese with English abstract).

Moore, S. S., Whan, V., Davis, G. P., Byrne, K., Hetzel, D. J. S. and Preston, N. (1999). The development and application of genetic markers for the Kuruma prawn *Penaeus japonicus*. *Aquaculture* **173**, 19–32.

Mora, L. (1994). Análisis de la pesquería del caracol pala (*Strombus gigas*) en Colombia. *In* "Queen Conch Biology, Fisheries and Mariculture" (R. S. Appel-doorn and B. Rodriguez, eds), pp. 137–144. Fundación Científica Los Roques, Caracas.

Morgan, G. R., Phillips, B. F. and Joll, L. M. (1982). Stock and recruitment relation-ships in *Panulirus cygnus*, the commercial rock (spiny) lobster of Western Austra-lia. *Fishery Bulletin, US* **80**, 475–486.

Morikawa, T. (1999). Status and prospects on the development and improvement of coastal fishing ground. *In* "Marine Ranching: Global Perspectives with Emphasis on the Japanese Experience. (unedited)", pp. 136–239. FAO Fisheries Circular No. 943, FAO, Rome.

Moriyasu, M., Landsburg, W. and Conan, G. Y. (1995). Sphyrion tag shedding and tag induced mortality of the American lobster, *Homarus americanus* H. Milne Edwards, 1837 (Decapoda, Nephropidae). *Crustaceana* **68**, 184–192.

Mortensen, S. (1999). Scallop introductions and transfers from an animal health point of view. *12th International Pectinid Workshop 5–11 May 1999, Bergen, Norway*, 21–22.

Mulley, J. C. and Latter, B. (1981). Geographic differentiation of tropical Australian penaeid prawn populations. *Australian Journal of Marine and Freshwater Research* **32**, 897–906.

Muliani (1993). Effect of different supplemental feeds and stocking densities on the growth rate and survival of sea cucumber, *Holothuria scabra* in Tallo river mouth, South Sulawesi. *Journal Penelitian Budidaya Pantai* **9**(4), 15–22.

Munro, J. L. (1989). Fisheries for giant clams (Tridacnidae: Bivalvia) and pro-spects for stock enhancement. *In* "Marine Invertebrate Fisheries: Their Assessment and Management" (J. F. Caddy, ed.), pp. 541–558. John Wiley & Sons, New York.

Munro, J. L. (1993a). Giant clams. *In* "Nearshore Marine Resources of the South Pacific" (A. Wright and L. Hill, eds), pp. 431–449. Forum Fisheries Agency, Honiara and Institute of Pacific Studies, Suva.

Munro, J. L. (1993b). Strategies for re-establishment of wild giant clam stocks. *In* "Genetic Aspects of Conservation and Cultivation of Giant Clams" (P. E.

Munro, ed.), pp. 17–20. ICLARM Conference Proceedings No. 39, Manila, Philippines.

Munro, J. L. and Bell, J. D. (1997). Enhancement of marine fisheries resources. *Reviews in Fisheries Science* **5**, 185–222.

Munro, P. E. (1993). "Genetic Aspects of Conservation and Cultivation of Giant Clams", p. 47. ICLARM Conference Proceedings No. 39, ICLARM, Manila, Philippines.

Murakoshi, M. (1986). Farming of the boring clam *Tridacna crocea*. *Galaxea* **5**, 239–254.

Myers, R. A. and Worm, B. (2003). Rapid worldwide depletion of predatory fish communities. *Nature* **423**, 280–283.

Mykles, D. L., Cotton, J. L. S., Taniguchi, H., Sana, K.-I. and Maeda, Y. (1998). Cloning of tropomyosins from lobster (*Homarus americanus*) striated muscles: Fast and slow isoforms may be generated from the same transcript. *Journal of Muscle Research and Cell Motility* **19**, 105–115.

Nadeau, M. and Cliche, G. (1998a). Predation of juvenile sea scallops (*Placopecten magellanicus*) by crabs (*Cancer irroratus* and *Hyas* sp.) and starfish (*Astrea vulgaris, Leptasterias polaris* and *Crossaster papposus*). *Journal of Shellfish Research* **17**, 905–910.

Nadeau, M. and Cliche, G. (1998b). Evaluation of the recapture rate of seeded scallops (*Placopecten magellanicus*) during commercial fishing activity in Iles-de-la-Madeleines, Quebec. *Bulletin of the Aquaculture Association of Canada* **98**, 79–81.

Nagasawa, K., Bresciani, J. and Lutzen, J. (1988). Morphology of *Pectinophilus ornatus*, new genus, new species, a copepod parasite of the Japanese scallop *Patinopecten yessoensis*. *Journal of Crustacean Biology* **8**, 31–42.

Nakai, K., Musashi, T., Inoguchi, N., Saido, T., Kishida, T. and Matsuda, H. (1993). Estimation of survival rate in the early stage of released artificial young abalone by the capture–recapture data. *Bulletin of the Japanese Society of Scientific Fisheries* **59**, 1845–1850 (in Japanese with English abstract, tables and figures).

Nakatsugawa, T. (1993). Relationship between water temperature and infectious nature of the agent of 'amyotrophia' in abalone. *Bulletin of the Kyoto Institute of Ocean and Fisheries Science* **16**, 35–38 (in Japanese, English abstract).

Nash, W. J. (1988). Hatchery rearing of trochus as a management tool. *Australian Fisheries* **47**, 36–39.

Nash, W. (1989). New size limit for west coast abalone. *Fishing Today* (Aug–Sep 1989) **2**(4), 38–39.

Nash, W. J. (1992). An evaluation of egg-per-recruit analysis as a means of assessing size limits for blacklip abalone (*Haliotis rubra*) in Tasmania. *In* "Abalone of the World: Biology, Fisheries and Culture" (S. A. Shepherd, M. J. Tegner and S. A. Guzmán del Próo, eds), pp. 318–338. Blackwell, Oxford.

Nash, W. J. (1993). Trochus. *In* "Nearshore Marine Resources of the South Pacific" (A. Wright and L. Hill, eds), pp. 452–495. Forum Fisheries Agency, Honiara and Institute of Pacific Studies, Suva.

Nash, W. J. (1996). Stock assessment report: Abalone. *Tasmanian Marine Resources Division, unpublished report,* 77.

Nash, W. J. and Catchpole, E. (in press). A population modeling approach to evaluating the role of restocking in fishery management. *4th World Fisheries Congress Proceedings*, Vancouver, Canada.

Nash, W. J., Sellers, T. L., Talbot, S. R., Cawthorn, A. J. and Ford, W. B. (1994). The population biology of abalone (*Haliotis* species) in Tasmania. I. Blacklip abalone from the north coast and the islands of Bass Strait. *Tasmanian Department of Primary Industry and Fisheries, Sea Fisheries Division Technical Report* **48**, 1–69.

Nash, W. J., Sanderson, J. C., Bridley, J., Dickson, S. and Hislop, B. (1995). Post-larval recruitment of blacklip abalone (*Haliotis rubra*) on artificial collectors in southern Tasmania. *Marine and Freshwater Research* **46**, 531–538.

Nash, W., Adams, T., Tuara, P., Terekia, O., Munro, D., Amos, M., Legata, J., Mataiti, N., Teopenga, M. and Whitford, J. (1995). "The Aitutaki Trochus Fishery: A Case Study", p. 72. South Pacific Commission, Noumea, New Caledonia.

National Fisheries Research and Development Institute (2000). "Annual Seedling Production at National Fisheries". NFRDI Press, Seoul, South Korea.

Natsukari, Y., Tanaka, N., Chung, S.-C. and Hirayama, K. (1995). A genetic comparison among three group (wild populations, artificial seed production, and mixed populations) of sea urchin *Pseudocentrotus depressus*: A preliminary report. *UJNR Technical Report* **22**, 17–25.

Neal, R. A. (1969). Methods for marking shrimp. pp. 1149–1165. FAO Fisheries Report 57. FAO, Rome.

Nei, M. (1973). Analysis of gene diversity in subdivided populations. *Proceedings of the National Academy of Sciences USA* **70**, 3321–3323.

Neigel, J. E. (1994). Analysis of rapidly evolving molecules and DNA sequence variants: Alternative approaches for detecting genetic structure in marine populations. *California Cooperative Oceanic Fisheries Investigations Reports* **35**, 82–89.

Nelson, M. M., Mooney, B. D., Nichols, P. D., Phleger, C. F., Smith, G. G., Hart, P. and Ritar, A. J. (2002). The effect of diet on the biochemical composition of juvenile *Artemia*: Potential formulations for rock lobster aquaculture. *Journal of the World Aquaculture Society* **33**, 146–157.

Nevo, E., Beiles, A. and Ben-Shlomo, R. (1984). The evolutionary significance of genetic diversity: Ecological, demographic and life-history correlates. *In* "Evolutionary Dynamics of Genetic Diversity" (G. S. Mani, ed.), pp. 13–213. Springer-Verlag, Berlin.

Newkirk, G. (1993). A discussion of genetic aspects of broodstock establishment and management. *In* "Genetic Aspects of Conservation and Cultivation of Giant Clams" (P. E. Munro, ed.), pp. 6–12. ICLARM Conference Proceedings No. 39, Manila, Philippines.

Newland, P. L. and Chapman, C. J. (1993). Locomotory behaviour and swimming performance of the Norway lobster, *Nephrops norvegicus*, in the presence of an acoustic tag. *Marine Biology* **115**, 33–37.

Nicolas, J. L., Basuyaux, O., Mazurié, J. and Thébault, A. (2002). *Vibrio carchariae*, a pathogen of *Haliotis tuberculata*. *Diseases of Aquatic Organisms* **50**, 35–43.

Nicosia, F. and Lavalli, K. (1999). Homarid lobster hatcheries: Their history and role in research, management, and aquaculture. *Marine Fisheries Review* **61**, 1–57.

Nielsen, E. E., Hansen, M. M. and Loeschcke, V. (1999). Analysis of DNA from old scale samples: Technical aspects, applications and perspectives for conservation. *Hereditas* **130**, 265–276.

Nielson, L. A. (1992). Methods of Marking Fish and Shellfish. *American Fisheries Society Special Publication* **23**, 208.

Nielsen, R. and Wakeley, J. (2001). Distinguishing migration from isolation. *Genetics* **158**, 885–896.

Nishimura, M. and Tsuji, H. (1980). Experimental studies on the release of the cultured seeds of abalone at artificial reef at Kamoiri area (in Japanese). *Scientific Report of the Kyoto Marine Centre* **3**, 1–17.

NMT (2003). Crustacean marking with coded wire tags. [http://www.nmt.us/Applications/CWT/Crustacean.pdf].

Nonaka, M., Fushimi, H. and Yamakawa, T. (2000). The spiny lobster fishery in Japan and restocking. *In* "Spiny Lobsters: Fisheries and Culture (2nd Edition)" (B. F. Phillips and J. Kittaka, eds), pp. 221–242. Fishing News Books, Oxford.

Norman, C. P., Yamakawa, H. and Yoshimura, T. (1994). Habitat selection, growth rate and density of juvenile *Panulirus japonicus* (von Siebold, 1824) (Decapoda, Palinuridae) at Banda, Chiba Prefecture, Japan. *Crustaceana* **66**, 366–383.

Norton, J. H., Shepherd, M. A., Long, H. M. and Prior, H. C. (1993a). Parasites of giant clams (Tridacnidae) *In* "Biology and Mariculture of Giant Clams (W. K. Fitt, ed.), pp. 18–23. ACIAR Proceedings No. 47, Australian Centre for International Agricultural Research, Canberra.

Norton, J. H., Braley, R. D. and Anderson, I. G. (1993b). A quarantine protocol to prevent the spread of parasitic diseases of giant clams (Tridacnidae) via translocation. *In* "Biology and Mariculture of Giant Clams" (W. K. Fitt, ed.), pp. 24–26. ACIAR Proceedings No. 47, Australian Centre for International Agricultural Research, Canberra.

Okaichi, T. and Yanagi, T. (eds) (1997). "The Preservation and Creation of Fisheries Grounds". Terra Scientific Publishing Company, Tokyo.

Oliver, M., Stewart, R., Gardner, C. and Mac Diarmid, A. (2001). Evaluating the release and survival of juvenile rock lobsters released for enhancement purposes. *In* "Developments in Rock Lobster Enhancement and Aquaculture III. Fisheries Research Development Corporation Rock Lobster Enhancement and Aquaculture Subprogram" (R. van Barneveld, ed.), p. 46. RLEAS Publication No. 6 [http://www.frdc.com.au/research/programs/rleas/pub/index.htm].

Olsen, D. A. (1968a). Banding patterns of *Haliotis rufescens* as indicators of botanical and animal succession. *Biological Bulletin* **134**, 139–147.

Olsen, D. A. (1968b). Banding patterns in *Haliotis*. 2—Some behavioural considerations and the effect of diet on shell coloration for *Haliotis rufescens, Haliotis corrugata, Haliotis sorenseni* and *Haliotis assimilis*. *The Veliger* **11**, 135–139.

Orensanz, J. M., Parma, A. M. and Iribarne, O. O. (1991). Population dynamics and management of natural stocks. *In* "Scallops: Biology, Ecology and Aquaculture" (S. E. Shumway, ed.), pp. 625–713. Elsevier, Amsterdam.

Ortiz-Quintanilla, M. (1980). Un sistema para inducir el desove de abulón en los campos pesqueros de Baja California, México. *Memorias del Segundo Simposio Latinoamericana de Acuacultura* **1**, 871–881.

Oshima, Y. (1976). A technique for resource propagation for the Japanese spiny lobster. *Fishery Engineering* **12**, 1–3.

Ovenden, J. R. and Brasher, D. J. (1994). Stock identity of the red (*Jasus edwardsii*) and green (*J. verreauxi*) rock lobsters inferred from mitochondrial DNA analysis. *In* "Spiny Lobster Management" (B. F. Phillips, J. S. Cobb and J. Kittaka, eds), pp. 230–249. Fishing News Books, Oxford.

Ovenden, J. R., Brasher, D. J. and White, R. W. G. (1992). Mitochondrial DNA analyses of the red rock lobster *Jasus edwardsii* support an apparent absence of population sub-division throughout Australasia. *Marine Biology* **112**, 319–326.

Owens, L. (1983). Bopyrid parasite *Epipenaeon ingens* Nobili as a biological marker for the banana prawn, *Penaeus merguiensis* de Man. *Australian Journal of Marine and Freshwater Research* **34**, 477–481.

Owens, L. (1985). *Polypocephalus* sp. (Cestoda: Lecanicephalidae) as a biological marker for banana prawns, *Penaeus merguiensis* de Man, in the Gulf of Carpentaria. *Australian Journal of Marine and Freshwater Research* **36**, 291–299.

Ozumi, K. (1998). Behavior and survival of juvenile paua (*Haliotis iris*) in the field. "Abstracts of the New Zealand Marine Science Society Conference., University of Otago, Dunedin".

Paille, N. and Gendron, L. (2001). "Conception et misé en place de recifs artificiels pour le homard (*Homarus americanus*): Revue des essais et recommandations", Institute of Maurice Lamontagne. Mont-Joli, Quebec, Ontario, Canada, no 261, pp. 50.

Palm, S., Dannewitz, J., Jarvi, T., Petersson, E., Prestegaard, T. and Ryman, N. (2003). Lack of molecular genetic divergence between sea-ranched and wild sea trout (*Salmo trutta*). *Molecular Ecology* **12**, 2057–2071.

Palma, A. T., Steneck, R. S. and Wilson, C. J. (1999). Settlement driven, multiscale demographic patterns of large benthic decapods in the Gulf of Maine. *Journal of Experimental Marine Biology and Ecology* **241**, 107–136.

Palumbi, S. R. (2003). Population genetics, demographic connectivity and the design of marine reserves. *Ecological Applications* **13** (Suppl. 5), 146–158.

Parrish, F. A. and Polovina, J. J. (1994). Habitat thresholds and bottlenecks in production of the spiny lobster (*Panulirus marginatus*) in the Northwestern Hawaiian Islands. *Bulletin of Marine Science* **54**, 151–163.

Pauly, D. (1987). Theory and practice of overfishing: A Southeast Asia perspective. "Proceedings of the 22nd Indo-Pacific Fisheries Commission, Darwin, Australia, 16–26 Feb, 1987", pp. 146–163.

Pauly, D. (1988). Some definitions of overfishing relevant to coastal zone management in Southeast Asia. *Tropical Coastal Area Management* **3**, 14–15.

Pauly, D., Christensen, V., Gúenette, S., Pitcher, T. J., Sumaila, U. R., Walters, C. J., Watson, R. and Zeller, D. (2002). Towards sustainability in world fisheries. *Nature* **418**, 689–695.

Penn, J. W. and Caputi, N. (1986). Spawning stock recruitment relationships and environmental influences on the tiger prawn (*Penaeus esculentus*) fishery in Exmouth Gulf, Western Australia. *Australian Journal of Marine and Freshwater Research* **37**, 491–505.

Penn, J. W., Caputi, N. and Hall, N. G. (1995). Stock–recruitment relationships for the tiger prawn (*Penaeus esculentus*) stocks in Western Australia. *ICES Marine Science Symposium* **199**, 320–333.

Penn, J. W., Watson, R., Caputi, N. and Hall, N. (1997). Protecting vulnerable stocks in multi-species prawn fisheries. *In* "Developing and Sustaining World Fisheries Resources: The State of Science and Management" (D. Hancock, D. Smith, A. Grant and J. P. Beumer, eds), pp. 122–129. *2nd World Fisheries Congress Proceedings*, Brisbane, Australia.

Pérez-Farfante, I. and Kensley, B. F. (1997). Penaeoid and sergestoid shrimps and prawns of the world: Keys and diagnoses for the families and genera. *Memoirs du Museum National d'Histoire Naturale, Paris* **175**, 1–233.

Peterson, C. H. (1990). On the role of ecological experimentation in resource management: Managing fisheries through mechanistic understanding of predator feeding behaviour. *In* "Behavioural Mechanisms of Food Selection" (R. N. Hughes, ed.), pp. 821–846. Springer-Verlag, Berlin.

Peterson, C. H. (2002). Recruitment overfishing in a bivalve mollusc fishery: Hard clams (*Mercenaria mercenaria*) in North Carolina. *Canadian Journal of Fisheries and Aquatic Sciences* **59**, 96–104.

Peterson, C. H., Ambrose, W. H. J. and Hunt, J. H. (1982). A field test of the swimming response of the bay scallop (*Argopecten irradians*) to changing biological factors. *Bulletin of Marine Science* **32**, 939–944.

Peterson, C. H., Summerson, H. C. and Huber, J. (1995). Replenishment of hard clam stocks using hatchery seed: Combined importance of bottom type, seed size, planting season and density. *Journal of Shellfish Research* **14**, 293–300.

Peterson, C. H., Summerson, H. C. and Luettich, R. A., Jr. (1996). Response of bay scallops to spawner transplants: A test of recruitment limitation. *Marine Ecology Progress Series* **132**, 93–107.

Phillips, B. F. (1972). A quantitative collector of the western rock lobster *Panulirus longipes cygnus* George (Decapoda: Palinuridae). *Crustaceana* **22**, 147–154.

Phillips, B. F. (1986). Prediction of commercial catches of the western rock lobster *Panulirus cygnus* George. *Canadian Journal of Fisheries and Aquatic Sciences* **43**, 2126–2130.

Phillips, B. F. and Booth, J. D. (1994). Design, use and effectiveness of collectors for catching the puerulus stage of spiny lobsters. *Reviews in Fisheries Sciences* **2**, 255–289.

Phillips, B. F. and Evans, L. H. (1997). Aquaculture and stock enhancement of lobsters: Report from a workshop. *Marine and Freshwater Research* **48**, 899–902.

Phillips, B. F. and Kittaka, J. (eds) (2000). "Spiny Lobsters: Fisheries and Culture" (2nd Edition), p. 679. Fishing News Books, Oxford.

Phillips, B. F. and Melville-Smith, R. (2001). Potential impacts of puerulus collection on the biological neutrality of the Western Australian rock lobster fishery and the relevance to other fisheries. *In* "Developments in Rock Lobster Enhancement and Aquaculture III" (R. van Barneveld, ed.), pp. 47–52. FRDC Rock Lobster Enhancement and Aquaculture Subprogram. RLEAS Publication No. 6. (http://www.frdc.com.au/research/programs/rleas/pub/index.htm).

Phillips, B. F., Joll, L. M., Sandland, R. I. and Wright, D. (1983). Longevity, reproductive condition and growth of the western rock lobster *Panulirus cygnus* George, reared in aquaria. *Australian Journal of Marine and Freshwater Research* **34**, 419–429.

Phillips, B. F., Joll, L. M. and Ramm, D. C. (1984). An electromagnetic tracking system for studying the movements of rock (spiny) lobsters. *Journal of Experimental Marine Biology and Ecology* **79**, 9–18.

Phillips, B. F., Cobb, J. S. and Kittaka, J. (eds) (1994)."Spiny Lobster Management" p. 550. Fishing News Books, Oxford.

Phillips, B. F., Cruz, R., Brown, R. S. and Caputi, N. (2000). Predicting the catch of spiny lobster fisheries. *In* "Spiny Lobsters: Fisheries and Culture (2nd Edition)" (B. F. Phillips and J. Kittaka, eds), pp. 285–301. Blackwell, Oxford.

Phillips, B. F., Melville-Smith, R., Cheng, Y. W. and Rossbach (2001). Testing collector designs for commercial harvesting of western rock lobster (*Panulirus cygnus*) puerulus. *Marine and Freshwater Research* **52**, 1465–1473.

Phleger, C. F., Nelson, M. M., Mooney, B. D., Nichols, P. D., Ritar, A. J., Smith, G. G., Hart, P. R. and Jeffs, A. G. (2001). Lipids and nutrition of the southern rock lobster, *Jasus edwardsii*, from hatch to puerulus. *Marine and Freshwater Research* **52**, 1475–1486.

Pitcher, C. R., Skewes, T. D. and Dennis, D. M. (1994). Research for management of the ornate tropical rock lobster, *Panulirus ornatus*, fishery in Torres Strait.

"Progress Report to Torres Strait Fisheries Scientific Advisory Council Meeting No. 21; April 1994" p. 10. CSIRO, Cleveland, Australia .

Pitcher, R., Skewes, T. and Dennis, D. (1995). Biology and fisheries ecology of the ornate rock lobster in Torres Strait. In "Recent Advances in Marine Science and Technology 1994" (O. Bellwood, O. H. Choat, and N. K. Saxena, eds), pp. 311–321. PACON International, Honolulu/James Cook University, Townsville, Australia 754 pp.

Pitt, R. (2001). Review of sandfish breeding and rearing methods. *Secretariat of the Pacific Community Bech-de-mer Information Bulletin* **14**, 14–21.

Pitt, R. and Duy, N. D. Q. (2004). Breeding and rearing of the sea cucumber *Holothuria scabra* in Viet Nam. In "Advances in Sea Cucumber Aquaculture and Management" (A. Lovatelli, C. Conand, S. Purcell, S. Uthicke, J. F. Hamel and A. Mercier, eds), pp. 333–346. FAO Fisheries Technical Paper No. 463, FAO, Rome.

Pitt, R., Duy, N. D. Q., Duy, T. V. and Long, H. T. C. (2004). Sandfish (*Holothuria scabra*) with shrimp (*Penaeus monodon*) co-culture tank trials. *Secretariat of the Pacific Community Beche-de-mer Information Bulletin* **20**, 12–22.

Poiner, I. R., Staples, D. J. and Kenyon, R. (1987). Seagrass communities of the Gulf of Carpentaria, Australia. *Australian Journal of Marine and Freshwater Research* **38**, 121–131.

Polovina, J. J. (1991). Fisheries applications and biological impacts of artificial habitats. In "Artificial Habitats for Marine and Freshwater Fisheries" (W. Seaman, Jr., and L. M. Sprague, eds), pp. 153–176. Academic Press, San Diego.

Posey, M. H., Alphin, T. D., Powell, C. M. and Townsend, E. (1999). Use of oyster reefs as a habitat for epibenthic fish and decapods. In "Oyster Reef Habitat Restoration: A Synopsis and Synthesis of Approaches" (M. W. Luckenbach, R. Mann and J. A. Wesson, eds), pp. 229–238. Virginia Institute of Marine Science, School of Marine Science, College of William and Mary, VIMS Press, Gloucester Point, Virginia.

Posgay, J. A. (1981). Movement of tagged sea scallops on Georges Bank. *Marine Fisheries Review* **43**, 19–25.

Potter, I. C., Manning, R. J. G. and Loneragan, N. R. (1991). Size, movements, distribution and gonadal stage of the western king prawn (*Penaeus latisulcatus*) in a temperate estuary and local marine waters. *Journal of Zoology, London* **223**, 419–445.

Preece, P. A., Shepherd, S. A., Clarke, S. M. and Keesing, J. K. (1997). Abalone stock enhancement by larval seeding: Effect of larval density on settlement and survival. *Molluscan Research* **18**, 265–273.

Prentice, E. F. (1990). A new internal telemetry tag for fish and crustaceans. In "Marine Farming and Enhancement: Proceedings of the 15th US-Japan Meeting on Aquaculture" (A. K. Sparks, ed.), pp. 1–9. NOAA Technical Report NMFS 85, Kyoto, Japan. US Department of Commerce, Washington.

Prentice, E. F. and Rensel, J. E. (1977). Tag retention of the spot prawn, *Pandalus platyceros,* injected with coded wire tags. *Journal of the Fisheries Research Board of Canada* **34**, 2199–2203.

Preston, G. L. (1993). Beche-de-mer. In "Nearshore Marine Resources of the South Pacific" (A. Wright and L. Hill, eds), pp. 371–407. Forum Fisheries Agency, Honiara and Institute of Pacific Studies, Suva.

Preston, G. (1997). Background to the workshop and narrative report. In "Workshop on Trochus Resource Assessment, Management and Development. Integrated Fisheries Management Project" (unedited), pp. 3–12. Technical Document No. 13, South Pacific Commission, Noumea, New Caledonia.

Preston, N. P., Smith, D. M., Kellaway, D. M. and Bunn, S. E. (1996). The use of enriched ^{15}N as an indicator of the assimilation of individual protein sources from compound diets for juvenile *Penaeus monodon*. *Aquaculture* **147**, 249–259.

Primavera, J. H. and Caballero, R. M. (1992). Effect of streamer tags on survival and growth of juvenile tiger prawns, *Penaeus monodon*, under laboratory conditions. *Australian Journal of Marine and Freshwater Research* **43**, 737–743.

Prince, J. D. and Hilborn, R. (1998). Concentration profiles and invertebrate fisheries management. *In* "Proceedings of the North Pacific Symposium on Invertebrate Stock Assessment and Management" (G. S. Jamieson and A. Campbell, eds), *Canadian Special Publication in Fisheries and Aquatic Sciences* **125**, 187–196.

Prince, J. D., Sellers, T. L., Ford, W. B. and Talbot, S. R. (1987). Experimental evidence for limited dispersal of haliotid larvae (genus *Haliotis*; Mollusca: Gastropoda). *Journal of Experimental Marine Biology and Ecology* **106**, 243–263.

Prince, J. D., Sellers, T. L., Ford, W. B. and Talbot, S. R. (1988a). Confirmation of a relationship between the localized abundance of breeding stock and recruitment for *Haliotis rubra* Leach (Mollusca: Gastropoda). *Journal of Experimental Marine Biology and Ecology* **122**, 91–104.

Prince, J. D., Sellers, T. L., Ford, W. B. and Talbot, S. R. (1988b). Recruitment, growth, mortality and population structure in a southern Australian population of *Haliotis rubra* (Mollusca: Gastropoda). *Marine Biology* **100**, 75–82.

Prince, J. D., Walters, C., Ruiz-Avila, R. and Sluczanowski, P. (1998). Territorial users' rights and the Australian abalone (*Haliotis* spp.) fishery. *In* "Proceedings of the North Pacific Symposium on Invertebrate Stock Assessment and Management" (G. S. Jamieson and A. Campbell, eds), *Canadian Special Publication in Fisheries and Aquatic Sciences* **125**, 367–375.

Purcell, S. (2001). Successful cage design for intermediate culture of trochus for restocking. *Secretariat of the Pacific Community Trochus Information Bulletin* **8**, 4–7.

Purcell, S. (2002). Cultured vs wild juvenile trochus: Disparate shell morphologies sends caution for seeding. *Secretariat of the Pacific Community Trochus Information Bulletin* **9**, 6–8.

Purcell, S. W. (2004a). Management options for restocked trochus fisheries. *In* "Stock Enhancement and Sea Ranching: Developments, Pitfalls and Opportunities" (K. M. Leber, S. Kitada, L. Blankenship and T. Svasand, eds), pp. 233–244. Blackwell, Oxford.

Purcell, S. W. (2004b). Criteria for release strategies and evaluating the restocking of sea cucumbers. *In* "Advances in Sea Cucumber Aquaculture and Management" (A. Lovatelli, C. Conand, S. Purcell, S. Uthicke, J. F. Hamel and A. Mercier, eds), pp. 181–191. FAO Fisheries Technical Paper No. 463, FAO, Rome.

Purcell, S. W. and Lee, C. L. (2001). Testing the efficacy of restocking trochus using broodstock transplantation and juvenile seeding—An ACIAR-funded project. *Secretariat of the Pacific Community Trochus Information Bulletin* **7**, 3–8.

Purcell, S., Gardner, D. and Bell, J. (2002). Developing optimal strategies for restocking sandfish: A collaborative project in New Caledonia. *Secretariat of the Pacific Community Beche-de-mer Information Bulletin* **16**, 2–4.

Purcell, S. W., Amos, M. J. and Pakoa, K. (2004). Releases of cultured sub-adult *Trochus niloticus* generate broodstock for fishery replenishment in Vanuatu. *Fisheries Research* **67**, 329–333.

Purcell, S., Blockmans, B. and Agudo, N. (2005). Transportation methods for restocking juvenile sea cucumber, *Holothuria scabra*. *Aquaculture* **249**, in press.

Radhakrishnan, E. V. and Vijayakumaran, M. (1995). Early larval development of the spiny lobster *Panulirus homarus* (Linnaeus, 1758) reared in the laboratory. *Crustaceana* **68**, 151–159.

Rahman, M. K. and Srikashnadhas, B. (1994). The potential for spiny lobster culture in India. *Infofish International* **1**, 51–53.

Ramirez-Rodriguez, M., Chávez, E. A. and Arreguin-Sánchez, F. (2000). Perspective of the pink shrimp (*Farfantepenaeus duorarum* Burkenroad) fishery of Campeche Bank, Mexico. *Ciencias Marinas* **26**, 97–112.

Ramm, D. C. (1980). Electromagnetic tracking of rock lobsters (*Jasus novaehollandiae*). *Australian Journal of Marine and Freshwater Research* **31**, 263–269.

Randall, J. E. (1964). Contributions to the biology of the queen conch, *Strombus gigas*. *Bulletin of Marine Science* **14**, 246–295.

Ray, M. and Stoner, A. W. (1994). Experimental analysis of growth and survivorship in a marine gastropod aggregation: Balancing growth with safety in numbers. *Marine Ecology Progress Series* **105**, 47–59.

Ray, M. and Stoner, A. W. (1995). Growth, survivorship, and habitat choice in a newly settled seagrass gastropod, *Strombus gigas*. *Marine Ecology Progress Series* **123**, 83–94.

Ray, M., Stoner, A. W. and O'Connell, S. M. (1994). Size-specific predation of juvenile queen conch, *Strombus gigas*: Implications for stock enhancement. *Aquaculture* **128**, 79–88.

Ray-Culp, M., Davis, M. and Stoner, A. W. (1997). The micropredators of settling and newly settled queen conch (*Strombus gigas* Linnaeus). *Journal of Shellfish Research* **16**, 423–428.

Ray-Culp, M., Davis, M. and Stoner, A. W. (1999). Predation by xanthid crabs on early post-settlement gastropods: The role of prey size, prey density, and habitat complexity. *Journal of Experimental Marine Biology and Ecology* **240**, 303–321.

Raymond, M. and Rousset, F. (2000). Updated version of GENEPOP (v. 1.2) as described in: Raymond, M. and Rousset, F. 1995. GENEPOP (v. 1.2): Population genetics software for exact tests and ecumenicism. *Journal of Heredity* **86**, 248–249.

Reboulet, A. M., Viallefont, A., Pradel, R. and Lebreton, J. D. (1999). Selection of survival and recruitment models with SURGE5.0. *Bird Study* **46**(Suppl.), 148–156 [ftp://ftp.cefe.cnrs-mop.fr/biom/Surge/].

Reeb, C. A. and Avise, J. C. (1990). A genetic discontinuity in a continuously distributed species: Mitochondrial DNA in the American oyster, *Crassostrea virginica*. *Genetics* **124**, 397–406.

Reisenbichler, R. R. and Rubin, S. P. (1999). Genetic changes from artificial propagation of Pacific salmon affect the productivity and viability of supplemented populations. *ICES Journal of Marine Science* **56**, 459–466.

Rice, M. A., Valliere, A. and Caporelli, A. (2000). A review of shellfish restoration and management projects in Rhode Island. *Journal of Shellfish Research* **19**, 401–408.

Richards, A. H., Bell, L. J. and Bell, J. D. (1994). Inshore fisheries resources of Solomon Islands. *Marine Pollution Bulletin* **29**, 90–98.

Richards, D. V. (2000). The status of rocky intertidal communities in Channel Islands National Park. *In* "Proceedings of 5th California Islands Symposium" (D. R. Browne, K. L. Mitchell and H. W. Chaney, eds), pp. 356–358. United States Department of the Interior Minerals Management Service.

Richards, D. V. and Davis, G. E. (1993). Early warnings of modern population collapse in black abalone *Haliotis cracherodii* Leach, 1814 at the California Channel Islands. *Journal of Shellfish Research* **12**, 189–194.

Richards, R. A. and Cobb, J. S. (1986). Competition for shelter between lobsters (*Homarus americanus*) and Jonah crabs (*Cancer borealis*): Effects of relative size. *Journal of the Fisheries Research Board of Canada* **43**, 2250–2255.

Richardson, B. J., Baverstock, P. R. and Adams, M. (1986). "Allozyme Electrophoresis: A Handbook for Animal Systematics and Population Studies", Academic Press, Sydney.

Richmond, R. (1997). Reproduction and recruitment in corals: Critical links in the persistence of reefs. *In* "Life and Death of Coral Reefs" (C. Birkeland, ed.), pp. 175–196. Chapman and Hall, New York.

Rickard, N. A. and Newman, R. A. (1988). Development of technology for harvesting and transplanting subtidal juvenile Pacific razor clams, *Siliqua patula* Dixon, along the coast of Washington State. *Journal of Shellfish Research* **7**, 131.

Ritar, A. J. (2001). The experimental culture of phyllosoma larvae of the southern rock lobster (*Jasus edwardsii*) in a flow through system. *Aquaculture Engineering* **24**, 149–156.

Ritar, A. J., Thomas, C. W. and Beech, A. R. (2002). Feeding *Artemia* and shellfish to phyllosoma larvae of the southern rock lobster (*Jasus edwardsii*). *Aquaculture* **212**, 179–190.

Roberts, R. D., Kawamura, T. and Takami, H. (1999). Abalone recruitment—An overview of some recent research in New Zealand, Australia and Japan. *Bulletin of the Tohoku National Fisheries Research Institute* **62**, 95–107.

Robertson, A. I. (1988). Abundance, diet and predators of juvenile banana prawns, *Penaeus merguiensis*, in a tropical mangrove estuary. *Australian Journal of Marine and Freshwater Research* **39**, 467–478.

Robertson, P. B. (1968). The complete larval development of the sand lobster *Scyllarus americanus* (Smith) (Decapoda, Scyllaridae) reared in the laboratory, with notes on larvae from the plankton. *Bulletin of Marine Science* **18**, 294–342.

Robinson, S. M. C. (1993). The potential for aquaculture technology to enhance wild scallop production in Atlantic Canada. *World Aquaculture* **24**(2), 61–67.

Rodda, K. R., Keesing, J. K. and Foureur, B. L. (1997). Variability in larval settlement of abalone on larval collectors. *Molluscan Research* **18**, 253–264.

Rodriguez, B. and Posada, J. M. (1994). Revisión histórica de la pesquería del botuto o guarura (*Strombus gigas*) y el alcance de su programa de manejo en al Parque Nacional Archipiélago de Los Roques, Venezuela. *In* "Queen Conch Biology, Fisheries and Mariculture" (R. S. Appeldoorn and B. Rodriguez, eds), pp. 13–24. Fundación Científica Los Roques, Caracas.

Rodriguez Gil, L. A. (1994). Análisis de la evolución de la pesquería del caracol en dos estados de la Península de Yucatán, México y una cooperativa de pescadores. *In* "Queen Conch Biology, Fisheries and Mariculture" (R. S. Appeldoorn and B. Rodriguez, eds), pp. 113–124. Fundación Científica Los Roques, Caracas.

Rogers-Bennett, L. and Pearse, J. S. (1998). Experimental seeding of hatchery-reared juvenile red abalone in Northern California. *Journal of Shellfish Research* **17**, 877–880.

Rothlisberg, P. C. (1998). Aspects of penaeid biology and ecology of relevance to aquaculture—A review. *Aquaculture* **164**, 49–65.

Rothlisberg, P. C. and Preston, N. P. (1992). Technical aspects of stocking: Batch marking and stock assessment. *In* "Recruitment Processes. Australian Society for Fish Biology Workshop, August 21, 1991" (D. A. Hancock, ed.), pp. 187–191. Bureau of Rural Resources Proceedings No. 16. Australian Government Printing Office, Canberra.

Rothlisberg, P. C., Craig, P. D. and Andrewartha, J. R. (1996). Modelling penaeid prawn larval advection in Albatross Bay, Australia: Defining the effective spawning population. *Australian Journal of Marine and Freshwater Research* **47**, 157–168.

Rothlisberg, P. C., Preston, N. P., Loneragan, N. R., Die, D. J. and Poiner, I. R. (1999). Approaches to reseeding penaeid prawns. *In* "Stock Enhancement and Sea Ranching" (B. R. Howell, E. Moskness and T. Svasand, eds), pp. 365–378. Fishing News Books, Blackwell, Oxford.

Rothschild, B. J., Ault, J. S., Goulletquer, P. and Heral, M. (1994). Decline of the Chesapeake Bay oyster population: A century of habitat destruction and overfishing. *Marine Ecology Progress Series* **111**, 29–39.

Rowe, F. W. E. and Gates, J. (1995). Echinodermata. *In* "Zoological Catalogue of Australia Volume 33" (A. Wells, ed.), pp. 294–295. CSIRO, Melbourne, Australia.

Ruiz, G. M., Fofonoft, P., Hines, A. H. and Grosholz, E. D. (1999). Non-indigenous species as stressors in estuarine and marine communities: Assessing invasion impacts and interactions. *Limnology and Oceanography* **44**, 950–972.

Rumrill, S. S. (1990). Natural mortality of marine invertebrate larvae. *Ophelia* **32**, 163–198.

Sahoo, D. and Ohno, M. (2000). Rebuilding the ocean floor—Construction of artificial reefs around the Japanese coast. *Current Science* **78**, 228–230.

Saila, S. B., Flowers, J. M. and Hughes, J. T. (1969). Fecundity of the American lobster, *Homarus americanus*. *Transactions of the American Fisheries Society* **98**, 537–539.

Sainte-Marie, B. and Chabot, D. (2002). Ontogenetic shifts in natural diet during benthic stages of American lobster (*Homarus americanus*), off the Magdalen Islands. *Fishery Bulletin, U.S.* **100**, 106–116.

Saito, K. (1979). Studies on propagation of Ezo abalone, *Haliotis discus hannai* Ino—I. Analysis of the relationship between transplantation and catch in Funka Bay coast. *Bulletin of the Japanese Society of Scientific Fisheries* **45**, 695–704.

Saito, K. (1981). The appearance and growth of 0-year-old Ezo abalone. *Bulletin of the Japanese Society of Scientific Fisheries* **47**, 1393–1400.

Saito, K. (1984). Ocean ranching of abalones and scallops in Northern Japan. *Aquaculture* **39**, 361–373.

Saito, K. (1992). Japan's sea urchin enhancement experience. *In* "Sea Urchins, Abalone, and Kelp. Their Biology, Enhancement and Management" (C. M. Dewees and L. T. Davies, eds), p. 21. California Sea Grant College, La Jolla, California.

Sakai, S. (1962). Ecological studies on the abalone, *Haliotis discus hannai* Ino. 2. Mutuality among the colored shell area, growth of the abalone and algal vegetation. *Bulletin of the Japanese Society of Scientific Fisheries* **28**, 780–783.

Sakai, S., Tajima, K. and Agatsuma, Y. (2004). Stock enhancement of the short-spined sea urchin *Strongylocentrotus intermedius* in Hokkaido, Japan. *In* "Stock Enhancement and Sea Ranching: Developments, Pitfalls and Opportunities" (K. M. Leber, S. Kitada, L. Blankenship and T. Svasand, eds), pp. 465–476. Blackwell, Oxford.

Sakamoto, J. K., Tanaka, K. and Kobayashi, K. (1984). On the utilization of the marine environment for the intermediate culture of laboratory reared abalone seed. *Chiba Prefectural Fisheries Research Laboratory Reports* **40**, 123 . Translated by M.G. Mottet in: "Summaries of Japanese Papers on Hatchery Technology and Intermediate Rearing Facilities for Clams, Scallops and Abalone", p. 26. Washington Department Fisheries Progress Report 203.

Salini, J. P., Blaber, S. J. M. and Brewer, D. T. (1990). Diets of piscivorous fishes in a tropical Australian estuary, with special reference to predation on penaeid prawns. *Marine Biology* **105**, 363–374.

Sandt, V. and Stoner, A. W. (1993). Ontogenetic shift in habitat by early juvenile queen conch, *Strombus gigas*: Patterns and potential mechanisms. *Fishery Bulletin, US* **91**, 516–525.

Sarver, S. K., Silberman, J. D. and Walsh, P. J. (1998). Mitochondrial DNA sequence evidence supporting the recognition of two species or subspecies of the Florida spiny lobster, *Panulirus argus*. *Journal of Crustacean Biology* **18**, 177–186.

Sasaki and Shepherd, M. A. (1995). Larval dispersal and recruitment of *Haliotis discus hannai* and *Tegula* spp. on Miyagi coasts, Japan. *Marine and Freshwater Research* **46**, 519–529.

Sasaki, R. and Shepherd, S. A. (2001). Ecology and post-settlement survival of the Ezo abalone, *Haliotis discus hannai,* on Miyagi coasts, Japan. *Journal of Shellfish Research* **20**, 619–626.

Saunders, N. C., Kessler, L. G. and Avise, J. C. (1986). Genetic variation and geographic differentiation in mitochondrial DNA of the horseshoe crab, *Limulus polyphemus*. *Genetics* **112**, 613–627.

Sawada, M., Miki, F. and Asuke, M. (1981). Aquatic afforestation with *Laminaria* seaweed beds. *Fisheries Science Series* **23**, 130–141 (in Japanese).

Sbordoni, V., De Matthaeis, E., Corolli Sbordoni, M., La Rosa, G. and Mattoccia, M. (1986). Bottleneck effects and the depression of genetic variability in hatchery stocks of *Penaeus japonicus* (Crustacea, Decapoda). *Aquaculture* **57**, 239–251.

Scarratt, D. J. (1968). An artificial reef for lobsters (*Homarus americanus*). *Journal of the Fisheries Research Board of Canada* **25**(12), 2683–2690.

Scarratt, D. J. (1970). Laboratory and field tests of modified sphyrion tags on lobsters (*Homarus americanus*). *Journal of the Fisheries Research Board of Canada* **27**, 257–264.

Scarratt, D. J. and Elson, P. F. (1965). Preliminary trials of a tag for salmon and lobsters. *Journal of the Fisheries Research Board of Canada* **22**, 421–423.

Schiel, D. R. (1992). The enhancement of paua (*Haliotis iris* Martyn) populations in New Zealand. *In* "Abalone of the World: Biology, Fishery and Culture" (S. A. Shepherd, M. J. Tegner and S. A. Guzmán del Próo, eds), pp. 474–484. Blackwell, Oxford.

Schiel, D. R. (1993). Experimental evaluation of commercial-scale enhancement of abalone *Haliotis iris* populations in New Zealand. *Marine Ecology Progress Series* **97**, 167–181.

Schiel, D. R. and Welden, B. C. (1987). Responses to predators of cultured and wild red abalone, *Haliotis rufescens,* in laboratory experiments. *Aquaculture* **60**, 173–188.

Schink, T. D., McGraw, K. A. and Chew, K. K. (1983). "Pacific Coast Clam Fisheries. Washington Sea Grant Technical Report 83–1", p. 72. University of Washington, Seattle.

Schneider, S., Rosseli, D. and Excoffier, L. (2000). "ARLEQUIN ver 2.00: A Software for Population Genetic Data Analysis", Genetics and Biometry Laboratory, University of Geneva, Switzerland.

Schwarz, C. J. and Arnason, A. N. (1990). Use of tag-recovery information in migration and movement studies. *American Fisheries Society Symposium* **7**, 588–603.

Schwarz, C. J., Schweigert, J. F. and Arnason, A. N. (1993). Estimating migration rates using tag-recovery data. *Biometrics* **49**, 177–193.

Scovacricchi, T. (1999). Preliminary studies on stocking depleted populations of the European lobster, *Homarus gammarus* (L.) (Decapoda, Nephropsidae), onto the natural beachrock outcrops in the northern Adriatic Sea. *In* "Stock Enhancement and Sea Ranching" (B. R. Howell, E. Moksness and T. Svasand, eds), pp. 393–400. Fishing News Books, Blackwell, Oxford.

Scovacricchi, T., Privileggi, N. and Ferrero, E. A. (1999). Pilot-scale production of hatchery-reared lobster juveniles *Homarus gammarus* (Linnaeus, 1758), for restocking actions over restricted marine areas. *Rivista Italiana Acquacoltura* **34**, 47–59.

Seaman, W., Jr and Sprague, L. M. (1991). "Artificial Habitats for Marine and Freshwater Fisheries", p. 285. Academic Press, San Diego.

Seijo, J., Salas, S., Arceo, P. and Fuentes, D. (1991). Análisis bioeconómico comparative de la pesquería de langosta *Panulirus argus* en la plataforma continental de Yucatán. FAO Fisheries Report 431(Suppl.), 39–58, FAO, Rome.

Seki, T. and Sano, M. (1998). An ecological basis for the restoration of abalone populations. *Bulletin of the Tohoku National Fisheries Research Institute* **60**, 23–40 (in Japanese with English abstract, tables and figures).

Seki, T. and Taniguchi, K. (2000). Rehabilitation of northern Japanese abalone, *Haliotis discus hannai*, populations by transplanting juveniles. *In* "Workshop on Rebuilding Abalone Stocks in British Columbia" (A. Campbell, ed.), *Canadian Special Publication in Fisheries and Aquatic Sciences* **130**, 72–83

Sekine, S., Suzuki, S., Shima, Y. and Nonaka, T. (2000). Larval rearing and molting in the Japanese spiny lobster *Panulirus japonicus* under laboratory conditions. *Fishery Science* **66**, 19–24.

Selvamani, M. J. P., Degnan, S. M. and Degnan, B. M. (2001). Microsatellite genotyping of individual abalone larvae: Parentage assignment in aquaculture. *Marine Biotechnology* **3**, 478–485.

Seyoum, S., Bert, T. M., Wilbur, A., Arnold, W. S. and Crawford, C. (2003). Development, evaluation and application of a mitochondrial DNA genetic tag for the bay scallop, *Argopecten irradians*. *Journal of Shellfish Research* **22**, 111–117.

Shaklee, J. B. and Bentzen, P. (1998). Genetic identification of stocks of marine fish and shellfish. *Bulletin of Marine Science* **62**, 589–621.

Sharp, W. C., Lellis, W. A., Butler, M. J., Herrnkind, W. F., Hunt, J. H., Pardee-Woodring, M. and Matthews, T. R. (2000). The use of coded microwire tags in mark–recapture studies of juvenile Caribbean spiny lobster, *Panulirus argus*. *Journal of Crustacean Biology* **20**, 510–521.

Shawl, A. L., Davis, M. and Corsaut, J. (2003). Captive breeding for the gastropod conch (*Strombus* spp.). *Proceedings of the Gulf and Caribbean Fisheries Institute* **54**, 427–536.

Sheehy, M. R. J., Bannister, R. C. A., Wickins, J. F. and Shelton, P. M. J. (1999). New perspectives on the growth and longevity of the European lobster (*Homarus gammarus*). *Canadian Journal of Fisheries and Aquatic Sciences* **56**, 1904–1915.

Shelley, C. C. (1986). The potential for re-introduction of a beche-de-mer "fishery" in Torres Strait. *In* "Torres Strait Fisheries Seminar, Port Moresby" (A. K. Haines, G. C. Williams and D. Coates, eds), pp. 140–150. Australian Government Publishing Service, Canberra.

Shelton, P. M. J. and Chapman, C. J. (1995). A moult recording tag for lobsters: Field trials. *ICES Marine Science Symposium* **199**, 222–230.

Shepherd, S. A. (1986). Movement of the Southern Australian abalone *Haliotis laevigata* in relation to crevice abundance. *Australian Journal of Ecology* **11**, 295–302.

Shepherd, S. A. (1990). Studies on southern Australian abalone (*genus Haliotis*). XII. Long-term recruitment and mortality dynamics of an unfished population. *Australian Journal of Marine and Freshwater Research* **41**, 475–492.

Shepherd, S. A. and Baker, J. L. (1998). Biological reference points in an abalone (*Haliotis laevigata*) fishery. *In* "Proceedings of the North Pacific Symposium on Invertebrate Stock Assessment and Management" (G. S. Jamieson and A. Campbell, eds), *Canadian Special Publication in Fisheries and Aquatic Sciences* **125**, 235–245.

Shepherd, S. A. and Rodda, K. R. (2001). Sustainability demands vigilance: Evidence for serial decline of the greenlip abalone fishery and a review of management. *Journal of Shellfish Research* **20**, 829–841.

Shepherd, S. A., Lowe, D. and Partington, D. (1992a). Studies on southern Australian abalone (*genus Haliotis*). XIII. Larval dispersal and recruitment. *Journal of Experimental Marine Biology and Ecology* **164**, 247–260.

Shepherd, S. A., Tegner, M. J. and Guzmán del Próo, S. A. (eds) (1992). "Abalone of the World: Biology, Fisheries and Culture". Blackwell, Oxford.

Shepherd, S. A., Preece, P. A. and White, R. W. G. (2000). Tired nature's sweet restorer? Ecology of abalone (*Haliotis* spp.) stock enhancement in Australia. *In* "Workshop on Rebuilding Abalone Stocks in British Columbia" (A. Campbell, ed.), *Canadian Special Publication in Fisheries and Aquatic Sciences* **130**, 84–97.

Shepherd, S. A., Rodda, K. R. and Vargas, K. M. (2001). A chronicle of collapse in two abalone stocks with proposals for precautionary management. *Journal of Shellfish Research* **20**, 843–856.

Shokita, S., Kakazu, K., Tomori, A. and Toma, T. (1991). "Aquaculture in Tropical Areas", p. 360. Midori Shobo Co., Ltd, Tokyo.

Shumway, S. E. (1991). Scallops: Biology, ecology and aquaculture. *In* "Developments in Aquaculture and Fisheries Science", pp. 715–751. Elsevier, New York.

Shumway, S. E. and Castagna, M. (1994). Scallop fisheries, culture and enhancements in the United States. *Memoirs of the Queensland Museum* **36**, 283–298.

Siddall, S. E. (1983). Biological and economic outlook for hatchery production of juvenile queen conch. *Proceedings of the Gulf and Caribbean Fisheries Institute* **35**, 46–52.

Silberman, J. D., Sarver, S. K. and Walsh, P. J. (1994). Mitochondrial DNA variation and population structure in the spiny lobster *Panulirus argus*. *Marine Biology* **120**, 601–608.

Sims, N. (1985). The abundance, distribution and exploitation of *Trochus niloticus* in the Cook Islands. *Proceedings 5th International. Coral Reef Symposium* **5**, 539–544.

Sims, N. (1988). Trochus research in the Cook Islands and its implications for management. Background Paper 37, Workshop on Inshore Fisheries Resources, p. 13. South Pacific Commission, Noumea, New Caledonia.

Sindermann, C. J. (1990). Principal Diseases of Marine Fish and Shellfish. (2nd Edition) Volume 2, Diseases of Marine Shellfish, p. 516. Academic Press, San Diego.

Sindermann, C. J. (1993). Disease risks associated with importation of nonindigenous marine animals. *Marine Fisheries Review* **54**, 1–10.

Skewes, T. D., Pitcher, C. R. and Dennis, D. M. (1997). Growth of ornate rock lobsters, *Panulirus ornatus*, in Torres Strait, Australia. *Marine and Freshwater Research* **48**, 497–501.

Smith, B. D. (1987). Growth rate, distribution and abundance of the introduced topshell, *Trochus niloticus*, Linnaeus on Guam, Mariana Islands. *Bulletin of Marine Science* **41**, 466–474.

Smith, P. J. and Conroy, A. M. (1992). Loss of genetic variation in hatchery-produced abalone, *Haliotis iris*. *New Zealand Journal of Marine and Freshwater Research* **26**, 81–85.

Smith, P. J., McKoy, J. L. and Machin, P. J. (1980). Genetic variation in the rock lobster *Jasus edwardsii* and *Jasus novaehollandiae*. *New Zealand Journal of Marine and Freshwater Research* **14**, 55–63.

Smith, I. P., Collins, K. J. and Jensen, A. C. (2000). Digital electromagnetic telemetry system for studying behaviour of decapod crustaceans. *Journal of Experimental Marine Biology and Ecology* **247**, 209–222.

Sosa-Cordero, E., Arce, A. M., Aguilar-Dávila, W. and Ramírez-González (1998). Artificial shelters for spiny lobster *Panulirus argus* (Latreille): An evaluation of occupancy in different benthic habitats. *Journal of Experimental Marine Biology and Ecology* **229**, 1–18.

Southworth, M. and Mann, R. (1998). Oyster reef broodstock enhancement in the Great Wicomico River, Virginia. *Journal of Shellfish Research* **17**, 1101–1114.

Spanier, E. (1994). What are the characteristics of a good artificial reef for lobster? *Crustaceana* **67**, 173–186.

Spanier, E. and Zimmer-Faust, R. K. (1988). Some physical properties of shelter that influence den preference in spiny lobsters. *Journal of Experimental Marine Biology and Ecology* **121**, 137–149.

Spanier, E., Tom, M., Pisanty, S. and Almog, G. (1988). Seasonality and shelter selection by the slipper lobster *Scyllarides latus* in southeastern Mediterranean. *Marine Ecology Progress Series* **42**, 247–255.

Spencer, B. E. (1992). Predators and methods of control in molluscan shellfish cultivation in north European waters. *In* "Aquaculture and the Environment" (N. D. Pauw and J. Joyce, eds), pp. 309–337. European Aquaculture Society Special Publication No. 16.

Sponaugle, S. and Lawton, P. (1990). *Kortunia* predation on juvenile hard clams: Effects of substrate type and prey density. *Marine Ecology Progress Series* **67**, 43–53.

Staples, D. J., Vance, D. J. and Heales, D. S. (1985). Habitat requirements of juvenile penaeid prawns and their offshore fisheries. *In* "Second Australian National Prawn Seminar, NPS2" (P. C. Rothlisberg, B. J. Hill and D. J. Staples, eds), pp. 47–54. Cleveland, Australia.

Star, B., Aptes, S. and Gardner, J. P. A. (2003). Genetic structuring among populations of the greenshell mussel *Perna canaliculus* revealed by analysis of randomly amplified polymorphic DNA. *Marine Ecology Progress Series* **249**, 171–182.

Steneck, R. S. (1991). The demographic consequences of intraspecific competition among lobsters (*Homarus americanus*). *Journal of Shellfish Research* **10**, 287.

Stickney, R. R. (2002). Issues associated with non-indigenous species in marine aquaculture. *In* "Responsible Marine Aquaculture" (R. R. Stickney and J. P. McVey, eds), pp. 205–220. CAB International, New York.

Stockhausen, W. T. and Lipcius, R. N. (2001). Single large or several small marine reserves for the Caribbean spiny lobster? *Marine and Freshwater Research* **52**, 1605–1614.

Stockhausen, W. T., Lipcius, R. N. and Hickey, B. H. (2000). Joint effects of larval dispersal, population regulation, marine reserve design and exploitation on production and recruitment in the Caribbean spiny lobster. *Bulletin of Marine Science* **66**, 957–990.

Stoner, A. W. (1994). Significance of habitat and stock pre-testing for enhancement of natural fisheries: Experimental analyses with queen conch *Strombus gigas*. *Journal of the World Aquaculture Society* **25**, 155–165.

Stoner, A. W. (1997a). The status of queen conch, *Strombus gigas,* research in the Caribbean. *Marine Fisheries Review* **59**, 14–22.

Stoner, A. W. (1997b). Shell middens as indicators of long-term distributional pattern in *Strombus gigas*, a heavily exploited marine gastropod. *Bulletin of Marine Science* **61**, 559–570.

Stoner, A. W. (1999). Queen conch stock enhancement: The need for an integrated approach. *Proceedings of the Gulf and Caribbean Fisheries Institute* **45**, 922–925.

Stoner, A. W. and Davis, M. (1994). Experimental outplanting of juvenile queen conch, *Strombus gigas*: Comparison of wild and hatchery reared stocks. *Fishery Bulletin, US* **92**, 390–411.

Stoner, A. W. and Davis, M. (1997). Abundance and distribution of queen conch veligers (*Strombus gigas* Linne) in the central Bahamas. II. Vertical patterns in nearshore and deep-water habitats. *Journal of Shellfish Research* **16**, 19–29.

Stoner, A. W. and Glazer, R. A. (1998). Variation in natural mortality: Implications for queen conch stock enhancement. *Bulletin of Marine Science* **62**, 427–442.

Stoner, A. W. and Ray, M. (1996). Queen conch (*Strombus gigas*) in fished and unfished locations of the Bahamas: Effects of a marine fishery reserve on adults, juveniles, and larval production. *Fishery Bulletin, US* **94**, 551–565.

Stoner, A. W. and Ray-Culp, M. (2000). Evidence for Allee effects in an over-harvested marine gastropod: Density-dependent mating and egg production. *Marine Ecology Progress Series* **202**, 297–302.

Stoner, A. W. and Sandt, V. (1991). Experimental analysis of habitat quality for juvenile queen conch in seagrass meadows. *Fishery Bulletin, US* **89**, 693–700.

Stoner, A. W. and Sandt, V. (1992). Transplanting as a test procedure before large-scale outplanting of juvenile queen conch. *Proceedings of the Gulf and Caribbean Fisheries Institute* **41**, 447–458.

Stoner, A. W. and Schwarte, K. C. (1994). Queen conch, *Strombus gigas,* reproductive stocks in the central Bahamas—Distribution and probable sources. *Fishery Bulletin, US* **92**, 171–179.

Stoner, A. W. and Waite, J. M. (1990). Distribution and behavior of queen conch, *Strombus gigas*, relative to seagrass standing crop. *Fishery Bulletin, US* **88**, 573–585.

Stoner, A. W., Sandt, V. and Boidron-Metairon, I. F. (1992). Seasonality of reproductive activity and abundance of veligers in queen conch, *Strombus gigas*. *Fishery Bulletin, US* **90**, 161–170.

Stoner, A. W., Hanisak, M., Smith, N. and Armstrong, R. (1994). Large-scale distribution of queen conch nursery habitats: Implications for stock enhancement. *In* "Queen Conch Biology, Fisheries and Mariculture" (R. Appeldoorn and B. Rodriguez, eds), pp. 169–189. Fundación Científica Los Roques, Caracas.

Stoner, A. W., Pitts, P. A. and Armstrong, R. A. (1996a). Interaction of physical and biological factors in the large-scale distribution of juvenile queen conch in seagrass meadows. *Bulletin of Marine Science* **58**, 217–233.

Stoner, A. W., Glazer, R. A. and Barile, P. (1996b). Larval supply to queen conch nurseries: Relationships with recruitment process and population size in Florida and the Bahamas. *Journal of Shellfish Research* **15**, 407–420.

Stoner, A. W., Ray-Culp, M. and O'Connell, S. M. (1998). Settlement and recruitment of queen conch, *Strombus gigas*, in seagrass meadows: Associations with habitat and micropredators. *Fishery Bulletin, US* **96**, 885–899.

Stoner, A. W., Mehta, N. and Ray-Culp, M. (1999). Mesoscale distribution patterns of queen conch (*Strombus gigas* Linné) in Exuma Sound, Bahamas: Links in recruitment from larvae to fishery yields. *Ocean and Coastal Management* **42**, 1069–1081.

Strand, O., Grefsrud, E. S., Haugum, G. A., Bakke, G., Helland, E. and Helland, T. E. (2004). Release strategies in scallop (*Pecten maximus*) sea ranching vulnerable to crab predation. In "Stock Enhancement and Sea Ranching: Developments, Pitfalls and Opportunities" (K. M. Leber, S. Kitada, L. Blankenship and T. Svasand, eds), pp. 544–555. Blackwell, Oxford.

Strathmann, R. R., Hughes, T. P., Kuris, A. M., Lindeman, K. C., Morgan, S. G., Pandolfi, J. M. and Warner, R. R. (2002). Evolution of local recruitment and its consequences for marine populations. *Bulletin of Marine Science* **70** (suppl.1), 377–396.

Strehlow, H. V. (2004). Economics and management strategies for restocking sandfish in Vietnam. *NAGA WorldFish Center Quarterly* **27**, 36–40.

Stuller, J. (1995). King of the clams. *Wildlife Conservation.* May/June 1995, pp. 48–53.

Su, M.-S. and Liao, I.-C. (1999). Research and development of prawn stock enhancement in Taiwan. In "Stock Enhancement and Sea Ranching" (B. R. Howell, E. Moskness and T. Svåsand, eds), pp. 379–392. Fishing News Books, Blackwell, Oxford.

Su, M.-S., Liao, I.-C. and Hirano, R. (1990). Restocking sub-adult grass prawn, *Penaeus monodon*, in the coastal waters of southwest Taiwan. In "Proceedings of the Second Asian Fisheries Forum" (R. Hirano and I. Hanyu, eds), pp. 99–102. Asian Fisheries Society, Manila.

Subasinghe, R. P., Bondad-Reantaso, M. G. and McGladdery, S. E. (2001). Aquaculture development, health and wealth. In "Aquaculture in the Third Millennium" (R. P. Subasinghe, P. Bueno, M. J. Phillips, C. Hough, S. E. McGladdery and J. R. Arthur, eds), pp. 167–191. NACA, Bangkok and FAO, Rome.

Svåsand, T., Skilbrei, G. I., van der Meeren, G. and Holm, M. (1998). Review of morphological and behavioural differences between reared and wild individuals: Implications for sea ranching of Atlantic Salmon, *Salmo salar* L., Atlantic cod, *Gadus morhua* L., and European lobster, *Homarus gammarus* L. *Fisheries Management and Ecology* **5**, 473–490.

Svåsand, T., Kristiansen, T., Pedersen, T. N., Salvanes, A., Engelsen, R., Nævdal, G. and Nødtvedt, M. (2000). The enhancement of cod stocks. *Fish and Fisheries* **1**, 173.

Swearer, S. E., Shima, J. S., Hellberg, M. E., Thorrold, S. R., Jones, G. P., Robertson, D. R., Morgan, S. G., Selkoe, K. A., Ruiz, G. M. and Warner, R. R. (2002). Evidence of self-recruitment in demersal marine populations. *Bulletin of Marine Science* **70**(Suppl.), 251–271.

Sweijd, N. (1999). Molecular markers and abalone seeding as tools for the conservation of the South African abalone (perlemoen), *Haliotis midae* Linn. resource. PhD thesis, Department of Zoology, University of Cape Town.

Sweijd, N., Snethlage, Q., Harvey, D. and Cook, P. (1998). Experimental abalone (*Haliotis midae*) seeding in South Africa. *Journal of Shellfish Research* **17**, 897–904.

Takahashi, M. and Saisho, T. (1978). The complete larval development of the scyllarid lobsters *Ibacus ciliatus* (von Siebold) and *Ibacus novemdentatus* Gibbes in the laboratory. *Memoirs of the Faculty of Fisheries, Kagoshima University* **27**, 305–363 (in Japanese).

Takeichi, M. (1981). Experimental studies of the cultured seeds of abalone, *Haliotis discus hannai*. Report of the 1980s Tohoku Block Aquaculture Conference pp. 27–31 (in Japanese).

Takeichi, M. (1988). Recapture and survival rate of mass released seed of abalone, *Haliotis discus hannai*. *Saibaigiken* **17**, 27–36 (in Japanese).

Takeichi, M., Tashiro, M. and Yada, T. (1978). Place of release and survival rate of artificially raised young abalone, *Haliotis discus*. *The Aquiculture* **26**(1), 1–5. Translated by M.G. Mottet. *In* "Summaries of Japanese Papers on Hatchery Technology and Intermediate Rearing Facilities for Clams, Scallops and Abalone". Washington Department Fisheries Progress Report 203, 26 pp.

Taki, J. and Higashida, I. (1964). Investigation and problem on introduction of rocks to fishing grounds to enhance the sea urchin *Hemicentrotus pulcherrimus* in Fukui Prefecture. *Aquaculture* **12**, 37–47 (in Japanese).

Tam, Y. K. and Kornfield, I. (1996). Characterisation of microsatellite markers in *Homarus* (Crustacea, Decapoda). *Molecular Marine Biology and Biotechnology* **5**, 230–238.

Tam, Y. K. and Kornfield, I. (1998). Phylogenetic relationships among clawed lobster genera (Decapoda, Nephropidae) based on mitochondrial 16S fRNA gene sequence. *Journal of Crustacean Biology* **18**, 138–146.

Tanaka, K. (1988). Study of the abalone culture in the coast of Awa region, Chiba Prefecture. *Bulletin of the Japanese Sea Regional Fisheries Research Laboratory* **38**, 21–132.

Tanaka, K. and Nakamura, T. (1971). Marking methods of kuruma shrimp. *Bulletin of the Chiba Fisheries Experimental Station* **24**, 145–150.

Tanaka, K., Tanaka, T., Ishida, O. and Oba, T. (1986). On the distribution of swimming and deposited larvae of nursery ground abalone at the southern coast off Chiba Prefecture. *Bulletin of the Japanese Society of Scientific Fisheries* **52**, 1525–1532.

Tanida, K., Ikewaki, Y., Aoyama, E., Okuyama, Y., Nozaka, M. and Fujiwarra, M. (2002). Investigation of stocking effectiveness for kuruma prawn (*Penaeus japonicus*) in the eastern Seto inland sea. Abstract, Second International Symposium on Stock Enhancement and Sea Ranching, Kobe, Japan 28th January–1st February 2002.

Taniguchi, K. (1991). Marine algal recolonization on the denuded sublittoral rock surface off Oshika Peninsula, Japan. *Bulletin of the Tohoku National Fisheries Research Institute* **53**, 1–5 (in Japanese with English abstract).

Taniguchi, K. (1996). Primary succession of marine algal communities in the sublittoral zone off Oshika Peninsula, Japan. *Nippon Suisan Gakkaishi* **62**, 765–771 (in Japanese with English abstract).

Tarbath, D., Mundy, C. and Haddon, M. (2003). Tasmanian abalone fishery 2002, p. 144. Tasmanian Aquaculture Fisheries Institute, Fisheries Assessment Report, Hobart, Australia.

Tarr, R. J. Q. (2000). The South African abalone (*Haliotis midae*) fishery: A decade of challenges and change. *In* "Workshop on Rebuilding Abalone Stocks in British Columbia" (A. Campbell, ed.), *Canadian Special Publication in Fisheries and Aquatic Sciences* **130**, 32–40.

Tarr, R. J. Q., Williams, P. V. G. and Mac Kenzie, A. J. (1996). Abalone, sea urchins and rock lobster: A possible ecological shift that may affect traditional fisheries. *South African Journal of Marine Science* **17**, 319–323.

Tautz, D. (1989). Hypervariability of simple sequences as a good general source for polymorphic DNA markers. *Nucleic Acids Research* **17**, 6463–6471.

Teboul, D. (1993). Internal monofilament tags used to identify juvenile *Penaeus monodon*: Tag retention and effects on growth and survival. *Aquaculture* **113**, 167–170.

Tegner, M. J. (1989). The California abalone fishery: Production, ecological interactions, and prospects for the future. *In* "Marine Invertebrate Fisheries: Their Assessment and Management" (J. F. Caddy, ed.), pp. 401–420. John Wiley & Sons, New York.

Tegner, M. J. (1989). The feasibility of enhancing red sea urchin *Strongylocentrotus franciscanus* stocks in California; an analysis of the options. *Marine Fisheries Review* **51**, 1–22.

Tegner, M. J. (1992). Brood-stock transplants as an approach to abalone stock enhancement. *In* "Abalone of the World: Biology, Fishery and Culture" (S. A. Shepherd, M. J. Tegner and S. A. Guzmán del Próo, eds), pp. 461–473. Blackwell, Oxford.

Tegner, M. J. (1993). Southern California abalones: Can stocks be rebuilt using marine harvest refugia? *Canadian Journal of Fisheries and Aquatic Sciences* **50**, 2010–2018.

Tegner, M. J. (2000). Abalone (*Haliotis* spp.) enhancement in California: What we've learned and where we go from here. *In* "Workshop on Rebuilding Abalone Stocks in British Columbia" (A. Campbell, ed.), *Canadian Special Publication in Fisheries and Aquatic Sciences* **130**, 61–71.

Tegner, M. J. and Butler, R. A. (1985). The survival and mortality of seeded and native red abalone, *Haliotis rufescens*, on the Palos Verdes Peninsula. *California Fish and Game* **71**, 150–163.

Tegner, M. J. and Butler, R. A. (1989). Abalone seeding. *In* "Handbook of Culture of Abalone and Other Marine Gastropods" (K. Hahn, ed.), pp. 157–182. CRC Press, Boca Raton, Florida.

Tegner, M. J. and Dayton, P. K. (1987). El Niño effects on southern California kelp forests. *Advances in Ecological Research* **17**, 243–279.

Tegner, M. J., Ebert, E. E. and Parker, D. O. (1986). "Abalone larval transplants as an approach to stock enhancement", p. 6. University of California, San Diego Report. R/NP–1–15D.

Tegner, M. J., Breen, P. A. and Lennert, C. (1989). Population biology of red abalones, *Haliotis rufescens*, in Southern California and management of the red and pink, *H. corrugata*, abalone fisheries. *Fishery Bulletin, US* **87**, 313–339.

Tegner, M. J., Dayton, P. K., Edwards, P. B. and Riser, K. L. (1996). Is there evidence for long-term climatic change in southern California kelp forest communities? *California Cooperative Oceanic Fisheries Investigations Reports* **37**, 111–125.

Templeton, A. R. (1998). Nested clade analyses of phylogeographic data: Testing hypotheses about gene flow and population history. *Molecular Ecology* **7**, 381–397.

Tettelbach, S. T. and Wenczel, P. (1993). Reseeding efforts and the status of bay scallop *Argopecten irradians* (Lamarck, 1819) populations in New York following the occurrence of "Brown Tide" algal blooms. *Journal of Shellfish Research* **12**, 423–431.

Tettelbach, S. T., Smith, C. F. and Wenczel, P. (1997). Bay scallop stock restoration efforts in Long Island, New York: Approaches and recommendations. *Journal of Shellfish Research* **16**, 276.

Tewfik, A. and Bene, C. (2000). Densities and age structure of fished versus protected populations of queen conch (*Strombus gigas L.*) in the Turks & Caicos Islands. *Proceedings of the Gulf and Caribbean Fisheries Institute* 51, 60–79.

Thomas, C. W., Carter, C. G. and Crear, B. J. (2003). Feed availability and its relationship to survival, growth, dominance and the agonistic behaviour of the southern rock lobster, *Jasus edwardsii* in captivity. *Aquaculture* 215, 45–65.

Thomson, J. D. (1992). Scallop enhancement—how, and is it worth the effort? *In* "Recruitment Processes, Australian Society for Fish Biology Workshop, August, 21, 1991" (D. A. Hancock, ed.), pp. 183–186. Bureau of Rural Resources Proceedings No. 16, Australian Government Printing Service, Canberra.

Thomson, J. D., Fujimoto, T., Moriya, H. and Ikeda, T. (1995). Enhancement of the scallop *Pecten fumatus* in Tasmania—Japanese technology transfer down under. *Journal of Shellfish Research* 14, 280.

Thorson, G. (1950). Reproductive and larval ecology of marine bottom invertebrates. *Biological Reviews of the Cambridge Philoshopical Society* 25, 1–45.

Tong, L. J., Moss, G. A. and Illingworth, J. (1987). Enhancement of a natural population of the abalone, *Haliotis iris*, using cultured larvae. *Aquaculture* 62, 67–72.

Tracey, M. L., Nelson, K., Hedgecock, D., Shleser, R. A. and Pressick, M. L. (1975). Biochemical genetics of lobster: Genetic variation and the structure of the American lobster (*Homarus americanus*) populations. *Journal of the Fisheries Research Board of Canada* 32, 2091–2101.

Travis, J., Coleman, F., Grimes, C., Conover, D., Bert, T. and Tringali, M. (1998). Critically assessing stock enhancement: An introduction to the Mote symposium. *Bulletin of Marine Science* 62, 305–311.

Trendall, J. T. (1988). Influence of tag type on the recapture rates in the tropical rock lobster, *Panulirus ornatus*. *In* "Tagging—Solution or Problem?" (D. A. Hancock, ed.), pp. 68–76. Australian Society for Fish Biology Tagging Workshop, 21–22 July 1988. Bureau of Rural Resources Proceedings No. 5, Australian Government Printing Office, Canberra.

Tringali, M. D. and Bert, T. M. (1998). Risk to genetic effective population size should be an important consideration in fish stock-enhancement programs. *Bulletin of Marine Science* 62, 641–659.

Tsuchida, K., Takeichi, M., Hirose, T., Kanazawa, T., Omura, R. and Sakashita, T. (1971). On the growth, recapture rate and distribution of released abalone, *Haliotis discus hannai*. *Bulletin of the Iwate Prefecture Fisheries Experimental Station* 1, 121–130 (in Japanese).

Tuan, L. A., Nho, N. T. and Hambrey, J. (2000). Status of cage mariculture in Vietnam. *In* "Cage Aquaculture in Asia: Proceedings of the First International Symposium on Cage Aquaculture in Asia" (I. C. Liao and C. K. Lin, eds), pp. 111–123. Asian Fisheries Society, Manila and World Aquaculture Society—Southeast Asian Chapter, Bangkok.

Tuara, P. N. (1997). Trochus resource assessment, development and management in the Cook Islands. *In* "Workshop on Trochus Resource Assessment, Management and Development, Integrated Fisheries Management Project", pp. 21–25. Technical Document No. 13, South Pacific Commission, Noumea, New Caledonia.

Tully, O. (1999). Restoration of lobster (*Homarus gammarus*) population egg production in depleted stocks. *Journal of Shellfish Research* 18, 729.

Tully, O., Roantree, V. and Robinson, M. (2001). Maturity, fecundity and reproductive potential of the European lobster (*Homarus gammarus*) in Ireland. *Journal of the Marine Biological Association of the United Kingdom* 81, 61–68.

Tveite, S. and Grimsen, S. (1995). Survival of one-year-old artificially raised lobsters (*Homarus gammarus*) released in southern Norway. *ICES Marine Science Symposia* **199**, 73–77.

Uchiba, S. and Futajima, K. (1980). Experimental studies on the release of the cultured seeds of abalone, *Haliotis discus*—III. *Report of the Fukuoka Prefecture Fisheries Experimental Station 1977*, pp. 105–110 (in Japanese).

Ueda, Y., Noguchi, T., Onoue, Y., Koyama, K., Kono, M. and Hashimoto, K. (1982). Occurrence of PSP-infested scallops in Ofunato Bay during 1976–1979 and investigation of responsible plankton. *Bulletin of the Japanese Society of Scientific Fisheries* **48**, 455–458.

Uglem, I. and Grimsen, S. (1995). Tag retention and survival of juvenile lobster, *Homarus gammarus* (L.), marked with coded wire tags. *Aquaculture Research* **26**, 837–841.

Uglem, I., Nœss, H., Farestveit, E. and Jørstad, K. E. (1996). Tagging of juvenile lobsters (*Homarus gammarus* (L.)) with visible implant fluorescent elastomer tags. *Aquacultural Engineering* **15**, 499–501.

Uki, N. (1981). Abalone culture in Japan. *In* "Proceedings of the 9th and 10th U.S.-Japan Meetings on Aquaculture" (C. J. Sinderman, ed.), pp. 83–88. NOAA Technical Report NMFS–16, U. S. Department of Commerce.

Uki, N. (1989). Abalone seeding production and its theory. *International Journal of Aquaculture Fisheries Technology* Part 1, pp. 3–15. Part 2, pp. 125–132; Part 3, pp. 224–231.

Underwood, A. J. (1992). Beyond BACI: The detection of environmental impact on populations in the real, but variable world. *Journal of Experimental Marine Biology and Ecology* **161**, 145–178.

Underwood, A. J. (1995). Detection and measurement of environmental impacts. *In* "Coastal Marine Ecology of Temperate Australia" (A. J. Underwood and M. G. Chapman, eds), pp. 311–324. University of New South Wales Press, Sydney.

Uno, Y. (1985). An ecological approach to mariculture of shrimp: Shrimp ranching fisheries. *In* "Proceedings of the First International Conference on the Culture of Penaeid Prawns/Shrimps" (Y. Taki, J. H. Primavera and J. A. Llobrera, eds), pp. 37–45. SEAFDEC, Iloilo, Philippines.

Uthicke, S. and Benzie, J. (1999). Allozyme variation as a tool for beche-de-mer fisheries management: A study on *Holothuria scabra* (sandfish). *Secretariat of the Pacific Community Beche-de-mer Information Bulletin* **12**, 18–23.

Uthicke, S. and Benzie, J. A. H. (2000). Allozyme electrophoresis indicates high gene flow between populations of *Holothuria* (*Microthele*) *nobilis* (Holothuria: Aspidochirotida) on the Great Barrier Reef. *Marine Biology* **137**, 819–825.

Uthicke, S. and Benzie, J. (2001). Restricted gene flow between *Holothuria scabra* (Echinodermata: Holothuroidea) populations along the north-east coast of Australia and the Solomon Islands. *Marine Ecology Progress Series* **216**, 109–117.

Uthicke, S and Purcell, S (2004). Preservation of genetic diversity in restocking of the sea cucumber *Holothuria scabra* planned through allozyme electrophoresis. *Canadian Journal of Fisheries and Aquatic Sciences* **61**, 519–528.

Uthicke, S., Conand, C. and Benzie, J. A. H. (2001). Population genetics of the fissiparous holothurians *Stichopus chloronotus* and *Holothuria atra* (Aspidochirotida): A comparison between the Torres Strait and La Reunion. *Marine Biology* **139**, 257–265.

Utter, F. (1998). Genetic problems of hatchery-reared progeny released into the wild, and how to deal with them. *Bulletin of Marine Science* **62**, 623–640.

Valdimarsson, G. and James, D. B. (2001). World fisheries—Utilization of catches. *Ocean and Coastal Management* **44**, 619–633.

van der Meeren, G. (1991a). Acclimatization techniques for lobster. *In* "Sea Ranching—Scientific Experiences and Challenges. Proceedings from the Symposium and Workshop 21–23 October 1990, Bergen, Norway" (T.N. Pedersen and E. Kjoersvikpp, eds.), pp. 95–97. Norwegian Society for Aquaculture Research, Bergen, Norway.

van der Meeren, G. (1991b). Out-of-water transportation effects on behaviour in newly released juvenile Atlantic lobsters *Homarus gammarus. Aquacultural Engineering* **10**, 55–64.

van der Meeren, G. (1993). Initial response to physical and biological conditions in naive juvenile lobsters *Homarus gammarus* L. *Marine Behavior and Physiology* **24**, 79–92.

van der Meeren, G. (1997). Preliminary acoustic tracking of native and transplanted European lobsters (*Homarus gammarus*) in an open sea lagoon. *Marine and Freshwater Research* **48**, 915–921.

van der Meeren, G. I. and Uksnoey, L. E. (2000). A comparison of claw morphology and dominance between wild and cultivated male European lobster. *Aquaculture International* **8**, 77–94.

van Montfrans, J., Capelli, J., Orth, R. J. and Ryer, C. H. (1986). Use of microwire tags for tagging juvenile blue crabs (*Callinectes sapidus* Rathbun). *Journal of Crustacean Biology* **6**, 370–376.

Velasquez, D. E. (1992). Shell fragility in juvenile geoducks (*Panopea abrupta*) and its implication for the geoduck enhancement program. *Journal of Shellfish Research* **11**, 208.

Ventilla, R. F. (1982). The scallop industry in Japan. *Advances in Marine Biology* **20**, 309–382.

Vermeij, G. J. (1976). Interoceanic differences in vulnerability of shelled prey to crab predation. *Nature* **260**, 135–136.

Waddy, S. L., Aiken, D. E. and De Klejin, D. P. V. (1995). Control of growth and reproduction. *In* "Biology of the Lobster, *Homarus americanus*" (J. R. Factor, ed.), pp. 217–266. Academic Press, San Diego.

Wahle, R. A. (1991). Implications for lobster fishery enhancement from natural benthic recruitment, hatchery-reared "blues", and experimental cobbles. *Journal of Shellfish Research* **10**, 287.

Wahle, R. A. and Incze, L. S. (1997). Pre- and post-settlement processes in recruitment of the American lobster. *Journal of Experimental Marine Biology and Ecology* **217**, 179–207.

Wahle, R. A. and Steneck, R. S. (1991). Recruitment habitats and nursery grounds of the America lobster (*Homarus americanus* Milne-Edwards): A demographic bottleneck. *Marine Ecology Progress Series* **69**, 231–243.

Wahle, R. A. and Steneck, R. S. (1992). Habitat restrictions in early benthic life: Experiments on habitat selection and *in situ* predation with the American lobster. *Journal of Experimental Marine Biology and Ecology* **157**, 91–114.

Walker, R. S. (1986). The first returns of tagged juvenile lobsters (*Homarus gammarus* (L.)) after release to the wild. *Aquaculture* **52**, 231–233.

Walton, W. C. and Walton, W. C. (2001). Problems, predators, and perceptions: Management of quahog (hardclam), *Mercenaria mercenaria*, stock enhancement programs in southern New England. *Journal of Shellfish Research* **20**, 127–134.

Wang, Q., Zhuang, Z. and Deng, J. (2002). The development and prospects of stock enhancement and sea ranching industry China. Abstract. Second International

Symposium on Stock Enhancement and Sea Ranching, Kobe, Japan, 28th January–1st February 2002.

Wang, X., Li, Y., Zhao, M., Xu, X. and Xing, X. (1999). Abalone raft-breeding technique on the north coast of China. *Transactions of Oceanology and Limnology* **3**, 72–77 (In Chinese, with English abstract.).

Wang, Y.-G. and Die, D. J. (1996). Stock-recruitment relationships of the tiger prawns (*Penaeus esculents* and *Penaeus semisulcatus*) in the Australian northern prawn fishery. *Marine and Freshwater Research* **47**, 87–95.

Wang, Y. G., Zhang, C. Y., Rong, X. J., Chen, J., Shi, C. Y., Sun, H. and Yan, J. P. (2004). Diseases of cultured sea cucumber, *Apostichopus japonicus*, in China. *In* "Advances in Sea Cucumber Aquaculture and Management" (A. Lovatelli, C. Conand, S. Purcell, S. Uthicke, J. F. Hamel and A. Mercier, eds), pp. 297–310. FAO Fisheries Technical Paper No. 463, FAO, Rome.

Waples, R. S. (1987). A multispecies approach to the analysis of gene flow in marine shore fishes. *Evolution* **41**, 385–400.

Ward, R. D. (2002). The genetics of fish populations. *In* "The Handbook of Fish and Fisheries, Vol. 1 Fish Biology" (P. J. B. Hart and J. D. Reynolds, eds), pp. 200–224. Blackwell, Oxford.

Ward, R. D. and Elliott, N. G. (2001). Genetic population structure of species in the south east fishery of Australia. *Marine and Freshwater Research* **52**, 563–573.

Ward, R. D. and Grewe, P. M. (1994). Appraisal of molecular genetic techniques in fisheries. *Reviews in Fish Biology and Fisheries* **4**, 300–325.

Ward, R. D., Skibinski, D. O. F. and Woodwark, M. (1992). Protein heterozygosity, protein structure and taxonomic differentiation. *Evolutionary Biology* **26**, 73–159.

Ward, R. D., Woodwark, M. and Skibinski, D. O. F. (1994). A comparison of genetic diversity levels in marine, freshwater and anadromous fish. *Journal of Fish Biology* **44**, 213–232.

Watts, R. J., Johnson, M. S. and Black, R. (1990). Effects of recruitment on genetic patchiness in the urchin *Echinometra mathaei* in Western Australia. *Marine Biology* **105**, 145–152.

Wells, S. M., Pyle, R. M. and Collins, N. M. (1983). Giant clams. *In* "The IUCN Invertebrate Red Data Book", pp. 97–107. Gland, Switzerland.

Wesson, J. A., Mann, R. and Luckenbach, M. (1999). Oyster restoration efforts in Virginia. *In* "Oyster Reef Habitat Restoration: A Synopsis and Synthesis of Approaches" (M. W. Luckenbach, R. Mann and J. A. Wesson, eds), pp. 117–130. Virginia Institute of Marine Science, School of Marine Science, College of William and Mary, VIMS Press, Gloucester Point, Virginia.

West, W. Q. B. and Chew, K. K. (1968). Application of the Bergman-Jefferts tag on the spot shrimp, *Pandalus platyceros* Brandt. *Proceedings of the National Shellfish Association* **58**, 93–100.

Westley, R. E., Rickard, N. A., Goodwin, C. L. and Scholz, A. J. (1990). Marine Farming and Enhancement. "Proceedings of the Fifteenth U.S.-Japan Meeting on Aquaculture Kyoto", Japan, October 22–23, 1986. pp. 49–52. NOAA Technical Report NMFS 85, U.S. Department of Commerce.

White, G. C. and Burnham, K. P. (1999). Program MARK: Survival estimation from populations of marked animals. *Bird Study* **46** (Suppl.), 120–138. (http://www.cnr.colostate.edu/~gwhite/mark/mark/htm).

Whitlatch, R. B. and Osman, R. W. (1999). Reefs as metapopulations: Approaches for restoring and managing spatially fragmented habitats. *In* "Oyster Reef Habitat Restoration: A Synopsis and Synthesis of Approaches" (M. W. Luckenbach, R.

Mann and J. A. Wesson, eds), pp. 199–212. Virginia Institute of Marine Science, School of Marine Science, College of William and Mary, VIMS Press, Gloucester Point, Virginia.

Wickens, J. F. and Lee, DO'C. (2002). "Crustacean Farming (2nd Edition)", p. 446. Blackwell, Oxford.

Wickins, J. F. and Barry, J. (1996). The effect of previous experience on the motivation to burrow in early benthic phase lobsters (*Homarus gammarus* (L)). *Marine and Freshwater Behaviour and Physiology* **28**, 211–228.

Wickins, J. F., Beard, T. W. and Jones, E. (1986). Microtagging cultured lobsters, *Homarus gammarus* (L.), for stock enhancement trials. *Aquaculture and Fisheries Management* **17**, 259–265.

Wickins, J. F., Roberts, J. C. and Heasman, M. S. (1996). Within-burrow behaviour of juvenile European lobsters *Homarus gammarus* (L.). *Marine and Freshwater Behaviour and Physiology* **28**, 229–253.

Wiedemeyer, W. L. (1992). Feeding behaviour of two tropical holothurians *Holothuria* (*Metriatyla) scabra* (Jager 1833) and *H.* (*Halodeima) atra* (Jager 1833), from Okinawa, Japan. *Proceedings of the Seventh International Coral Reef Symposium* **2**, 863–870.

Wiegardt, L. J. and Bourne, N. (1989). Introduction and transfer of molluscs in the northeast Pacific. *Journal of Shellfish Research* **8**, 467.

Wilding, C. M., Beaumont, A. R. and Latchford, J. W. (1997). Mitochondrial DNA variation in the scallop *Pecten maximus* (L.) assessed by a PCR-RFLP method. *Heredity* **79**, 178–189.

Williams, K. C. (ed.) (2004). "Spiny lobster ecology and exploitation in the South China Sea region. ACIAR Proceedings No. 120, p. 73. Australian Centre for International Agricultural Research, Canberra.

Williams, K. C., Pitcher, C. R. and Jones, C. (2002)."Interim Report on Rock Lobster Study Mission to Viet Nam, May 2002", p. 21. Australian Centre for International Agricultural Research, Canberra.

Williams, S. T., Jara, J., Gomez, E. and Knowlton, N. (2002). The marine Indo-Pacific break: Contrasting the resolving power of mitochondrial and nuclear genes. *Integrative and Comparative Biology* **42**, 941–952.

Wilson, A. C., Cann, R. L., Carr, S., George, M. J., Gyllensten, U. B., Helm-Bychowski, K. M., Higuchi, R. G., Palumbi, S. R., Prager, E. M., Sage, R. D. and Stoneking, M. (1985). Mitochondrial DNA and two perspectives on evolutionary genetics. *Biological Journal of the Linnean Society* **26**, 375–400.

Wilson, A. J. and Ferguson, M. M. (2002). Molecular pedigree analysis in natural populations of fishes: Approaches, applications, and practical considerations. *Canadian Journal of Fisheries and Aquatic Sciences* **59**, 696–707.

Withler, R. (2000). Genetic tools for identification and conservation of exploited abalone (*Haliotis* spp.) species. *In* "Workshop on Rebuilding Abalone Stocks in British Columbia" (A. Campbell, ed.), *Canadian Special Publication in Fisheries and Aquatic Sciences* **130**, 101–110.

Wittenberg, R. and Cock, M. J. W. (2001). "Invasive Alien Species: A Toolkit of Best Prevention and Management Practices", p. 228. CAB International, Oxford.

Wright, S. (1943). Isolation by distance. *Genetics* **28**, 114–138.

Xu, J., Xia, M., Ning, X. and Mathews, C. P. (1997). Stocking, enhancement, and mariculture of *Penaeus orientalis* and other species in Shanghai and Zhejiang Provinces, China. *Marine Fisheries Review* **59**, 8–14.

Xu, Z. K., Primavera, J. H., de la Pena, L. D., Pettit, P., Belak, J. and Alcivar-Warren, A. (2001). Genetic diversity of wild and cultured black tiger shrimp (*Penaeus monodon*) in the Philippines using microsatellites. *Aquaculture* **199**, 13–40.

Yamakawa, T., Nishimura, M., Matsuda, H., Tsujigado, A. and Kamiya, N. (1989). Complete larval rearing of the Japanese spiny lobster *Panulirus japonicus*. *Nippon Suisan Gakkaishi* **55**, 745.

Yanagisawa, T. (1996). Sea-cucumber ranching in Japan and some suggestions for the South Pacific. *In* "Proceedings of The International Workshop on Present and Future of Aquaculture Research and Development in Pacific Island Countries", 20–24 November 1995, Nuku'alofa, Tonga. (unedited), pp. 387–411. Japan International Co-operation Agency, Tokyo.

Yanagisawa, T. (1998). Aspects of the biology and culture of the sea cucumber. *In* "Tropical Mariculture" (S. S. De Silva, ed.), pp. 292–308. Academic Press, London.

Yanai, T., Yanai, T., Takagi, H. and Nihira, A. (1995). Infestation of the Japanese abalone *Haliotis discus hannai* Ino by the boring polychaete, *Polydora*, in the coast of Ibaraki Prefecture, east Japan. *Bulletin of Fisheries Experimental Station, Ibaraki Ken* **33**, 119–125 (in Japanese, English abstract).

Ye, Y. (2000). Is recruitment related to spawning stock in penaeid shrimp fisheries? *ICES Journal of Marine Science* **57**, 1103–1109.

Ye, Y., Loneragan, N. R. and Harch, B. (2003). The bio-economic model to evaluate the costs and benefits of tiger prawn stock enhancement in Exmouth Gulf. *In* "Developing Techniques for Enhancing Prawn Fisheries, with a Focus on Brown Tiger Prawns (*Penaeus esculentus*) in Exmouth Gulf" (N. Loneragan, ed.), pp. 246–272. Final Report Australian Fisheries Research and Development Corporation Project 1999/222. CSIRO, Cleveland, Australia.

Ye, Y., Loneragan, N. R., Die, D. J., Watson, R. A. and Harch, B. (2005). Bioeconomic modelling and risk assessment of tiger prawn (*Penaeus esculentus*) stock enhancement in Exmouth Gulf, Australia. *Fisheries Research* **73**, 231–249.

Yotsui, T. and Maesako, N. (1993). Restoration experiments of *Eisenia bicyclis* beds on barren grounds at Tsushima Islands. *Suisan Zoshoku* **41**, 67–70 (in Japanese with English abstract).

Yu, E. T., Juinio-Menez, M. and Monje, V. D. (2000). Sequence variation in the ribosomal DNA internal transcribed spacer of *Tridacna crocea*. *Marine Biotechnology* **2**, 511–516.

Zhao, B., Hirayama, N., Yamada, S. and Yamada, J. (1991a). A mathematical model for estimating the harvesting rate of released reared-abalone and its applications. *Journal of the Tokyo University of Fisheries* **78**, 207–215.

Zhao, B., Yamada, J., Hirayama, N. and Yamada, S. (1991b). The optimum size of released reared-abalone in southern fishing ground of Akita Prefecture. *Journal of the Tokyo University of Fisheries* **78**, 217–226 (in Japanese with English abstract, tables and figures).

Zimmerman, R. J., Minello, T. J. and Zamora, G. (1984). Selection of vegetated habitat by brown shrimp, *Penaeus aztecus*, in a Galveston Bay salt marsh. *Fishery Bulletin, US* **82**, 325–336.

Zuñiga, G., Guzmán del Próo, S. A., Cisnero, R. and Rodriguez, G. (2000). Population genetic analysis of the abalone *Haliotis fulgens* (Mollusca: Gastropoda) in Baja California, Mexico. *Journal of Shellfish Research* **19**, 853–859.

Appendix

Nomenclature of species mentioned in the text, tables and figures.

Group	Scientific name
Bivalves	
Giant clams	*Hippopus hippopus* L.
	Tridacna crocea Lamarck
	T. derasa (Roeding)
	T. gigas (L.)
	T. maxima (Roeding)
	T. squamosa Lamarck
Scallops	*Argopecten irradians* (Lamarck)
	Chlamys islandica (Mueller)
	Chlamys opercularis L.
	Patinopecten yessoensis (Jay)
	Pecten maximus L.
	P. novaezelandiae Reeve
	Placopecten magellanicus (Gmelin)
Other bivalves	*Crassostrea virginica* (Gmelin)
	Gemma gemma (Totten)
	Macoma balthica (L.)
	Mercenaria campechinensis (Gmelin)
	M. mercenaria (L.)
	Mya arenaria L.
	Mytilus edulis L.
	Panopea abrupta (Conrad)
	Perna canaliculus (Gmelin)
	Rangia cuneata (Gray)
	Ruditapes philippinarum (Adams et Reeve)
	Siliqua patula (Dixon)
	Venerupis japonica (Deshayes)
Gastropods	
Abalone	*Haliotis asinina* L.
	H. assimilis Dall
	H. corrugata W. Wood
	H. cracherodii Leach
	H. discus discus Reeve
	H. discus hannai Ino

(Continued)

ADVANCES IN MARINE BIOLOGY VOL 49
© 2005 Elsevier Ltd. All rights reserved

0065-2881/05 $35.00
DOI: 10.1016/S0065-2881(05)49013-6

Appendix (Continued)

Group	Scientific name
	H. diversicolor Reeve
	H. fulgens Philippi
	H. gigantea Gmelin
	H. iris Martyn
	H. kamtschatkana Jonas
	H. laevigata Donovan
	H. midae L.
	H. rubra Leach
	H. rufescens Swainson
	H. sieboldi Reeve
	H. sorenseni Bartsch
	H. tuberculata L.
Topshell	*Trochus niloticus* L.
Queen conch	*Strombus gigas* L.
Other gastropods	*Crepidula fornicata* (L.)
	Turbo setosus Gmelin
	Urosalpinx cinerea (Say)
Crustaceans	
Alpheid shrimp	*Alpheus lottini* Guerin
Mantis shrimp	*Haptosquilla pulchella* Miers
Shrimp	*Litopenaeus stylirostris* Stimpson
	Metapenaeus ensis (De Haan)
	Pandalus platyceros Brandt
	Penaeus aztecus Ives
	P. chinensis (Osbeck)
	P. duorarum Burkenroad
	P. esculentus Haswell
	P. indicus Milne Edwards
	P. japonicus Bate
	P. latisulcatus Kishinouye
	P. merguiensis De Man
	P. monodon Fabricius
	P. orientalis Kishinouye
	P. penicillatus Alcock
	P. semisulcatus De Haan
	P. setiferus (L.)
	P. vannamei Boone
Spiny lobsters	*Jasus edwardsii* (Hutton)
	J. frontialis (H. Milne-Edwards)
	J. lalandii (H. Milne-Edwards)
	J. paulensis (Heller)
	J. verreauxi (H. Milne-Edwards)
	Palinurus elephas (Fabricius)
	Panulirus argus (Latreille)
	P. cygnus George
	P. homarus Linnaeus
	P. interruptus (Randall)
	P. japonicus (von Siebold)
	P. longipes (A. Milne-Edwards)

Appendix (Continued)

Group	Scientific name
	P. marginatus (Quoy et Gaimard)
	P. ornatus (Fabricius)
	P. penicillatus (Olivier)
	P. polyphagus (Herbst)
	P. versicolor (Latreille)
Slipper lobsters	*Ibacus ciliatus* (von Siebold)
	I. novemdentatus Gibes
	Scyllarides latus (Latreille)
	Scyllarus americanus (S. I. Smith)
	S. demani Holthuis
	Thenus orientalis (Lund)
Lobsters	*Homarus americanus* H. Milne-Edwards
	H. gammarus (L.)
	Nephrops norvegicus (L.)
Crabs	*Birgus latro* Rumphius
	Callinectes sapidus Rathbun
	Cancer magister Dana
	C. pagurus L.
	Carcinus maenas (L.)
	Neopanope texana (Stimpson)
	Paralithodes brevipes (Milne-Edwards et Lucas)
	Portunus trituberculatus (Miers)
	Pugettia quadridens (De Haan)
	Scylla serrata (Forskal)
	Telmessus cheiragonus (Tilesius)
Horseshoe crab	*Limulus polyphemus* (L.)
Barnacles	*Balanus glandula* Darwin
Echinoderms	
Sea cucumbers	*Apostichopus japonicus* (Selenka)
	Cucumaria miniata (Brandt)
	C. pseudocurata Deichmann
	Holothuria atra Jaeger
	H. nobilis Selenka
	H. scabra Jaeger
	Stichopus chloronotus Brandt
Sea urchins	*Anthocidaris crassispina* Lutken
	Centrostephanus rodgersii (Agassiz)
	Diadema antillarum Philippi
	Echinometra mathaei (Blainville)
	Echinothrix diadema (L.)
	Evechinus chloroticus Valenciennes
	Hemicentrotus pulcherrimus Mortensen
	Parechinus angulosus (Leske)
	Pseudocentrotus depressus Mortensen
	Strongylocentrotus franciscanus (Agassiz)
	S. intermedius (Agassiz)
	S. nudus (Agassiz)
	Tripneustes gratilla L.

(Continued)

Appendix   (Continued)

Group	Scientific name
Starfish	*Astrostole scabra* (Hutton)
	Linckia laevigata (L.)
Fish	*Ciliata mustela* (L.)
	Gadus morhua L.
	Hippoglossoides platessoides (Fabricius)
	Lates calcarifer (Bloch)
	Myxocephalus octodecemspinosus (Mitchill)
	M. scorpius (L.)
	Neopanope texana (Sayi)
	Pomatoschistus minutus (Pallas)
	Pleuronectes americanus Walbaum
	Scomberoides commersonianus Lacepede
	Tautogolabrus adspersus (Walbaum)
	Urophycis tenuis (Mitchill)
Turtles	*Caretta caretta* (L.)
Micro-algae	*Dinophysis fortii* Pavillard
	Gonyaulax catenella [=*Alexandrium catenella* (Whedon et Kofoid) Balech]
	Protogonyaulax tamarensis (Lebour) Taylor
Macro-algae	*Laurencia poitei* (Lamouroux) Howe
	Ulvella lens P. Crouan et H. Crouan
Seagrasses and other plants	*Cymodocea serrulata* (R. Brown) Ascherson et Magnus
	Halodule uninervis (Forskal) Aschers
	Thalassia testudinum Banks
	Reynoutria sachalinensis (F. Schmidt) Nakai
	Zostera capricorni Aschers
Parasites, pathogens and pests (excluding toxic dinoflagellates)	*Aerococcus viridianus* Williams *et al.*
	Aureococcus anophagefferens Hargraves et Sieburth
	Haplosporidium nelsoni (Haskin, Stauber et Mackin)
	Labyrinthuloides haliotidis (Bower) [= *Aplanochytridium haliotidis* (Bower) Leander et Porter]
	Perkinsus karlssoni McGladdery, Cawthorne et Bradford
	Perkinsus marinus Levine
	Polydora ciliata (Johnston)
	Vibrio anguillarum Bergeman

TAXONOMIC INDEX

SUBJECT INDEX

Series Contents for Last Ten Years*

*The full list of contents for volumes 1–37 can be found in volume 38.

—